Oxford Resources for IB
Diploma Programme

IB PREPARED ✓

2023 EDITION

BIOLOGY

Debora M. Primrose

OXFORD
UNIVERSITY PRESS

OXFORD
UNIVERSITY PRESS

Great Clarendon Street, Oxford, OX2 6DP, United Kingdom

Oxford University Press is a department of the University of Oxford. It furthers the University's objective of excellence in research, scholarship, and education by publishing worldwide. Oxford is a registered trade mark of Oxford University Press in the UK and in certain other countries.

© Oxford University Press 2024

The moral rights of the author have been asserted

First published in 2024

All rights reserved. No part of this publication may be reproduced, stored in a retrieval system, transmitted, used for text and data mining, or used for training artificial intelligence, in any form or by any means, without the prior permission in writing of Oxford University Press, or as expressly permitted by law, by licence or under terms agreed with the appropriate reprographics rights organization. Enquiries concerning reproduction outside the scope of the above should be sent to the Rights Department, Oxford University Press, at the address above.

You must not circulate this work in any other form and you must impose this same condition on any acquirer

British Library Cataloguing in Publication Data
Data available

9781382058315

9781382058346 (ebook)

10 9 8 7 6 5 4 3 2 1

Paper used in the production of this book is a natural, recyclable product made from wood grown in sustainable forests.

The manufacturing process conforms to the environmental regulations of the country of origin.

Printed in China by Shanghai Offset Printing Products Ltd

Acknowledgements
The "In cooperation with IB" logo signifies the content in this textbook has been reviewed by the IB to ensure if fully aligns with current IB curriculum and offers high-quality guidance and support for IB teaching and learning.

The publisher wishes to thank the International Baccalaureate Organization for permission to reproduce their intellectual property.

The publisher and author would like to thank the following for permissions to use photographs and other copyright material:

Cover: Alexius Sutandio / 500PX Plus / Getty Images.

Photos: ppiii-iv CLAUDE NURIDSANY & MARIA PERENNOU / Science Photo Library; Alexius Sutandio / 500PX Plus / Getty Images. Photos: p2: Sebastian Kaulitzki / Shutterstock; p6(t): Toni Genes / Shutterstock; p6(b): Sergey Uryadnikov / Shutterstock; p18: Peter Hermes Furian / Alamy Stock Photo; p19(t): Sebastian Kaulitzki / Shutterstock; p19(b): ERIC GRAVE / SCIENCE PHOTO LIBRARY; p20: Miami University / International Baccalaureate; p22: creativeneko / Shutterstock; p41: Liliya Butenko / Shutterstock; p46: AustralianCamera / Shutterstock; p47(t): MARTYN F. CHILLMAID / SCIENCE PHOTO LIBRARY; p47(b): MARTYN F. CHILLMAID / SCIENCE PHOTO LIBRARY; p64: Aldona Griskeviciene / Shutterstock; p71: Jose Luis Calvo / Shutterstock; p78: Science History Images / Alamy Stock Photo; p81: Science Photo Library / Alamy Stock Photo; p93: Praisaeng / Shutterstock; p98: Alex Mustard / Nature Picture Library / Alamy Stock Photo; p99: Goinyk Production / Shutterstock; p101(t): Rick & Nora Bowers / Alamy Stock Photo; p101(b): robertharding / Alamy Stock Photo; p105(a): David Bokuchava / Alamy Stock Photo; p105(b): MichaelGrantPlants / Alamy Stock Photo; p108: Vittorio Ricci - Italy / Getty Images; p123: SINCLAIR STAMMERS / SCIENCE PHOTO LIBRARY; p133: DENNIS KUNKEL MICROSCOPY / SCIENCE PHOTO LIBRARY; p146: DAVID M. PHILLIPS / SCIENCE PHOTO LIBRARY; p159: Samib123 / Shutterstock; p172: Christopher Hope-Fitch / Getty Images; p190(t): Rattiya Thongdumhyu / Shutterstock; p190(b): Science History Images / Alamy Stock Photo; p226: Emperor Bird of Paradise Paradisaea guilielmi by William T Cooper (from The Birds of Paradise and Bowerbirds by William T Cooper & Joseph M Forshaw); p243(l): Mircea Costina / Shutterstock; p243(r): Design Pics / Alamy Stock Photo; p244, p246, p251: CLAUDE NURIDSANY & MARIA PERENNOU / Science Photo Lib p256: JOSE CALVO / SCIENCE PHOTO LIBRARY; p257: Science History Images / Alamy Stock Photo; p258: Sergey Spritnyuk / Shutterstock.

Artwork by Alison Allot, Q2A Media, Aptara Corp., Greengate Publishing Services, and Oxford University Press.

Molecular graphics and analyses performed with UCSF Chimera, developed by the Resource for Biocomputing, Visualization, and Informatics at the University of California, San Francisco, with support from NIH P41-GM103311.

Every effort has been made to contact copyright holders of material reproduced in this book. Any omissions will be rectified in subsequent printings if notice is given to the publisher.

Links to third party websites are provided by Oxford in good faith and for information only. Oxford disclaims any responsibility for the materials contained in any third party website referenced in this work.

MIX
Paper | Supporting responsible forestry
FSC® C109093

Contents

Introduction	iv

Theme A—Unity and diversity — 2

A1.1 Water	2
A1.2 Nucleic acids	7
A2.1 Origins of cells	12
A2.2 Cell structure	16
A2.3 Viruses	22
A3.1 Diversity of organisms	27
A3.2 Classification and cladistics	31
A4.1 Evolution and speciation	34
A4.2 Conservation of biodiversity	40

Theme B—Form and function — 46

B1.1 Carbohydrates and lipids	46
B1.2 Proteins	52
B2.1 Membranes and membrane transport	57
B2.2 Organelles and compartmentalization	63
B2.3 Cell specialization	68
B3.1 Gas exchange	72
B3.2 Transport	80
B3.3 Muscle and motility	90
B4.1 Adaptation to environment	96
B4.2 Ecological niches	102

Theme C—Interaction and interdependence — 108

C1.1 Enzymes and metabolism	108
C1.2 Cell respiration	115
C1.3 Photosynthesis	122
C2.1 Chemical signalling	132
C2.2 Neural signalling	138
C3.1 Integration of body systems	145
C3.2 Defence against disease	152
C4.1 Populations and communities	158
C4.2 Transfers of energy and matter	167

Theme D—Continuity and change — 172

D1.1 DNA replication	172
D1.2 Protein synthesis	176
D1.3 Mutation and gene editing	184
D2.1 Cell and nuclear division	188
D2.2 Gene expression	195
D2.3 Water potential	199
D3.1 Reproduction	203
D3.2 Inheritance	211
D3.3 Homeostasis	219
D4.1 Natural selection	224
D4.2 Stability and change	229
D4.3 Climate change	236

Internal Assessment	246
Practice Exam Papers	251
Index	262

Answers
www.oxfordsecondary.com/ib-prepared-2e

Introduction

This book provides coverage of the IB diploma syllabus in biology and offers support to students preparing for their examinations. The book will help you to revise the study material, learn the essential terms and concepts, strengthen your problem-solving skills and improve your approach to IB examinations. The book is packed with worked examples and exam tips that demonstrate best practices and warn against common errors. All topics are illustrated by annotated student answers to questions from past examinations, which explain why marks may be scored or missed.

Practice problems and a complete set of IB-style examination papers provide further opportunities to check your knowledge and skills, boost your confidence and monitor the progress of your studies. Full solutions to all problems and examination papers are given online at **www.oxfordsecondary.com/ib-prepared-2e**.

As any study guide, this book is not intended to replace your course materials, such as textbooks, laboratory manuals, past papers and markschemes, the IB Biology syllabus and your own notes. To succeed in the examination, you will need to use a broad range of resources, many of which are available online. This book will help you to navigate through this critical part of your studies, making your preparation for the exam less stressful and more efficient.

DP Biology assessment

All standard level (SL) and higher level (HL) students must complete the internal assessment and take two papers as part of their external assessment. Paper 1 is separated into two sections: in paper 1A you answer a range of multiple-choice questions and in paper 1B you answer data-based questions related to experimental work and the syllabus. Papers 1A and 1B are completed together without interruptions. Paper 2 is separated into Section A with data-based questions and short-answer questions and Section B with extended-response questions. The internal and external assessment marks are combined as shown in the table at the top of page v to give your overall DP Biology grade, from 1 (lowest) to 7 (highest).

Overview of the book structure

The book is divided into several sections that cover the internal assessment, core SL and additional higher level (AHL) topics, data-based and practical questions, and a complete set of practice examination papers.

The main section of the book follows the structure of the IB diploma biology syllabus (for first assessment 2025) and covers all understandings and applications and skills assessment statements. The nature of science concepts are also discussed where applicable.

The internal assessment section outlines the nature of the investigation that you will have to carry out and explains how to select a suitable topic, collect and process experimental data, draw conclusions and present your report in a suitable format to satisfy the marking criteria and achieve the highest grade.

The final section contains IB-style practice examination papers 1 and 2, written exclusively for this book. These papers will give you an opportunity to test yourself before the exam and at the same time provide additional practice problems for every topic.

The answers and solutions to all practice problems and examination papers are given online at **www.oxfordsecondary.com/ib-prepared-2e**.

Assessment overview

Assessment	Description	SL Marks	SL Weight	HL Marks	HL Weight
Internal	Experimental work with a written report	24	20%	24	20%
Paper 1	A: Multiple-choice	30	36%	40	36%
	B: Data-based questions	25		35	
Paper 2	Short- and extended-response questions	50	44%	80	44%

Command terms

Command terms are pre-defined words and phrases used in all IB Biology questions and problems. Each command term specifies the type and depth of the response expected from you in a particular question. For example, the command terms state, outline, explain and discuss require answers with increasingly higher levels of detail, from a single word, short sentence or numerical value ("state") to comprehensive analysis ("discuss"), as shown in the next table.

Question	Possible answer
State the effect of increasing temperature on the reaction rate.	Rate increases.
Outline how an increase in temperature affects the reaction rate.	For most reactions, the rate approximately doubles when temperature increases by 10 degrees.
Explain why an increase in temperature increases the reaction rate.	As temperature increases, the average speed and thus kinetic energy of particles also increases. The particles collide with one another more frequently and with a greater force. As a result, the frequency of successful collisions increases, so the rate increases.
Discuss the effects of increasing temperature and the presence of an enzyme on the reaction rate.	Both factors increase the rate by increasing the frequency of successful collisions. However, an increase in temperature increases the frequency and intensity of all collisions (successful and unsuccessful) but has no effect on the activation energy. In contrast, an enzyme has no effect on the frequency or intensity of collisions but lowers the activation energy by providing an alternative reaction pathway and thus allowing slow-moving particles to collide successfully. Thus, the same macroscopic effect is achieved by different microscopic changes.

A list of commonly used command terms in biology examination questions is given in the following table. Understanding the exact meaning of frequently used command terms is essential for your success in the examination. Therefore, you should explore this table and use it regularly as a reference when answering questions in this book.

Command term	Definition
Analyse	Break down in order to bring out the essential elements or structure.
Annotate	Add brief notes to a diagram or graph.
Calculate	Obtain a numerical answer showing your working.
Comment	Give a judgement based on a given statement or result of a calculation.
Compare	Give an account of the similarities between two or more items.
Compare and contrast	Give an account of similarities and differences between two or more items.
Construct	Display information in a diagrammatic or logical form.
Deduce	Reach a conclusion from the information given.
Define	Give the precise meaning of a word, phrase, concept or physical quantity.
Describe	Give a detailed account.
Design	Produce a plan, simulation or model.
Determine	Obtain the only possible answer.
Discuss	Offer a considered and balanced review that includes a range of arguments, factors or hypotheses.
Distinguish	Make clear the differences between two or more concepts or items.
Draw	Represent by a labelled, accurate diagram or graph, drawn to scale, with plotted points (if appropriate) joined by a straight line or smooth curve.

Command term	Definition
Estimate	Obtain an approximate value.
Evaluate	Make an appraisal by weighing up the strengths and limitations.
Explain	Give a detailed account including reasons or causes.
Identify	Provide an answer from a number of possibilities.
Justify	Give a valid reason or evidence to support an answer or conclusion.
Label	Add labels to a diagram.
List	Give a sequence of brief answers with no explanation.
Measure	Obtain a value or quantity.
Outline	Give a brief account or summary.
Predict	Give an expected result.
Sketch	Represent by means of a diagram or graph (labelled as appropriate), giving a general idea of the required shape or relationship.
State	Give a specific name, value or other brief answer without explanation or calculation.
Suggest	Propose a solution, hypothesis or other possible answer.

A complete list of command terms is available in the subject guide.

Preparation and exam strategies

In addition to the above suggestions, there are some simple rules you should follow during your preparation study and the exam itself.

1. **Get ready for study.** Have enough sleep, eat well, drink plenty of water and reduce your stress by positive thinking and physical exercise. A good night's sleep is particularly important before the exam day, as it can improve your score.
2. **Organize your study environment.** Find a comfortable place with adequate lighting, temperature and ventilation. Avoid distractions. Keep your papers and computer files organized. Bookmark useful online and offline material.
3. **Plan your studies.** Make a list of your tasks and arrange them by importance. Break up large tasks into smaller, easily manageable parts. Create a schedule for your studying time and make sure that you can complete each task before the deadline.
4. **Use this book as your first point of reference.** Work your way through the topics systematically and identify the gaps in your understanding and skills. Spend extra time on the topics where improvement is required. Check your textbook and online resources for more information.
5. **Read actively.** Focus on understanding rather than memorizing. Recite key points and definitions using your own words. Try to solve every worked example and practice problem before looking at the answer. Make notes for future reference.
6. **Get ready for the exams.** Practise answering exam-style questions under a time constraint. Solve as many problems from past papers as you can. Take a trial exam using the papers at the end of this book.
7. **Optimize your exam approach.** Read all questions carefully, paying extra attention to command terms. Keep your answers as short and clear as possible. Double-check all numerical values and units. Label axes in graphs and annotate diagrams. Use exam tips from this book.
8. **Do not panic.** Take a positive attitude and concentrate on things you can improve. Set realistic goals and work systematically to achieve these goals. Be prepared to reflect on your performance and learn from your errors in order to improve your future results.

Key features of the book

Each chapter typically covers one topic, and starts with "**You should know**" and "**You should be able to**" checklists. These outline the understandings and applications and skills sections of the IB diploma biology syllabus. Some assessment statements have been reworded or combined together to make them more accessible and simplify the navigation. These changes do not affect the coverage of key syllabus material, which is always explained within the chapter.

Chapters contain the features outlined on this page.

Key definitions are discussed at a level sufficient for answering typical examination questions. Most definitions are given in a grey side box like this one, and explained in the text.

Example

Examples offer solutions to typical problems and demonstrate common problem-solving techniques. Many examples provide alternative answers and explain how the marks are awarded.

Assessment tip

This feature highlights the essential terms and statements that have appeared in past markschemes and warns against common errors.

Nature of science

Nature of science relates a biology concept to the overarching principles of the scientific approach.

Links provide a reference to relevant material, within another part of this book, that relates to the text in question. **Linking questions** from the guide are given at the end of each subtopic.

Sample student answers show typical student responses to IB-style questions (most of which are taken from past examination papers). In each response, the correct points are often highlighted in blue while incorrect or incomplete answers are highlighted in orange. Positive or negative feedback on the student's response is given in the blue and orange pull-out boxes. An example is given below.

The marks the student may have earned based on their response

Number of marks available

You will see an exam paper icon on the right when the question has been adapted from a past IB paper

Positive feedback

Examination question

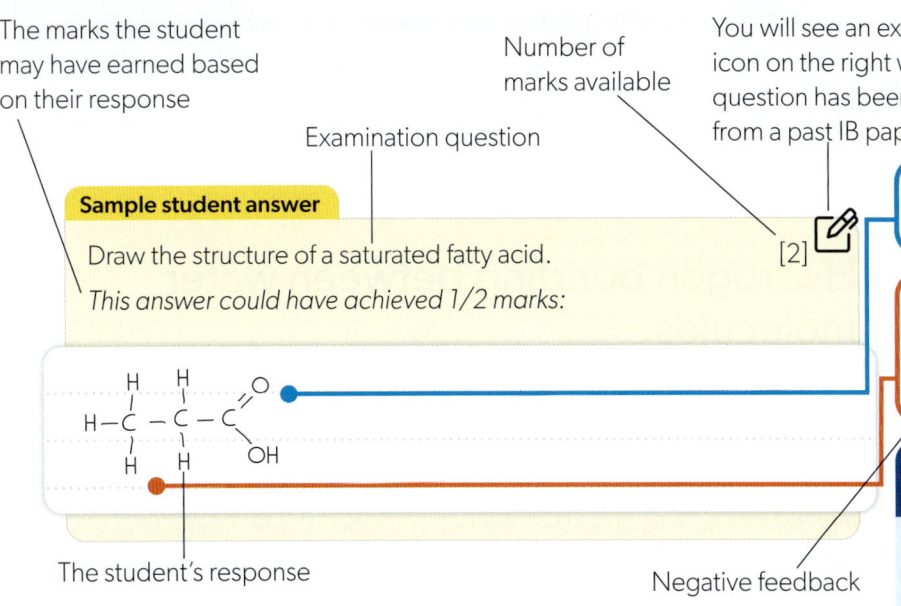

The student's response

Negative feedback

▲ One mark was given for the acid group and hydrogens shown.

▼ The structure only has three carbons and the shortest fatty acid known is butyric acid, with four carbons.

Examiner tip

These sit alongside sample student answers to show you how to optimize your approach to particular questions.

Questions not taken from past IB examinations will not have the exam paper icon.

Practice problems

Practice problems are given at the end of each chapter. These are IB-style questions that provide you with an opportunity to test yourself and improve your problem-solving skills. Some questions introduce factual or theoretical material from the syllabus that can be studied independently.

A Unity and diversity

A1.1 Water

You should know:
- ✔ water is the medium for life.
- ✔ water molecules have covalent bonds between oxygen and hydrogen atoms.
- ✔ water molecules are polar—they have a positive end and a negative end.
- ✔ hydrophilic substances are attracted to water (water-loving) and hydrophobic substances are repelled by water (water-fearing).

Additional higher level:
- ✔ water originated in asteroids.
- ✔ water is retained on Earth due to gravity and low temperature.

You should be able to:
- ✔ describe water's cohesive, adhesive, thermal and solvent properties.
- ✔ explain these properties in terms of hydrogen bonding and dipolarity.
- ✔ contrast physical properties of water and air in terms of buoyancy, viscosity, thermal conductivity and specific heat capacity.
- ✔ explain how organisms are adapted to live in water, air or land.

Additional higher level:
- ✔ explain why water is liquid on Earth.

Water as the medium for life

A water molecule hardly ever changes chemically, but without it most chemical reactions within organisms would not take place. Scientists think that the first cells originated in water. Water remains the medium in which most life processes occur. Water remains liquid at a wide range of temperatures. This allows it to keep body temperature constant, act as a lubricant, and dissolve and transport substances. Water affects chemical reactions in organisms—it is a substrate in photosynthesis and takes part in the enzymatic reactions of hydrolysis. Many macromolecules such as enzymes become active only in the presence of water. Water also forms part of cells and is therefore an essential part of living tissues.

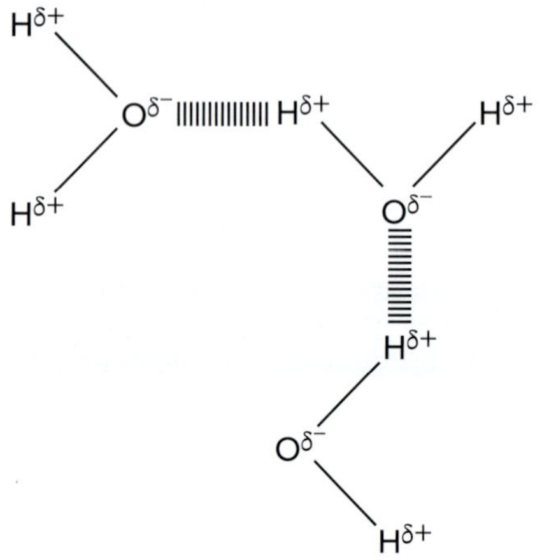

Hydrogen bonding between water molecules

The unique properties of water are due to strong attractive forces among water molecules. A molecule of water contains two hydrogen atoms which each bond to an oxygen atom with a polar covalent bond. The atom of oxygen shares one electron with each hydrogen atom, therefore only two of the four electrons it has in its outer shell are being shared. The two remaining electrons give the oxygen its electronegative character. The hydrogen atoms are electropositive because they share their electron with the oxygen atom and have no extra electrons. The V-like form of a water molecule and the polarized nature of the O–H bond result in a negative charge close to the oxygen atom and positive charge in the hydrogen atoms.

▲ Figure 1 Hydrogen bonding between two water molecules

- **Hydrogen bonds** are electromagnetic bonds between water molecules.

Water can engage in multiple hydrogen bonding on a three-dimensional basis. The negative part of one water molecule is attracted to the positive part of another water molecule, forming electromagnetic bonds called **hydrogen bonds**. Compared with covalent bonds hydrogen bonds are weak, but many hydrogen bonds together form a strong attraction between water molecules.

Cohesion of water molecules

Hydrogen bonds allow water molecules to stick together. This "stickiness" between water molecules is called **cohesion**. Cohesion between water molecules allows water to be transported under tension in xylem. Cohesion causes surface tension, which allows organisms to walk on water and to use water surfaces as habitats.

- **Cohesion** is the attraction of water molecules to other water molecules.

Example 1

Which diagram best illustrates the interaction between water molecules?

A. 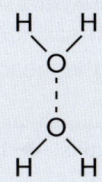 B.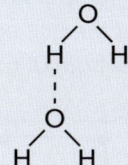

C. 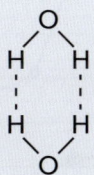 D.

Solution

The correct answer is **B**, as it shows the hydrogen bonding between the oxygen (O) of one water molecule and the hydrogen (H) of the other. In **A**, the bonding is shown between the oxygen atoms and in **C** and **D** between two hydrogen atoms, which is incorrect.

Water molecules are attracted to other kinds of molecules or compounds that are polar or charged. Water **adheres** to cellulose and other **hydrophilic** wall components in the xylem vessels. Transpiration pull, which uses capillary action and the surface tension of water, is the main mechanism of water movement in plants.

- **Adhesion** is the attraction of water molecules to other molecules or compounds.
- **Hydrophilic** describes water-soluble molecules or compounds.

Solvent properties of water

The functions of some molecules depend on them being hydrophilic (soluble in water) or **hydrophobic** (insoluble in water). Water has a high solvent capacity, making it an ideal medium for metabolic processes, including those that involve enzyme-catalysed reactions. Water-soluble substances include ionic salts, such as sodium chloride, amino acids and carbohydrates with many hydroxyl groups, such as glucose. In plants and animals, these substances are transported dissolved in water. Lipids, such as cholesterol and triglycerides, are non-polar (insoluble in water) so in animals they need to be transported associated with proteins. In cell membranes, hydrophilic portions of proteins and phospholipids face outwards and shield internal hydrophobic components.

- **Hydrophobic** molecules or compounds are insoluble in water.

Unity and diversity

Examiner tip

This is a paper 2A question—a data-based question based on an unfamiliar context.

Sample student answer

During aerobic cell respiration, oxygen is consumed and carbon dioxide is produced inside cells. This generates concentration gradients between respiring cells and the environment, which cause diffusion of oxygen and carbon dioxide. Both oxygen and carbon dioxide are soluble in water. As the temperature rises, water becomes saturated at a lower concentration of the gas.

Laternula elliptica is a mollusc that lives on the sea bed in Antarctica. Its body temperature is always similar to that of the environment around it. To investigate the effect of temperature on *Laternula elliptica*, specimens were kept in temperature-controlled aquaria. The oxygen concentrations of water near the gills and in the body fluids were measured, at a range of temperatures from 0 °C to 9 °C. The graph shows the mean results.

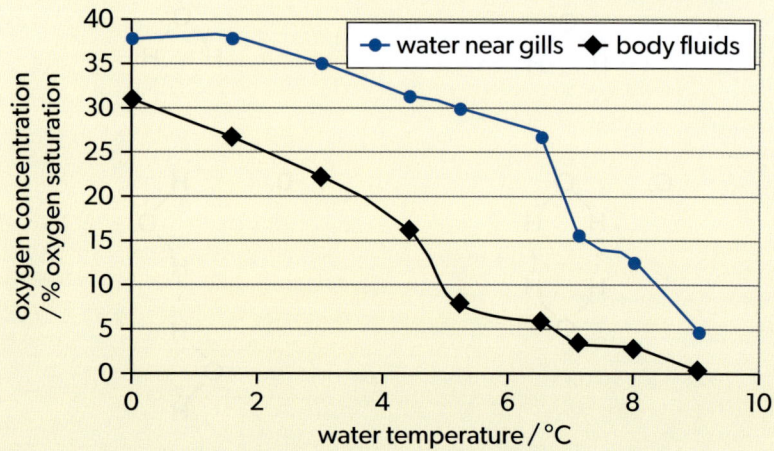

Source: Pörtner, H.O., et al. (2006). Hyperoxia alleviates thermal stress in the Antarctic bivalve, *Laternula elliptica*: evidence for oxygen limited thermal tolerance. *Polar Biology*, **29**(8), 688–693. www.doi.org/10.1007/s00300-005-0106-1

a) (i) Outline the relationship between temperature and oxygen concentration in the body fluids in *Laternula elliptica*. [2]

This answer could have achieved 1/2 marks:

> As the temperature increases, the oxygen concentration decreases.

▼ The negative correlation is correct, but the fact that this occurs for both fluids should have been mentioned. The oxygen concentration decreases faster in body fluids than in water near the gills up to 6.5 °C, where it decreases abruptly in water near the gills.

(ii) Suggest **two** reasons for the relationship. [2]

This answer could have achieved 0/2 marks:

> Oxygen concentration depends on temperature.

▼ The relationship was already stated in the answer to part (i), so does not need to be repeated. The water in body fluids contains less oxygen because of respiration. This process requires oxygen and produces carbon dioxide. In the water near the gills, there is more oxygen at low temperatures, as this is oxygen dissolved in the Antarctic Ocean. As the temperature increases, in both locations, oxygen is less soluble in water because it has more motion and can evaporate, so is lost as gas.

Physical properties of water

The physical properties of water are important for animals in aquatic habitats. **Buoyancy** occurs as water exerts an upwards pressure on the floating organism that is equal to the mass of the displaced water. If the organism is lighter than this force, it will be able to float in water, otherwise it will sink. Some organisms have adaptations that allow them to float or sink when needed, for example swim bladders in bony fish. The buoyant force exerted by air on an object depends on the volume of air displaced and on the pressure and temperature of the air.

The **viscosity** of water is the measure of its resistance to deformation at a given rate. Viscosity is measured as a force multiplied by a time divided by an area. Water is approximately 50 times more viscous than air.

Thermal conductivity is the property that describes the ability to conduct heat. The thermal conductivity of water is low, but it is much higher (almost 20 times) than the thermal conductivity of air. The more solute and ions that are present in a sample of water, the higher the thermal conductivity.

An average energy of 20 kJ mol^{-1} is required to break each hydrogen bond. There are many hydrogen bonds in water, so a lot of heat is required to increase its temperature (high specific **heat capacity**). It also requires a lot of heat to change state, causing the boiling and melting points to be high. This helps aquatic environments to remain at constant temperatures despite changing atmospheric conditions. Water acts as an insulator by resisting the flow of heat through it because it can store excess heat without releasing it to the environment. The specific heat capacity of water is around four times greater than that of air.

> - **Buoyancy** occurs as water (or air) exerts an upwards pressure on the floating organism that is equal to the mass of the displaced water (or air).
> - **Viscosity** is the measure of flow resistance.
> - **Thermal conductivity** is the ability to conduct heat.
> - **Heat capacity** is the energy required to raise the temperature of 1 g of substance by 1 °C.

Example 2

Explain how buoyancy allows aquatic organisms to swim.

Solution

Buoyancy is the net upwards force on an organism equal to the water displaced by the organism. This allows organisms to float in water and move along using their limbs.

Example 3

What property of water is valuable to organisms?
A. Its high thermal conductivity causes mineral ions to deposit on ocean beds.
B. Its high specific heat capacity keeps their bodies cool.
C. Its viscosity allows greater speed of movement than in air.
D. Its solubility decreases thermal conductivity.

Solution

The correct answer is **B**, as a high specific heat capacity means that water remains at a lower temperature than the atmosphere in warm areas. A high thermal conductivity is caused by mineral ions dissolved in water. Water's high viscosity produces more resistance and therefore slows down movement. Increasing the solubility increases the dissolved minerals. Thermal conductivity increases with more minerals dissolved.

▲ Figure 2 Black-throated loon (*Gavia arctica*)

▲ Figure 3 Ringed seal (*Pusa hispida*)

Black-throated loons (*Gavia arctica*) are aquatic birds found in the northern hemisphere. They live on land and spend winters on ice-free coasts of the north-east Atlantic Ocean and the eastern and western Pacific Ocean. They are excellent swimmers and divers. They have a long, pointed beak which helps them catch their prey when diving into water. Although they are energetic flyers, they have short legs so they need to run on the water surface before they can take off for flight.

The ringed seal (*Pusa hispida*) can be found in the north Pacific, Atlantic and Arctic Oceans. They feed in water but rest and breed on floating ice (they are unable to reproduce on land). The ringed seal maintains a breathing hole in the ice allowing it to use ice habitats that other seals cannot. They can hold their breath for a long time, allowing them to dive frequently. With declines in ice packs due to global warming, survival has become tougher for them. They are the main prey for polar bears and seal pups are being forced into the water sooner due to lack of ice, so their population is decreasing. Nevertheless, the number of polar bears will probably also decrease with global warming, which might mean an eventual increase in the ringed seal population rather than a downfall.

Extraplanetary origin of water on Earth

Earth is the only planet in the solar system to contain oceans of liquid water, which is essential for life. The abundance of water over billions of years of the Earth's history has allowed life to evolve. Scientists think that water originally came from the outer solar system through the impact of asteroids or small planets containing water, especially during a period called the Late Heavy Bombardment about 4×10^9 years ago. This hypothesis is supported by similarities in the abundance and isotope ratios of water between the Earth and the oldest known meteorites that originate from the solar system's asteroid belt.

Goldilocks zone

- **Goldilocks zone** is the distance from the Sun where life could exist on planets.

Earth has been able to conserve its water because of gravitational forces and because of its distance from the Sun: it is far enough from the Sun to avoid vaporization due to extreme heat and close enough not to have frozen due to extreme cold. The distance Earth orbits the Sun is just right for water to remain a liquid. This distance from the Sun is called the habitable zone, or the **Goldilocks zone** (in reference to the story of *Goldilocks and the Three Bears*, where Goldilocks chooses items that are not found in the extremes—for example, not too cold or not too hot). Planets outside the solar system found in the habitable zones of their stars are more likely targets for detecting liquid water on their surfaces.

You will study adaptations of organisms to different environments in subtopic B4.1 and adaptations for swimming in marine mammals in subtopic B3.3 "Reasons for locomotion".

Example 4

What is a reason the planet Earth is considered to be in the Goldilocks zone?

A. There is no impact of meteorites.
B. It is close enough to the Sun to vaporize water.
C. The amount of oxygen in the atmosphere allows living organisms to survive.
D. The gravitational forces and temperatures allow water to remain liquid.

Solution

The correct answer is **D**. Water appeared through the impact of meteorites and the Earth is at a distance from the Sun to avoid water vaporization. It is true that the amount of oxygen is adequate, but this is not the reason it is considered to be in the Goldilocks zone.

- Could there be life without water?
- How are organisms adapted to living in water?

A1.2 Nucleic acids

You should know:
- the nucleic acids DNA and RNA are polymers of nucleotides.
- DNA is the genetic material of all organisms.
- viruses are non-living and use DNA or RNA as their genetic material.
- complementary base pairing allows genetic information to be replicated and expressed.
- DNA has an enormous capacity for storing data with great economy.
- conservation of the genetic code is evidence of universal common ancestry.

Additional higher level:
- RNA and DNA have their 5' end in a phosphate and 3' end in a sugar.
- the amount of pyrimidine bases is equal to purine bases across diverse life forms.
- the structure of DNA is suited to its function.
- nucleosomes help to supercoil DNA.

You should be able to:
- draw simple diagrams of the structure of single nucleotides of DNA and RNA, pentoses and bases.
- draw the structure of DNA showing two antiparallel strands of nucleotides linked by hydrogen bonding between complementary base pairs.
- compare DNA and RNA in terms of the number of strands present, the base composition and the type of pentose.

Additional higher level:
- describe the structure of nucleosomes.
- analyse the association between protein and DNA within a nucleosome using molecular visualization software.
- analyse results of the Hershey and Chase experiment providing evidence that DNA is the genetic material.
- explain that diversity in any length of DNA molecule and any base sequence is possible.

Structure of DNA and RNA

DNA is the genetic material of all living organisms. Although viruses are not considered to be living, they use **RNA** or DNA as their genetic material.

Nucleotides are monomers that are joined together in a condensation reaction to form DNA and RNA polymers. Each nucleotide is formed by a pentose sugar with a phosphate joined at carbon 5 and a **nitrogenous base** at carbon 1. At carbon 3 of this sugar, a new nucleotide is joined by its phosphate. This sugar–phosphate bonding makes a continuous chain of covalently bonded atoms in each strand of DNA or RNA nucleotides, forming the "backbone" of the molecule.

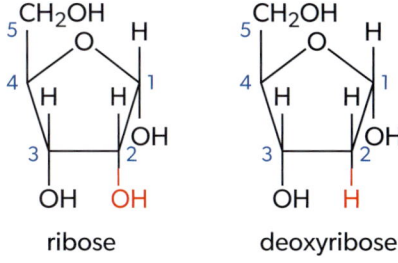

▲ Figure 4 Pentose sugars showing the carbon number

In DNA, the sugar is deoxyribose while, in RNA, it is ribose. These pentose sugars differ only in the hydroxyl group in carbon 2, which is only present in ribose.

- **DNA** (deoxyribonucleic acid) is a polymer of nucleotides containing the genetic instructions responsible for inheritance.
- **RNA** (ribonucleic acid) is a polymer of nucleotides containing the genetic instructions for protein synthesis.
- A **nucleotide** is a molecule containing a phosphate, a sugar and a nitrogenous base.
- The **nitrogenous bases** adenine (A), guanine (G), cytosine (C) and thymine (T) are found in DNA. Uracil (U) replaces thymine in RNA.

Unity and diversity

Nature of science

Francis Crick and James Watson created models to discover the structure of DNA. They used information from Rosalind Franklin's X-ray diffraction images of DNA and Erwin Chargaff's measurements of the proportions of each of the four bases.

You will study the advantages of diversity in DNA in subtopic A4.1.

- **Complementary base pairs** are linked by hydrogen bonds on opposite strands of DNA.

Assessment tip

Use circles, pentagons and rectangles to represent phosphates, sugars and bases, respectively. When drawing DNA, show the two DNA strands as antiparallel.

Complementary base pairing and the genetic code

DNA has a wide range of possible combinations in its nucleotide sequence. It is organized and as it is folded it does not occupy much space. The combination of the four nitrogenous bases in different orders produces different codons. These code for different amino acids and consequently the human DNA molecule can produce around 25,000 different proteins. This allows variation within a population, which at the same time allows adaptation to change in the environment. Natural selection will favour beneficial traits. The conservation of the genetic code across all life forms is evidence that there must have been a universal common ancestor to all organisms.

DNA is a double helix made of two antiparallel strands of nucleotides. The two strands are linked by hydrogen bonding between **complementary base pairs**: adenine (A) is paired with thymine (T), and guanine (G) is paired with cytosine (C).

Example 5

a) Draw the labelled structure of one nucleotide.
b) Explain how nucleotides join to form a DNA molecule.

Solution

a)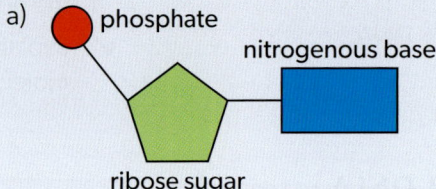

b) The phosphate of the deoxyribose sugar of the new nucleotide is joined by a covalent bond to the deoxyribose sugar of the previous nucleotide. This sugar–phosphate bonding makes a continuous chain of covalently bonded atoms in each strand of DNA. At the same time, complementary nitrogenous bases form hydrogen bonds between nucleotides from antiparallel DNA strands, forming a double helix.

Sample student answer

Label the parts of two paired nucleotides in the polynucleotide of DNA. [3]

This answer could have achieved 2/3 marks:

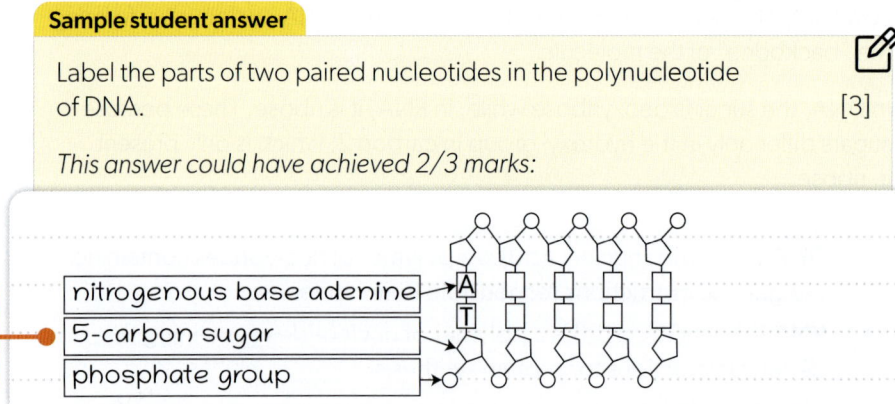

▼ "5-carbon sugar" is too vague for a mark. The correct answer is deoxyribose.

Example 6

Distinguish between DNA and RNA.

Solution

	DNA	RNA
Number of strands present	2	1
Pentose sugar	deoxyribose	ribose
Types of nitrogenous bases	adenine, thymine, cytosine, guanine	adenine, uracil, cytosine, guanine

Assessment tip

You do not need to use a table in your answer, but it can help to lay out your answer to distinguish between DNA and RNA. If the question is worth 3 marks you need to write at least three differences.

Directionality of RNA and DNA

The 5' end of a nucleic acid corresponds to the phosphate joined at carbon 5 of the ribose or deoxyribose.

During replication, DNA polymerase can only add nucleotides in the 5' to 3' direction. This means it adds a new nucleotide by attaching it to the 3' end of the growing DNA. In transcription, RNA polymerase adds the 5' end of the free RNA nucleotide to the 3' end of the growing mRNA molecule. In translation, the nucleotide of the first codon to be read (start codon) is close to the 5' end of the mRNA.

You will study the processes of replication, transcription and translation in detail in subtopics D1.1 and D1.2.

Example 7

The percentages of each nitrogenous base in the DNA and mRNA in a neuron of the giant octopus (*Enteroctopus dofleini*) are given in the table.

	Bases / %				
	Purines		Pyrimidines		
	Guanine	Adenine	Cytosine	Thymine	Uracil
DNA	20	30		32	0
mRNA	30	35	16	0	19

a) Determine the percentage of cytosine in the DNA of the neuron of the giant octopus.
b) Explain the percentages of bases found in DNA and mRNA.
c) State the location of DNA and mRNA in the neurons of the giant octopus.

Solution

a) $100 - (20 + 30 + 32 + 0) = 18\%$
b) In DNA the percentages of adenine and thymine are very similar (30% or 32%). The same is true for guanine and cytosine (20% and 18%). However, in RNA all bases are found in different percentages. This is because DNA forms a double strand of complementary base pairs between C and G, and between A and T, and therefore these pairs must have similar values (slight differences may be due to experimental inaccuracies). In RNA, as they are in single strands, the amounts of each nucleotide are variable.
c) DNA is found in the nucleus of the cell and in mitochondria. The mRNA is found in the nucleus, in the cytoplasm, attached to ribosomes, in rough endoplasmic reticulum and in mitochondria.

- Adenine and guanine are **purines** (larger nucleotides) and the rest of the nucleotides (cytosine, thymine and uracil) are **pyrimidines**.

Nature of science

Rosalind Franklin's X-ray diffraction provided crucial evidence that DNA is a double helix. Rosalind Franklin and Maurice Wilkins used a high-resolution X-ray camera to obtain very clear images of diffraction patterns from DNA. They were able to deduce that the DNA molecule was helical in shape. This helped James Watson and Francis Crick to elucidate the structure of DNA.

DNA structure and replication

DNA consists of a double strand of nucleotides in the form of a helix. Purines (A and G), which are larger, bind to pyrimidines (C and T), which are smaller, so the complementary pairs have equal length. Consequently, the DNA helix has the same three-dimensional structure, regardless of the base sequence.

In eukaryotes, DNA is associated with histone proteins forming nucleosomes. This allows the DNA to be condensed in the nucleus of the cell. A nucleosome consists of a core of four pairs of histone proteins with around 150 base pairs of DNA coiled around the proteins. A short section of naked DNA connects one nucleosome to the next. An additional histone protein molecule binds the DNA to the core particle.

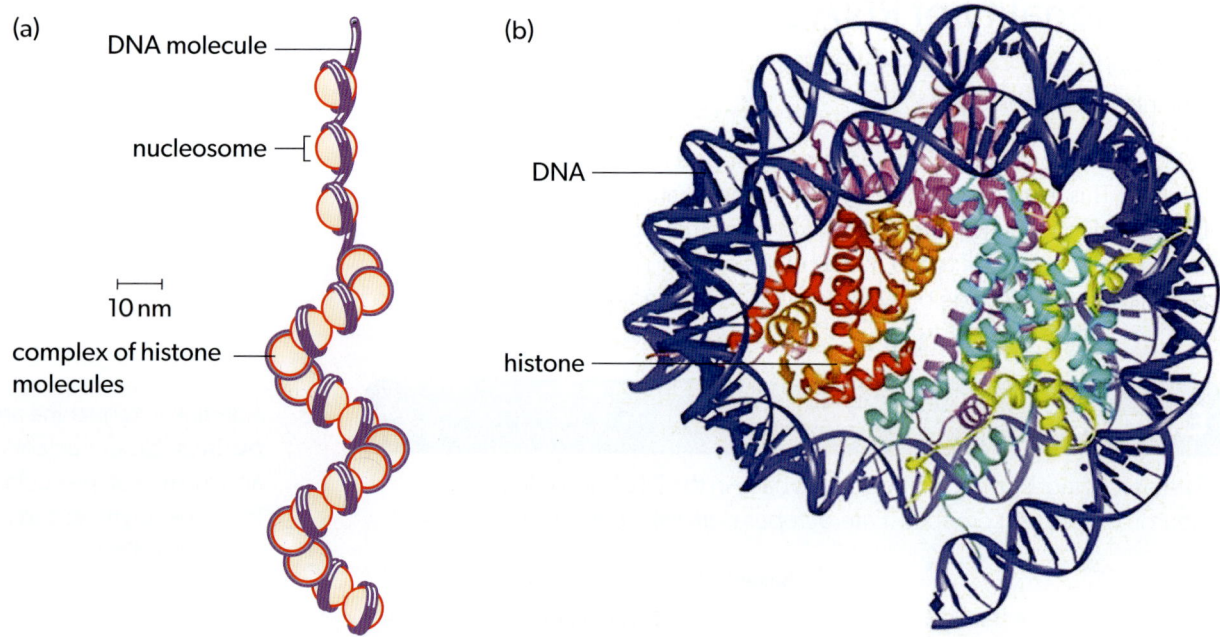

▲ Figure 5 (a) Structure of human nucleosome. (b) Detail of one nucleosome as seen using molecular visualization software

Alfred Hershey and Martha Chase devised an experiment that provided evidence that DNA is the genetic material and not proteins as it was believed at the time. In their experiment, they cultured viruses (**bacteriophages**) that contained proteins with radioactive (^{35}S) sulfur and they separately cultured viruses that contained DNA with radioactive (^{32}P) phosphorus. They infected bacteria separately with the two types of viruses. They separated the non-genetic component and measured the radioactivity in the pellet (where the infected bacteria were) and the supernatant (where the coats were). They found that most radioactivity was found in the infected bacteria when using radiolabelled DNA.

In subtopic B2.1, you will study the structure of channel proteins.

- **Bacteriophages** are viruses that create a channel across the bacterial cell membrane and wall to transfer the viral genome DNA to the host cell cytoplasm, initiating infection in the bacterium.

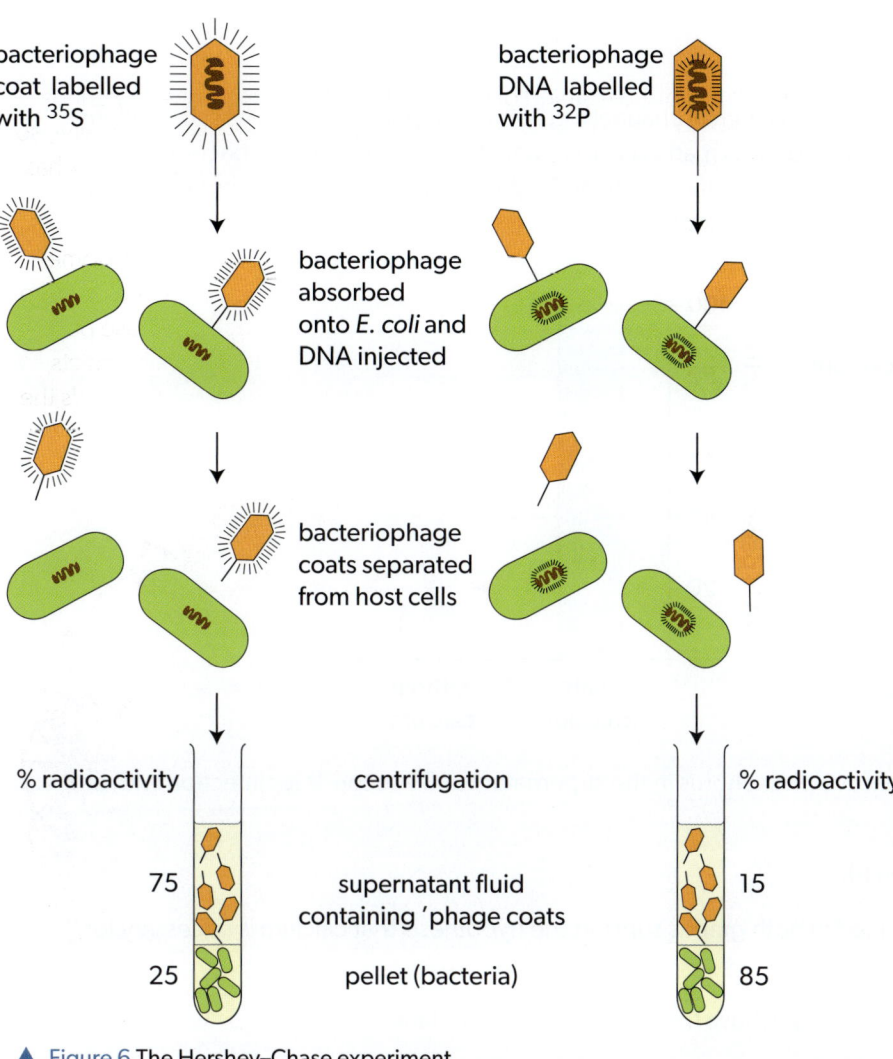

▲ Figure 6 The Hershey–Chase experiment

Nature of science

The Hershey–Chase experiment only became possible when radioisotopes were made available to scientists as research tools.

Example 8

The table shows the ratio of nucleotides in the DNA of different organisms.

Source	A to G	A to T	C to T	C to G
human	1.29	1.04	0.57	1.00
salmon	1.43	1.02	0.70	0.98
wheat	1.22	1.00	0.85	1.03
E. coli	1.05	1.09	1.05	1.01

a) State the two nucleotide ratios that are closest to 1 for most organisms.

b) Discuss whether this data supports the hypothesis that adenine (A) always pairs with thymine (T), and cytosine (C) pairs with guanine (G) by complementary base pairing in the DNA structure.

Solution

a) A to T and C to G.

b) DNA is a double-stranded helix. Each base in one strand is paired with its complementary base in the second strand—there are no unpaired nucleotides. The ratios A to G and C to T are not close to 1, whereas A to T and C to G are. This means that the total amount of A in the DNA of these organisms is similar to the total amount of T. The same happens with C and G. This supports the idea that they are joining by complementary base pairing.

Nature of science

Chargaff's data on the relative amounts of pyrimidine and purine bases across diverse life forms falsified the tetranucleotide hypothesis that there was a repeating sequence of the four bases in DNA.

Unity and diversity

Practice problems

Scientists wanted to study the importance of calcium in channelling of bacteriophage DNA to host bacterium *Bacillus subtilis*. The DNA of the bacteriophage was marked with radioactive phosphorus and bacteria were infected with and without calcium present. The amount of radioactivity was recorded in the supernatant (left) and the percentage of infected bacteria calculated (right).

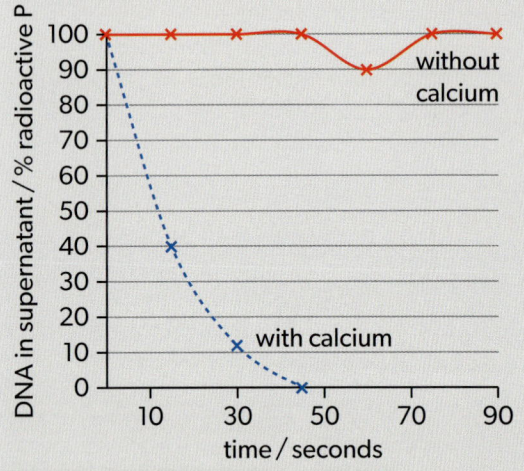
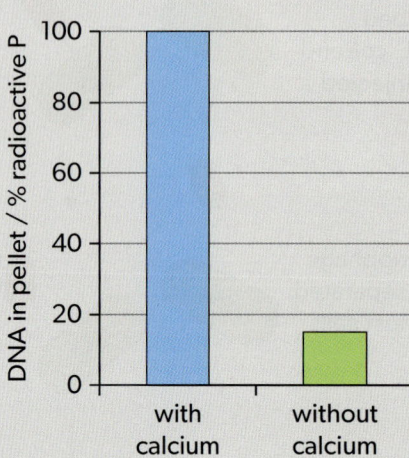

a) (i) State the percentage of radioactive phosphorus in the supernatant at 30 seconds for infection with and without calcium.

(ii) Explain the results obtained in **(i)**.

b) Suggest whether the results obtained in both graphs support the hypothesis that calcium is necessary for infection.

c) Discuss the results in terms of the Hershey–Chase theory of genetic inheritance.

- What makes RNA more likely to have been the first genetic material, rather than DNA?
- How can polymerization result in emergent properties?

A2.1 Origins of cells

Additional higher level:

You should know:

✔ the conditions on early Earth.
✔ carbon compounds formed spontaneously by chemical processes.
✔ cells are the smallest unit of life.
✔ cells can be formed only by division of pre-existing cells.
✔ the first cells must have arisen from non-living material.
✔ RNA is presumed to be the first genetic material.

You should be able to:

✔ analyse evidence that spontaneous generation of cells and organisms does not now occur on Earth.
✔ evaluate data from Miller and Urey's experiment.
✔ describe how vesicles are formed spontaneously by coalescence of fatty acids into spherical bilayers.
✔ explain the concept of the last universal common ancestor (LUCA).
✔ analyse evidence for LUCA.

A2.1 Origins of cells

Conditions on early Earth

Scientists think that the conditions on early Earth (in its first 1 billion years) included lack of free oxygen and ozone, which resulted in high temperatures and high ultraviolet light penetration. These conditions may have caused a variety of carbon compounds to form spontaneously by chemical processes.

Cells as the smallest units of self-sustaining life

Cells are the smallest units of life that can survive on their own. All living organisms are formed by cells. Viruses resemble living organisms as they have genetic material and can reproduce, but they are not formed by cells so are not considered to be living.

1. Pasteur poured nutrients into two flasks like this.

Spontaneous origin of cells

In the past, many people believed that living things could spring up from non-living materials. Their belief was based on observations they had made, such as maggots appearing in rotten meat. This idea was called spontaneous generation. Pasteur's experiments with broth in swan-necked flasks were carried out to prove whether microbes could be spontaneously generated or whether they could come only from pre-existing cells.

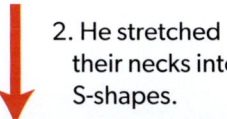

2. He stretched their necks into S-shapes.

Louis Pasteur investigated how broths turned bad in the following way. Since he knew that excess heat could kill living things, he boiled some broth in flasks to kill anything that might be living in it at the start. He then heated the necks of the glass flasks until they were soft, and pulled them out into a long, thin, curving tube called a swan-neck. The broths in the flasks did not go bad. Then Pasteur broke open one of the flasks and exposed the broth to the open air and this time he noticed that the broth did go bad. This made Pasteur conclude that there was something in the air that was affecting the broth. In the swan-necked flasks where the broth was clear, whatever was affecting the broth might have settled in the bend of the neck and therefore not reached the broth. To test his idea, he tipped a swan-necked flask so that some of the broth went into the bend where dust and other particles may have collected, and then he tipped it back again. The broth in this second flask then went bad. Pasteur concluded that whatever was causing the change could be carried by air currents, but it must be heavier than air as it settled in the bend in the swan-neck.

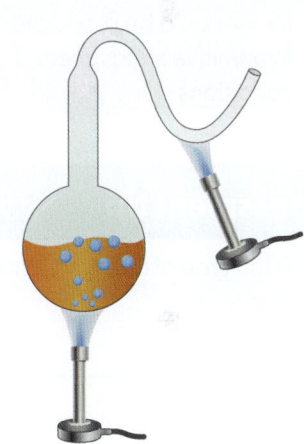

3. He boiled the nutrients.

▲ Figure 1 Pasteur's experiments

Example 1

What theory did Pasteur falsify with his experiments?

A. Independent assortment
B. Spontaneous generation
C. Endosymbiosis
D. Evolution

Solution

The correct answer is **B**, as with his experiments Pasteur showed that microorganisms did not appear spontaneously, as they could not grow in a broth unless dust particles (covered in microorganisms) were allowed into the flask. Independent assortment is related to meiosis and not to Pasteur's experiments. Endosymbiosis is the theory that explains the formation of mitochondria and chloroplasts. Evolution is the theory described by Darwin.

Nature of science

Hypotheses and theories need to be testable, but in some cases, this is very difficult. The exact conditions on pre-biotic Earth are not well known and there are no fossils of the first protocells.

Unity and diversity

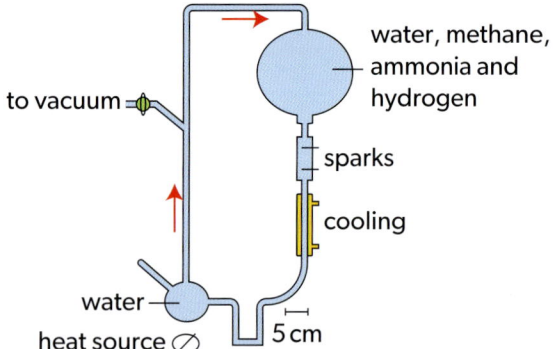

▲ Figure 2 The production of amino acids under possible primitive Earth conditions

Source: Miller, S. L. (1953). A Production of Amino Acids Under Possible Primitive Earth Conditions. Science, 117(3046), 528–529. www.doi.org/10.1126/science.117.3046.528

- **Miller and Urey** found that amino acids and other carbon compounds needed for life could be produced in primitive atmosphere conditions.

Assessment tip

This is the type of question found in paper 1B.

You will study separation by chromatography in subtopic C1.3 "Chromatography to separate photosynthetic pigments".

Evidence for the origin of carbon compounds

Stanley Miller and Harold Urey carried out experiments to show how the first cells might have arisen from non-living material. They passed steam through a mixture of gases representing the early Earth atmosphere. Electrical discharges were used to simulate lightning. They found that amino acids and other carbon compounds needed for life were produced.

Example 2

In 1953, Stanley Miller wanted to test whether organic compounds were formed when the Earth had an atmosphere of methane, ammonia, water and hydrogen, instead of the current atmosphere containing nitrogen, oxygen, carbon dioxide and water.

He designed an apparatus to circulate these gases past an electric discharge. He tested the resulting mixture for amino acids using two-dimensional paper chromatography. He first ran it in butanol-acetic acid and then turned this result in a 90° direction and ran it again but in phenol.

Miller then used the R_f value (ratio of the distance travelled by the amino acid to the distance travelled by the solvent front) to identify the amino acids.

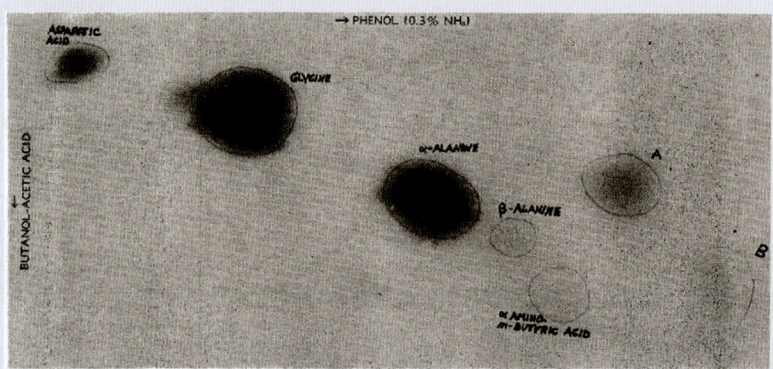

Source: Miller, S. L. (1953). 'A production of amino acids under possible primitive Earth conditions'. Science, 117(3046), 528–529. www.doi.org/10.1126/science.117.3046.528

a) State the use of the electrical discharges in this experiment.
b) Deduce, with a reason, whether glycine or aspartic acid had a larger R_f.
c) Justify whether α-alanine or β-alanine is more soluble in the solvents used in this chromatography.
d) Comment on Miller's results on the production of organic compounds in the primitive Earth atmosphere.

Solution

a) Simulated lightning.
b) Glycine had a larger R_f because it moved farther from the origin.
c) β-alanine is more soluble, as it moved farther from the origin. It will have a larger R_f.
d) Miller proved that organic compounds could be formed in a reducing atmosphere of methane, ammonia, water and hydrogen after the striking of lightning (simulated using sparks). Amino acids are organic compounds, and they could be identified in the paper chromatography.

Formation of vesicles and RNA as the first genetic material

A protocell is a spherical collection of lipids proposed as a stepping-stone to the origin of life. Compartmentalization of primitive biochemical reactions within membrane-bound water vesicles is considered an essential step in the origin of life. Spontaneous self-organization of lipid and fatty acid molecules into spherical bilayer vesicles by **coalescence** is a simple form of membrane-based compartmentalization. The assembly of vesicles was crucial for primitive cells to form. These vesicles produced boundaries or compartments, separating the internal part from the exterior. This allowed molecules such as primitive RNA (thought to be the first genetic material) to be retained, as well as isolating chemical reactions and energy storage.

RNA may have acted initially as both the genetic material (can be replicated) and the enzymes (has some catalytic activity) of the earliest cells. Ribozymes are RNA molecules that catalyse peptide bond formation during protein synthesis in the ribosome.

- **Coalescence** is the process by which two or more particles merge during contact to form a single particle.

Evidence for a last universal common ancestor

There is little evidence of the origin of life. Fossilized bacteria which resemble modern bacteria have been found in rocks in hydrothermal vents in northern Canada. In Greenland, graphite carbon isotopes were recorded in sedimentary rocks probably derived from mineralization associated with fossilized bacteria. Life was found in iron-containing sandstone in Australia, mainly in sedimentary formations produced by the build-up of photosynthetic microorganisms, such as cyanobacteria and sulfate-reducing bacteria. This provided evidence of some of the most ancient life. Sulfur isotope data suggest that sulfur-metabolizing bacteria existed almost 3.5×10^9 years ago, leading to suggestions that the earliest microbial ecosystems were sulfur based. The microstructures identified showed indicators of organization of multiple cells, giving origin to the **last universal common ancestor (LUCA)**, which is believed to have lived nearly 4.2×10^9 years ago.

- The **last universal common ancestor (LUCA)** is the most recent population of organisms from which all actual living organisms descend.

In genomic analysis, information is obtained by comparing sequences of nucleic acids (DNA and RNA) or proteins. Conserved sequences are identical or similar sequences across species or within a genome. Conservation indicates that the sequence has been conserved through natural selection and is unchanged since far back in geological time. As there is no fossil evidence for LUCA, it is studied by comparing the genetic sequences of hundreds of genes. The universal genetic code and shared genes across all organisms provide evidence that other forms of life evolved but became extinct due to competition from LUCA and its descendants.

- For what reasons is heredity an essential feature of living things?
- What is needed for structures to be able to evolve by natural selection?

Unity and diversity

A2.2 Cell structure

You should know:
- cells are the basic structural unit of all living organisms.
- unicellular organisms consist of only one cell that carries out all functions of life in that organism.
- the advantages of techniques in microscopy.
- the structures found in typical cells in living organisms.
- the cell components found in prokaryotes (unicellular organisms) and eukaryotes (multicellular organisms).
- the life processes of unicellular organisms include homeostasis, metabolism, nutrition, movement, excretion, growth, response to stimuli and reproduction.
- the differences in cell structure between animals, fungi and plants.
- examples of atypical cell structure in terms of numbers of nuclei.

Additional higher level:
- multicellular organisms have properties that emerge from the interaction of their cellular components.
- interactions between cells in groups have given them emergent properties that have favoured the evolution of multicellular organisms.

You should be able to:
- make temporary mounts of cells and tissues, stain them, measure size using an eyepiece graticule, take photographs and add a scale bar to the image.
- calculate the magnification and actual size of structures and ultrastructures shown in drawings or micrographs.
- identify cells in light and electron micrographs as prokaryote, plant or animal.
- identify in electron micrographs the nucleoid region, prokaryotic cell wall, nucleus, mitochondrion, chloroplast, sap vacuole, Golgi apparatus, rough and smooth endoplasmic reticulum, chromosomes, ribosomes, cell wall, plasma membrane and microvilli.
- draw and annotate diagrams, based on electron micrographs, of organelles (nucleus, mitochondria, chloroplasts, sap vacuole, Golgi apparatus, rough and smooth endoplasmic reticulum, and chromosomes) and other cell structures (cell wall, plasma membrane, secretory vesicles and microvilli).

Additional higher level:
- analyse the evidence for the endosymbiotic theory to explain the origin of eukaryotic cells.
- describe differentiation as the process for developing specialized tissues in multicellular organisms, driven by patterns of gene expression.

Nature of science

Scientists use deductive reasoning to make predictions from theories. Based on the cell theory, scientists can predict whether a newly discovered organism consists of one or more cells.

Cell theory

The cell theory states that all living organisms are formed of **cells**.

- **Cells** are the basic structural unit of all living organisms.

Microscopy

Example 3

The images of the radiolarian, a single-celled marine organism, were produced using a light microscope (left) and a scanning electron microscope (right). Use the images to answer the questions.

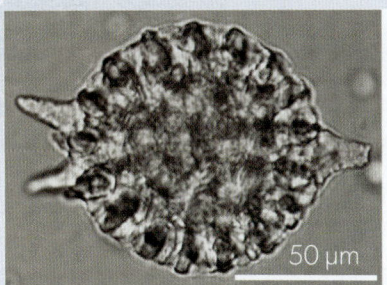 

Source: Averill B. and Eldredge P. 2011. Principles of General Chemistry. Chapter 6.4 figure 6.17. Saylor Academy (CC BY-NC-SA 3.0)

a) What is a reason for the difference in quality of these images?
   A. Light cannot pass through the specimen.
   B. Higher magnification can be achieved with the electron microscope.
   C. The resolution of the electron microscope is higher.
   D. Samples are stained with methylene blue when viewed with the light microscope.

b) The diameter of the radiolarian is 100 µm. What is the magnification of the scanning electron micrograph?

Solution

a) The correct answer is **C**, as an image as small as 50 µm can only be seen with a higher resolution under an electron microscope. Light is only required under a light microscope. The magnification is the same, as the bar on the bottom right shows the same size. Methylene blue does modify the cells slightly, but not enough to explain the differences in the image.

b) The magnification is the size of the image divided by the actual size of the organism.

$$\frac{3\,cm}{100\,\mu m} = \frac{30{,}000\,\mu m}{100\,\mu m} = \times 300$$

Developments in microscopy led to a greater understanding of cell structure. Electron microscopes have a greater power of **magnification** and a greater **resolution** than light microscopes. The maximum magnification of a light microscope is usually lower than ×2,000 and the maximum resolution is 0.2 µm. Beams of electrons have a much shorter wavelength compared with light waves, so electron microscopes have a much higher resolution. The maximum magnification of modern electron microscopes is around ×10,000,000 and the maximum resolution is less than 0.0001 µm.

Freeze fracture is an electron microscope technique that uses a fractured frozen biological sample, therefore exposing structural details. This technique has allowed scientists to decipher the structure of the cell membrane. Cryogenic electron microscopy involves the freezing of samples. It allows scientists to see the structure of proteins without the need for crystallization. Researchers use fluorescent dyes in light microscopy to visualize cell structures. Atoms absorb the light from the microscope at one wavelength and then emit it at a longer wavelength, creating a bright excitation light. However, fluorescent dyes cannot penetrate deeply enough and can sometimes be unspecific. Furthermore, they are damaged by prolonged exposure to light, leading to loss of fluorescence signal over time.

Nature of science

Using instruments to make measurements is a form of quantitative observation.

- **Magnification** is how much an image has been enlarged.
- **Resolution** is the minimum distance at which two points that are close together can be distinguished.

Unity and diversity

Example 4

The micrograph shows onion epidermal cells seen under the light microscope with a magnification of ×400.

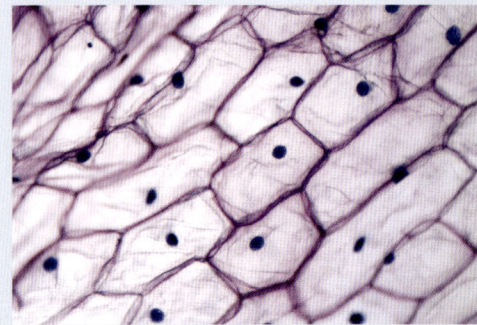

a) (i) Label the nucleus of one cell.
   (ii) Calculate the actual width of one of the cells. Show your working.
b) Suggest how the surface area-to-volume ratio of a cell can affect its function.

Solution

a) (i) Any dark circle labelled.
   (ii) The magnification is the size of the image divided by the actual size of the cell. Therefore the formula for actual size is:

   $$\text{actual size of cell} = \frac{\text{size of image}}{\text{magnification}}$$

   First, use a ruler to measure the width of the chosen cell (size of image).

   For example, measured cell width = 10 mm

   $$\text{actual width of cell} = \frac{10\,\text{mm}}{400}$$

   $$= 0.025\,\text{mm}$$

b) If the ratio is too small the exchange of substances will be too slow, waste substances will accumulate and heat will not be lost efficiently.

Cell structure

The cells of all living organisms have DNA as their genetic material and a cytoplasm composed mainly of water, which is enclosed by a plasma membrane made of lipids.

Prokaryotic cells do not have a nucleus or membrane-bound organelles. Instead, their nuclear material is found in the nucleoid or nuclear region and their DNA is naked, not bound to proteins. Prokaryotic cells have a cell wall, a plasma membrane, cytoplasm, DNA in a loop and 70S **ribosomes** (according to their sedimentation coefficients, in Svedberg units).

Example 5

The diagram shows a prokaryotic cell.

What are the structures labelled Y and Z?

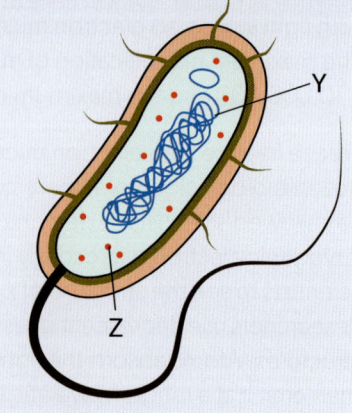

	Y	Z
A.	nucleus	70S ribosome
B.	nucleoid	80S ribosome
C.	nucleus	80S ribosome
D.	nucleoid	70S ribosome

Solution

The correct answer is **D**, as only eukaryotes have a nucleus and 80S ribosomes.

Eukaryotic cells have a plasma membrane enclosing a compartmentalized cytoplasm that contains 80S ribosomes and other membrane-bound organelles, such as mitochondria, endoplasmic reticulum, Golgi apparatus and a variety of vesicles or vacuoles including lysosomes. The cytoplasm also contains a cytoskeleton of microtubules and microfilaments. The nucleus in eukaryotic cells is surrounded by a double membrane with pores and contains DNA bound to histones. Some eukaryotic cells—such as plant cells—have a cell wall. Plant cells also contain **chloroplasts**, which are not present in animal cells.

- **Ribosomes** are 70S in prokaryotes and 80S in eukaryotes.
- **Chloroplasts** and large **vacuoles** are only found in plant cells.

Life processes in unicellular organisms

Unicellular organisms are formed of only one cell that performs all the functions of life (homeostasis, metabolism, nutrition, movement, excretion, growth, response to stimuli and reproduction).

Comparing eukaryotic cell structure in animals, fungi and plants

Example 6

Compare and contrast animal, plant and fungi cells using a table.

Solution

Characteristic	Type of eukaryotic cell		
	Animal	Plant	Fungus
cell wall	absent	made of cellulose	made of chitin, glucans or glycoproteins
vacuole	small	large	small
plastid	absent	chloroplast	absent
centriole	present	absent	absent
cilia/flagella	present in some cells	present in some cells	absent

Atypical cell structure in eukaryotes

There are some atypical cells. For example, in aseptate fungal hyphae and skeletal muscle there are no divisions between cells, so cells are multinucleate. However, in some atypical cells there is no nucleus, to leave space for transport. Examples are red blood cells, which carry oxygen, and phloem sieve tube elements, which carry nutrients.

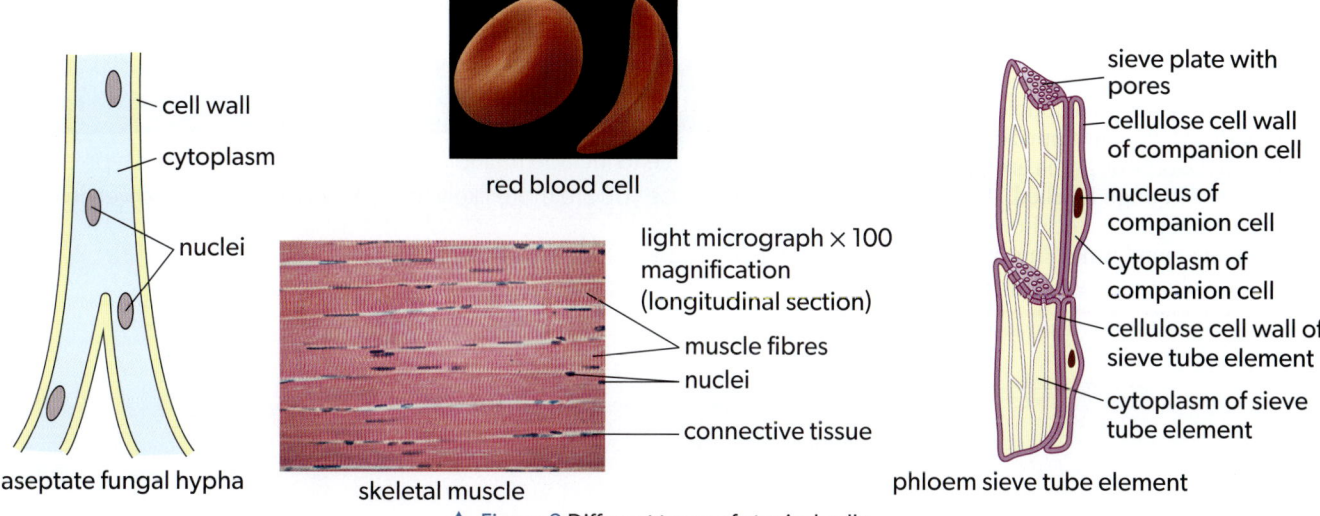

▲ Figure 3 Different types of atypical cells

Unity and diversity

Cell structures viewed under a microscope

Example 7

What can be identified in the electron micrograph?

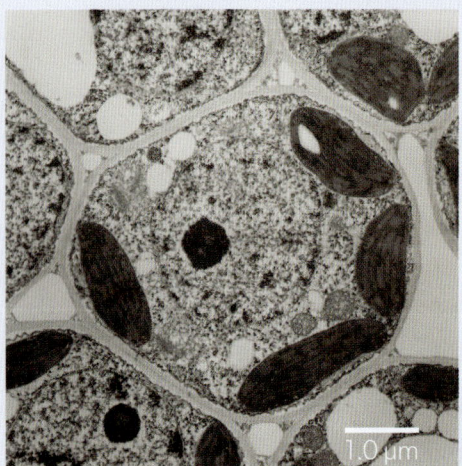

A. Plant cells with cell walls, nuclei and chloroplasts
B. Plant cells with chloroplasts and large central vacuoles containing nuclei
C. Several prokaryotes with cell walls and cytoplasm
D. Animal cells with prominent mitochondria

Solution

The correct answer is **A**. It is clearly a plant cell as the shape is well defined due to the presence of a cell wall. Although large vacuoles can be seen, these do not contain nuclei. Mitochondria can also be seen. They are smaller than the chloroplasts and rounder in shape (for example, there is one on the left of the vacuole next to the magnification bar).

Drawing from electron micrographs

Sample student answer

Draw a labelled diagram of a eukaryotic plant cell as seen in an electron micrograph. [4]

This answer could have achieved 2/4 marks:

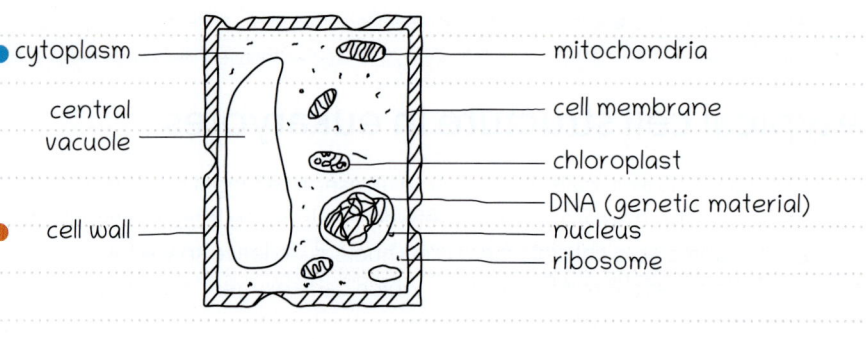

▲ The first mark is given because the cell membrane is shown as a single continuous line, as an inner line of the cell wall and is clearly labelled. The second mark is for the vacuole shown as a single continuous line.

▼ Although the cell wall is shown, it is not continuous and does not have two lines to indicate the thickness (the second line is shown as the membrane). The nuclear membrane is not shown with a double membrane with nuclear pores, therefore did not score a mark. The chloroplast is not shown with a double line to indicate the envelope and the thylakoids or grana are not shown. Although labelled, the mitochondrion is not shown with a double membrane. The cristae are not clear enough for a mark.

Examiner tip

It is good practice to include multiple labels when answering this type of question because you are more likely to identify all the answers given in the mark scheme.

Theory of endosymbiosis

The **endosymbiotic** theory explains the evolution of eukaryotic cells from prokaryotic cells. All eukaryotes evolved from a **common unicellular ancestor** that had a nucleus and reproduced sexually. Mitochondria are thought to be aerobic prokaryotes that were engulfed by other prokaryotes and remained inside the cells. Likewise, chloroplasts are thought to have been photosynthetic prokaryotes engulfed by other prokaryotes. Both these organelles maintain many of the features of prokaryotic cells, such as their own naked circular DNA and 70S ribosomes, and they also have double membranes produced by the endocytic mechanism. They are both still able to replicate.

> **Nature of science**
>
> The strength of a theory comes from the observations the theory explains and the predictions it supports. The theory of endosymbiosis is supported by a wide variety of evidence.

- An **endosymbiont** is one organism that lives inside another.
- Evidence for **endosymbiosis** in the evolution of chloroplasts and mitochondria includes the presence of 70S ribosomes, naked circular DNA and the ability to replicate.
- All eukaryotes evolved from a **common unicellular ancestor** that had a nucleus and reproduced sexually.

Differentiation by patterns of gene expression

Multicellular organisms are composed of many cells that become specialized by **differentiation**, forming different tissues. These tissues form organs which together make up organ systems. To differentiate, cells must express different genes and therefore produce different proteins. This is often triggered by changes in the environment. All cells in an organism have the same genetic material, but if some genes are expressed and others are not, the resulting cells will be different.

- Cell **differentiation** is the process for developing specialized tissues in multicellular organisms.

Evolution of multicellularity

Unicellular organisms evolved methods of cell–cell adhesion that allowed them to remain together after cell division. Cells in these aggregates specialized due to selective pressures, resulting in the evolution of differentiated tissues. Much of the diversity of multicellular organisms lies in how cells are arranged and epigenetic modification of patterns of gene expression. Multicellular organisms can partition complementary tasks among different cells, increasing division of labour and metabolic cooperation. This allows for larger body size and cell specialization. Many eukaryotic algae and fungi, and all plants and animals, are multicellular.

In subtopic D2.2 "Epigenetic changes affect phenotype not genotype" you will study the importance of epigenetics in the development of patterns of differentiation in the cells of a multicellular organism.

- What explains the use of certain molecular building blocks in all living cells?
- What are the features of a compelling theory?
- Describe one theory for the origin of life on Earth.

A2.3 Viruses

Additional higher level:

You should know:
- some viruses are enveloped in host cell membrane.
- viruses are highly diverse in their shape and structure.
- bacteriophages have a lytic and a lysogenic life cycle.
- viruses share an extreme form of obligate parasitism as a mode of existence.
- the genetic code is shared between viruses and living organisms.
- viruses evolve rapidly.

You should be able to:
- describe the structural features common to all viruses—small, fixed size; double- or single-stranded DNA or RNA as genetic material; a protein capsid; no cytoplasm; few or no enzymes.
- identify viral structures from electron micrographs or diagrams.
- identify stages in bacteriophage life cycles.
- explain theories of the origin of viruses.

Structure of viruses

Viruses can cause diseases, but are not considered living organisms because they can only multiply inside an infected cell. Viruses rely on a host cell for energy supply, nutrition, protein synthesis and other life functions.

Viruses have a segment of genetic material surrounded by a protein cover or capsid. Viruses do not have a cytoplasm as they are not formed by cells, and they have very few or no enzymes. Viruses must use the components of the host cell to replicate their own genetic material. Eventually the infecting virus might destroy the host cell, causing damage to the organism.

Some examples of diseases caused by viruses infecting humans are chicken pox caused by the varicella zoster virus, acquired immune deficiency syndrome (AIDS) caused by human immunodeficiency virus (HIV) and coronavirus disease 2019 (COVID-19) caused by severe acute respiratory syndrome coronavirus 2 (SARS-CoV-2).

There are millions of different types of viruses. These include viruses with different shapes, that can contain single-stranded DNA, double-stranded DNA or RNA. Viruses are usually small (between 10 and 300 nanometres in diameter)—approximately 100 times smaller than bacteria—and have a fixed size. In some cases, the virus contains glycoproteins exposed on the outer surface, forming spikes or an outer lipid layer. Some viruses are surrounded by host cell membrane and others are non-enveloped.

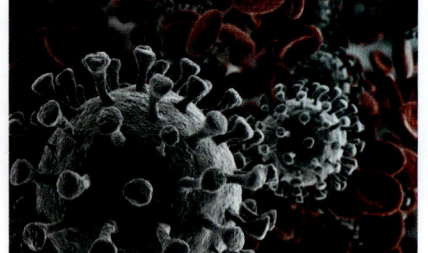

▲ Figure 4 **COVID-19 virus**

Lytic and lysogenic cycles of a virus

Bacteriophage lambda is a bacteria-infecting virus with a complex structure. The head contains double-stranded DNA associated with proteins. The tail is a tube through which the DNA is injected into the host cell. The tail ends in a hexagonal plate with spikes in each angle connected to six long, thin fibres that help the bacteriophage to adhere to the host (**Figure 5**).

Once the viral DNA is injected into the host, it starts to multiply, eventually destroying the host cell to be liberated. This is called the **lytic cycle**.

As well as the lytic cycle, bacteriophages can also have a **lysogenic cycle**. In this case, the DNA of the virus persists inside the host cell without the production of new bacteriophages. When the bacterium reproduces, the viral DNA is also passed on to the offspring in the bacterial chromosome. When activated, the dormant viral DNA in lysogenic phase can start multiplying, generating mature bacteriophages.

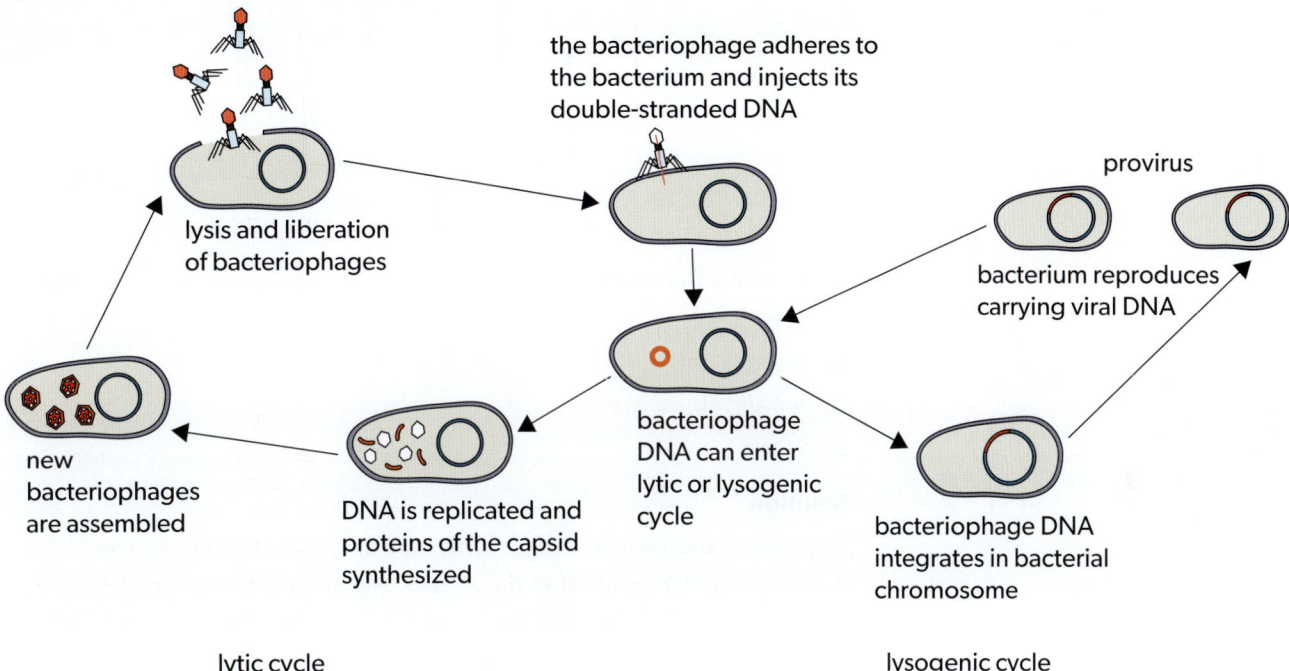

▲ Figure 5 Lytic and lysogenic cycles in bacteriophage lambda

- In the **lytic cycle**, the host cell (or bacterium) is destroyed after infection.
- In the **lysogenic cycle**, viral DNA integrates into the host DNA without destroying the host.

Origins of viruses from other organisms

The genetic code is shared between viruses and living organisms, so the viral nucleic acid uses the host's cellular machinery to synthesize its own enzymes and structural proteins. Viral genomes can code for between 4 and over 100 proteins. Mutations in viral DNA can cause a change in the amino acid sequence of a protein or in some cases not affect the protein at all (silent mutations). Mutations can accumulate in viral genomes and undergo natural selection. Repetitive selection of the favoured traits has led to viruses sharing common structural features, for example, all are extreme obligate parasites. This adaptive evolution can be regarded as convergent evolution because they have similar characteristics although they generated from a different ancestor.

 You will study convergent evolution in subtopic A4.1.

Example 8

The diagram shows the changes at a single nitrogenous base on a gene tree in viruses.

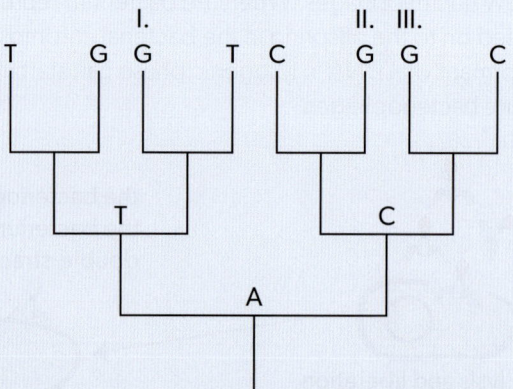

Which of the three labelled nitrogenous bases show convergent evolution?

A. I and II only
B. II and III only
C. I, II and III
D. None

Solution

The correct answer is **A**, as both I and II are analogous as they evolved from a different ancestor (T and C, respectively). II and III are from the same ancestor (C). I and III would also be a correct combination (but is not given as an answer).

There are several hypotheses for the origin of viruses. One hypothesis is that viruses appeared before cells, by self-assembly of nucleic acids and proteins in the pre-biotic Earth. Another hypothesis is that small parasitic cells lost most of their genes and eventually could only reproduce in hosts. The third hypothesis is that viruses evolved from fragments of DNA or RNA that escaped from the genes of their hosts. Given that there are no viral fossils, these hypotheses are difficult to prove. Nevertheless, there are viral sequences integrated into the genomes of organisms that could be used to study the evolution of viruses. For example, the presence of viral sequences in insect fossils suggests that viruses already infected insects more than 3×10^8 years ago.

Rapid evolution in viruses

The genomic sequences of viruses that are highly mutable and cause chronic infection tend to evolve along a distinct path over time (diverging evolution). However, in the absence of immune selective pressure, there is reversion towards ancestral characteristics in those strains that do not have the mutation (converging evolution).

Example 9

The HIV virus has a high rate of mutation. Scientists studied the evolution of this virus in a small group of men with a slow rate of disease progression for a period of 12 years. The viral divergence from the original viral population and the diversity within the populations were recorded over the years.

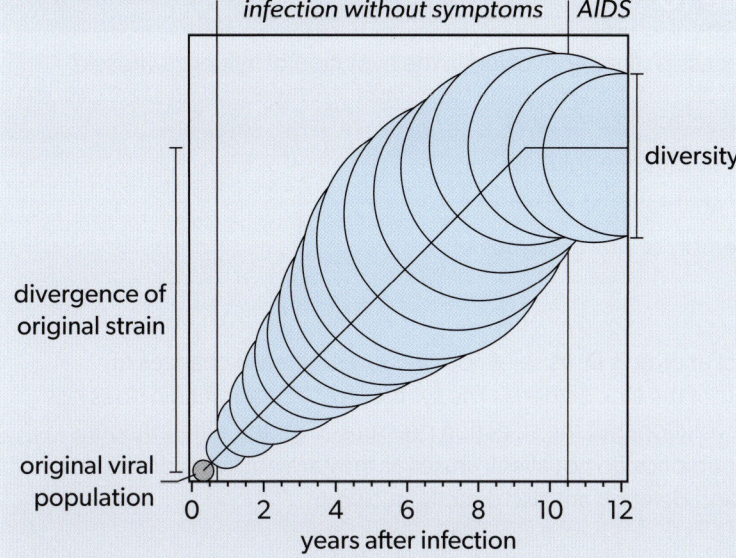

Source: Shankarappa, R., et al. (1999). Consistent viral evolutionary changes associated with the progression of human immunodeficiency virus type 1 infection. *Journal of Virology*, 73(12), 10489–10502. www.doi.org/10.1128/jvi.73.12.10489-10502.1999.

a) (i) Identify the pattern of divergence of the viral sequence from the original viral population.

   (ii) Identify the change in the pattern of diversity of the populations.

b) Estimate the percentage increase of the population diversity from the stage of the initial infection to 12 years later.

c) Using the data provided, suggest how the change in divergence and diversity of viral RNA may be of evolutionary benefit to the HIV virus.

Solution

a) (i) The divergence increases for the first 9 years and then levels off.

   (ii) The diversity increases (greatly) up to about 6 to 7 years and then decreases.

b) This is calculated as the difference between the diameter at 12 years and the initial diameter, divided by the initial diameter and multiplied by 100. The initial diameter is 7 mm and the diameter at 12 years is 24 mm.

$$\text{percentage increase in population} = \frac{24 - 7}{7} \times 100 = 243\%$$

c) During the first years there is a very rapid increase in both divergence of viral RNA and diversity. A greater divergence leads to greater diversity from the initial strain. With the increased diversity from the initial strain there is more chance of survival, which makes it more difficult for the infected person to produce new T-lymphocytes, making the reaction of the immune system weaker. As the most successful forms of HIV are naturally selected, after 6.5 years of natural selection only the most adapted strains survive.

The influenza A virus is a flu-causing pathogen in birds and some mammals, including humans. It is divided into subtypes depending on two surface proteins, hemagglutinin (HA) and neuraminidase (NA). Mutations in the HA and NA genes happen continually over time as influenza viruses replicate. Furthermore, the reassortment of HA and NA from different strains occurs during assembly. The host cell then forms new viruses that combine their surface proteins, for example H3N2 and H5N1 can form H5N2. In 2022, the most common subtypes in humans were H1N1 and H3N2. As this can change over time, vaccines are designed to target one or more of the surface proteins of influenza viruses.

Something similar has happened with the virus causing AIDS. While HIV has an extremely fast rate of evolution, the virus must have circulated within human populations for many years before it was first recognized for the extent of diversity that accumulated.

Example 10

Which process is directly involved in the evolution of influenza viruses?

A. Lack of selection pressures
B. Synthesis of hemagglutinin
C. Use of antibiotics to kill viruses
D. Reassembly of viral genomes

Solution

The correct answer is **D**, as selection pressures increase chances of evolution (not the lack of them). The synthesis of hemagglutinin does not produce evolution, but the selection pressures over the different forms of it do. Antibiotics do not affect viruses as they are not living organisms. Mutation and deletion are also directly involved.

> In subtopic A3.2 "Clades" you will study clades as groups of organisms with common ancestry and shared characteristics.

Practice problems

1. Describe how the structure of a cell can be observed using different fluorescent stains.

2. Scientists studied viruses infecting archaea from extreme thermal acidic environments (70–92 °C, pH 1.0–4.5) found in three sites of Yellowstone National Park, USA. They extracted total DNA and amplified the coat protein gene by polymerase chain reaction (PCR). They compared the amino acid sequences produced with those of viruses found in Japan, Iceland and Russia.

Site of study		Amino acid sequence similarity / %					
		Japan	Iceland	Russia	Amphitheater Springs	Nymph Lake	Norris Basin
Country	Japan	100	66	36	69	68	64
	Iceland		100	48	57	66	65
	Russia			100	37	39	39
Yellowstone National Park	Amphitheater Springs				100	64	71
	Nymph Lake					100	91
	Norris Basin						100

Source: modified from Rice, G., et al. (2001). Viruses from extreme thermal environments. *PNAS*, **98**(23), 13341–13345.

a) State the similarity between Japanese and Russian virus coat protein gene sequences.

b) Phylogenetic comparison of the coat protein gene indicates that isolates from Nymph Lake and Norris Basin cluster together whereas Amphitheater Springs is more closely related to the virus from Japan. Explain how the viruses in Yellowstone National Park could have evolved.

- What mechanisms contribute to convergent evolution?
- To what extent is the natural history of life characterized by increasing complexity or simplicity?

A3.1 Diversity of organisms

You should know:

- species are groups of organisms with shared traits that show variation.
- species are named and classified using an internationally agreed system—the binomial system.
- the biological species concept defines a species as a group of organisms that can breed and produce fertile offspring.
- speciation is the splitting of one species into two or more.
- it is difficult to distinguish between populations during speciation.
- organisms in the same species share most of their genome except for single-nucleotide polymorphisms.

Additional higher level:
- the biological species concept does not work well with groups of organisms that do not breed sexually or in bacteria with horizontal gene transfer.
- species share chromosome number.
- cross-breeding between closely related species is unlikely to produce fertile offspring.

You should be able to:

- analyse karyograms and classify chromosomes by size and banding.
- evaluate evidence on the origin of human chromosome 2.
- compare genome size from different species.
- evaluate the current and potential use of whole genome sequencing.

Additional higher level:
- determine the classification of one plant and one animal species from domain to species level.
- construct a dichotomous key for use in identifying specimens.
- describe how to identify species in a habitat using barcodes from environmental DNA.

Classification of species

All organisms within a species have similarities but also differences—no two individuals are identical. Natural classifications help scientists to identify species and predict characteristics shared by **species** within a group. Sometimes taxonomists reclassify groups of species when new evidence shows that a previous taxon contains species that have evolved from different ancestral species. For example, sequencing of the 16S rRNA bases (part of the 30S subunit of the 70S ribosome of prokaryotes) has resulted in the reclassification of many bacterial species. All organisms in a species share common morphological characteristics. The **binomial system**, designed by Carl Linnaeus in 1735, is used to avoid the confusion which often arises from the use of common names. The genus and species names are normally underlined or written in italics. The genus name begins with an uppercase (capital) letter and the species name with a lowercase (small) letter, for example *Homo sapiens* is the scientific name for humans.

- A **species** is a group of organisms that can breed and produce fertile offspring.
- The **binomial system** of classification uses "*Genus species*" to name organisms.
- **Speciation** is the splitting of one species into two or more.

Biological species concept

There are challenges associated with the definition of a species as many different definitions exist. The biological species concept defines a species as a group of organisms that can interbreed and produce fertile offspring. A species is maintained by interbreeding within a species as well as by having reproductive barriers between organisms of other species. **Speciation** occurs gradually after two populations are separated by geographic, behavioural or temporal reproductive isolation. While speciation occurs, it is hard to distinguish between non-interbreeding populations and species.

Unity and diversity

Nature of science

Some hypotheses are testable, such as the origin of chromosome 2, but not all statements in science are testable, for example the origin of water on Earth (see subtopic A1.1).

- The **karyotype** is the number and type of chromosomes present in the nucleus.
- A **karyogram** is a photograph of chromosomes ordered by size.

Diversity in chromosome numbers of plant and animal species

There is a great diversity in the number of chromosomes in plant and animal species. All members of a species have the same number of chromosomes (with a few exceptions, for example in cases such as Down syndrome). Prokaryotes have one chromosome consisting of a circular DNA molecule. Eukaryote chromosomes are found inside the nucleus of the cell. Diploid nuclei have pairs of homologous chromosomes, therefore an even number of chromosomes. Haploid nuclei have half the number of chromosomes as diploid nuclei—they have one chromosome of each pair.

The number of chromosomes is not an indication of evolution. Humans have 46 chromosomes in diploid cells (23 in haploid), chimpanzees have 48 (24 in haploid), while dogs have 78 chromosomes (39 in haploid).

Karyotyping and karyograms

A comparative analysis of chromosomes from orangutan, gorilla, chimpanzee and human suggests that chromosome 2 in humans was formed from the fusion of chromosomes 12 and 13 of their ancestors (**Figure 1**).

Example 1

A phylogenetic tree shows estimated timing of duplications, inversions and relocations of the current human chromosome 2 based on sequence identity measures. It shows an estimate of the time the most recent common ancestor (MRCA) of all modern humans appeared. The average percentage and standard deviation (SD) sequence differences between human and the hominids chimpanzee, gorilla and orangutan are given on the right.

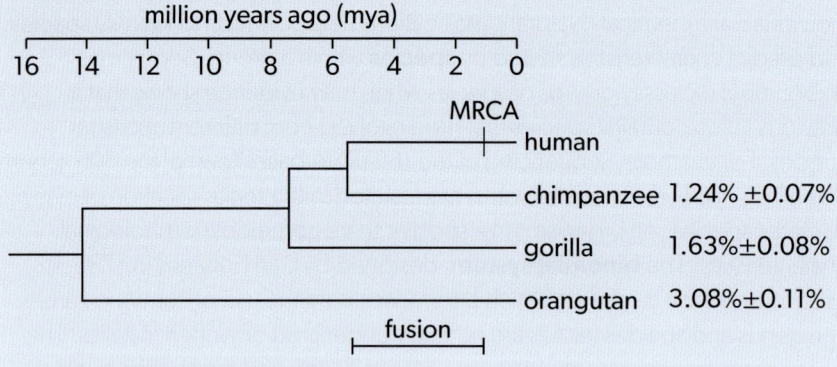

Source: Yunis, J. J. & Prakash, O. (1982).

a) State the percentage difference between human and gorilla sequences.
b) Identify the species that is most distantly related to humans.
c) Explain how speciation could have occurred after fusion of chromosomes.

Solution

a) 1.63%
b) Orangutan

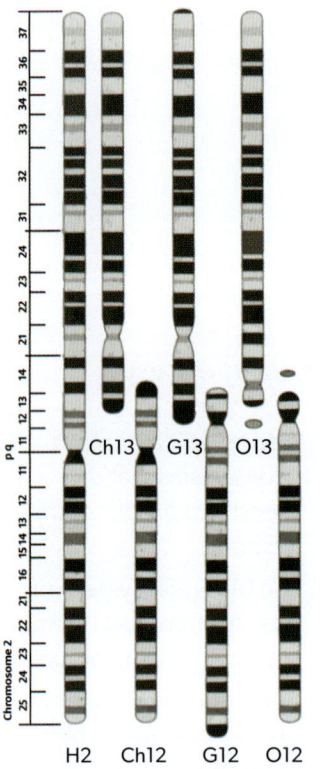

▲ Figure 1 Chromosome 2 in humans (H2) compared with chromosomes 12 and 13 in chimpanzees (CH12 and CH13), gorillas (G12 and G13) and orangutans (O12 and O13)

Source: Yunis, J. J. & Prakash, O. (1982). The origin of man: a chromosomal pictorial legacy. *Science*, **215**(4539), 1525–1530. www.doi.org/10.1126/science.7063861

c) Fusion of chromosomes (12 and 13) of primates produced chromosome 2 in humans. Consequently, humans and primates had a different number of chromosomes, so could not breed. The chromosomes could not align in meiosis as there were differences in the homologous pairs. If they did manage to breed, the offspring were infertile hybrids. Organisms were isolated in reproduction, so they split into two species.

Example 2

What is used to order the chromosomes in a karyogram?
I. Size of chromosome
II. Banding pattern after staining
III. Length of chromatid arm according to centromere

A. I only
B. I and II only
C. II and III only
D. I, II and III

Solution

The correct answer is **D**, as in a karyogram each chromosome is cut from a micrograph picture (using scissors) or using digital sources and ordered by homologous pairs according to their size and the length of the arm of each chromatid, and by the banding pattern produced from staining.

Diversity of genomes

The **genome** is all the genetic information of an organism. It includes the total amount of DNA, and therefore includes genes that are expressed and those that are not expressed. Organisms of the same species share most of their genome but variations such as **single-nucleotide polymorphisms** (SNPs) give some diversity. These SNPs can be found in coding and non-coding regions of the genome such as introns or promoters. Genomes vary in base sequence more between species than within a species.

The genome size does not give an idea of how complex in form the organism is. Humans have a genome of around 3×10^9 base pairs, while *Paris japonica*, a woodland plant, has 1.5×10^{11} base pairs.

- **Genome** is the whole genetic information of an organism.
- **Single-nucleotide polymorphisms** are substitutions (or point mutations) of nucleotides in a genome that are present in a large proportion of the population.

Whole genome sequencing

Whole genome sequencing determines the entire DNA sequence in a genome, all chromosomal DNA as well as mitochondrial DNA and, in plants, chloroplast DNA. In 2004, the Human Genome Project published a version of the human genome, which has been updated ever since. The genome sequence can be an important tool for researchers in the study of evolution, drug response, personalized medicine, early detection of disease and many related scientific uses.

Challenges to the biological species concept with asexually reproducing species and bacteria with horizontal gene transfer

Bacteria and viruses can transfer genetic material through horizontal transfer of genes. This means nucleic acid (DNA or RNA) is transferred from one organism to another and not from parents to offspring. The transfer can be from one species to another, which increases genetic diversity as genes from one species can appear in a different species. This will increase the chances of divergent evolution. It is believed that viruses played an important role in early evolution of bacteria and archaea before the divergence from LUCA through this horizontal gene transfer. Horizontal gene transfer is the primary mechanism for the spread of antibiotic resistance in bacteria. It is hard to apply the biological concept of species in organisms that do not breed sexually or in those where there is horizontal transfer of genes.

Chromosome number as a shared trait within a species

- **Hybridization** is cross-breeding (interspecific breeding) between closely related species.

Cross-breeding between closely related species (**hybridization**) is not frequent and usually has negative consequences. A growing amount of whole genome sequence data produced in recent years has shown that a broad range of species have experienced hybridization events. It has been estimated that around 25% of plant species and 10% of animal species have been affected by hybridization. Given that the number of chromosomes is a shared trait within species, hybridization is unlikely to produce fertile offspring if the number of chromosomes differs between these species.

Dichotomous keys

Example 3

The following is a dichotomous key for a group of beetles.

1. Wing cases with spots ---------------------------- Ladybird
   Wing cases without spots ---------------------------- 2
2. Elongated head ---------------------------- Weevil
   Head not elongated ---------------------------- 3
3. Long antennae ---------------------------- Longhorn beetle
   Short antennae ---------------------------- 4
4. Wing cases shorter than body ---------------------------- Oil beetle
   Wing cases as long as body ---------------------------- Colorado beetle

Using the key, identify the following beetles.

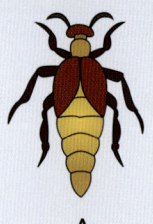    

A B C D E

Solution

A: oil beetle; B: longhorn beetle; C: Colorado beetle; D: weevil; E: ladybird.

Identifying species using DNA barcodes

- **DNA barcoding** allows identification of many species in an environment by PCR.

DNA barcoding is a fast and simple process based upon amplification and sequencing of all DNA sequences in a sample of multiple individuals or species. PCR primers are designed to join (anneal) to highly conserved DNA sequences to amplify a variety of species. High-throughput next-generation sequencing is used to determine the species composition. This sequencing results in millions of DNA sequence reads which are classified according to similarity of their DNA sequences to those of known species. DNA sequence tends to vary relatively little within species and much between species. A DNA barcode is chosen when it has low intraspecific and high interspecific variability.

- What might cause a species to persist or go extinct?
- How do species exemplify both continuous and discontinuous patterns of variation?

A3.2 Classification and cladistics

Additional higher level:

You should know:

- a clade is a group of organisms that have evolved from a common ancestor.
- base or amino acid sequences are evidence for which species are part of a clade.
- in classification according to evolutionary relationships all members of a taxonomic group evolved from a common ancestor.
- sequence differences accumulate gradually—there is a positive correlation between the number of differences between two species and the time since they diverged from a common ancestor.
- evidence from cladistics has shown that classifications of some groups based on structure did not correspond with the evolutionary origins of a group or species.

You should be able to:

- explain the importance of classification.
- discuss the advantages of classification by evolutionary relationships over the traditional hierarchy of taxa.
- construct cladograms—tree diagrams that show the most probable sequence of divergence in clades.
- analyse cladograms to deduce evolutionary relationships.
- analyse cladistic evidence that led to the reclassification of the figwort family.
- explain that the domains archaea, bacteria and eukarya are classified according to 16S ribosomal RNA.

Classification of organisms

Classifying the more than 2 million known species is important in facilitating meaningful communication among scientists and society in general. Traditional taxonomic identification based on observable traits is rapidly being replaced by molecular approaches. Classification based on genetic or protein sequences has two advantages over observation of morphology. Genetically distinct species that cannot be visually distinguished are easily recognized by differences in their DNA or protein sequences and there is no human error when identifying species by their DNA. However, because it is difficult to obtain genetic sequences during a field trip, classifying organisms using external features is the most advantageous approach when not in a laboratory.

Clades

Evolutionary relationships are used in classification. This means all the members of a **clade** have evolved from a common ancestor. Evidence for which species are part of a clade can be obtained from the base sequence of a gene or the corresponding amino acid sequence of a protein.

Nature of science

There has been a paradigm shift in the classification of organisms. The traditional hierarchy with fixed ranking of taxa (kingdom, phylum, class, order, family, genus and species) is arbitrary because it does not reflect the gradation of variation. Cladistics offers an alternative approach to classification using unranked clades.

- A **clade** is a group of organisms with common ancestry and shared characteristics.

Unity and diversity

- **Cladograms** are constructed using base sequences of genes or amino acid sequences of proteins.
- A **molecular clock** shows the gradual accumulation of sequence differences until clades diverged from a common ancestor.
- **Mutation rates** are affected by the length of the generation time, the size of a population and the intensity of selective pressure.

A **cladogram** is a tree diagram used in cladistics to show relationships between organisms. A cladogram uses lines that branch off in different directions ending at a clade. A clade is a group of organisms with a last common ancestor. At each node a splitting event occurs. The node therefore represents the end of the ancestral taxonomic group (taxon), and the stems represent the newly formed species that split from the ancestor.

The **molecular clock** is a form of estimating when a clade diverged from a common ancestor by the gradual accumulation of sequence differences. The time is not exact as **mutation rates** are affected by the length of the generation time, the size of a population and the intensity of selective pressure, among other factors.

Example 4

The cladogram shows the evolutionary relationship between some animals.

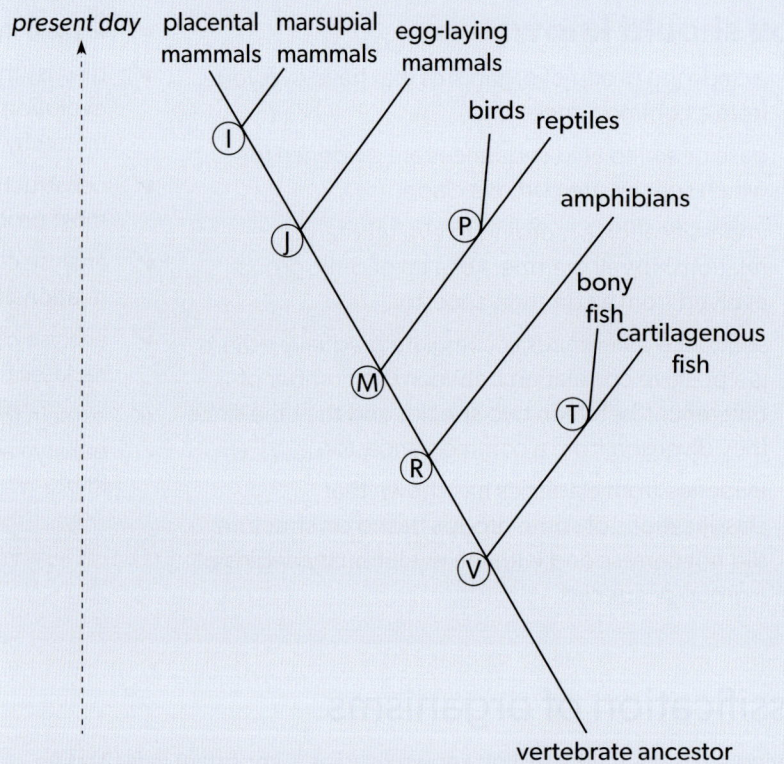

a) State the organisms that are closely related to marsupials.
b) Identify the group of organisms that is most distantly related to amphibians.
c) Identify the node where birds diverged.

Solution

a) Placental mammals
b) Mammals
c) P

Nature of science

Parsimony analysis is used to select the most probable cladogram. This means that the hypothesis of relationships that requires the smallest number of changes is preferred.

Constructing cladograms

DNA and protein sequences are usually shown in a format called FASTA. The similarities in DNA sequences can be shown using the software BLASTn and protein sequences using BLASTp. Sequence alignments can be performed in BLAST or ClustalW.

Example 5

The rate of nucleotide substitution is used as an indication of evolutionary change. A study was carried out on mitochondrial DNA and nuclear DNA of different species of *Acropora*, a Pacific Ocean coral.

The red bars represent the range of genetic sequence divergence between the species for specific mitochondrial DNA sequences (mtDNA) and the blue bars represent specific nuclear DNA sequences (nDNA).

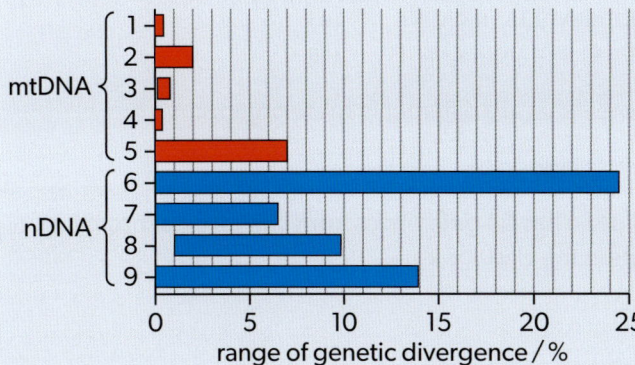

Source: Shearer, T.L, (2002). Slow mitochondrial DNA sequence evolution in the Anthozoa (Cnidaria). *Molecular Ecology* 11 2475–2487. www.doi.org/10.1046/j.1365-294x.2002.01652.x

a) Measure the percentage of maximum genetic divergence for the mtDNA2 sequence.
b) Calculate the maximum difference in genetic divergence between the mtDNA5 sequence and nDNA6 sequence.
c) Compare and contrast the variations in the range of genetic divergence of the mtDNA and nDNA sequences.
d) Discuss the significance of this data in terms of possible evolutionary changes in these genes.

Solution

a) 2%
b) 24.5 − 7 = 17.5%
c) The average genetic divergence of mtDNA is much less than the average nDNA. There is a greater range of genetic variation in nDNA than in mtDNA. Three of the mtDNA have less than 1% genetic divergence while none of nDNA have less than 5%. The highest divergence of mtDNA is similar to the lowest of nDNA.
d) The mtDNA is more stable due to less genetic diversity and has fewer genes, which could be a limit on the accumulation of mutations. The mDNA has genes of indispensable proteins, so natural selection would exert less pressure. The problem with using mutations as an evolutionary clock is that there is a different genetic divergence and different rates of mutation depending on the genes examined. The high rates of nDNA6 divergence could be neutral substitutions which have no effect (silent mutations). There is insufficient data to know the effects of these mutations.

Analysing cladograms

Plant families have been reclassified because of evidence from cladistics. In 2001, Richard Olmstead and Pat Reeve studied the DNA sequences of three chloroplast genes through PCR sequencing of 65 taxa of the aquatic angiosperms in *Scrophulariaceae*, commonly known as the figwort family. They found that the groups previously included in this family were not all descendants of a common ancestor—they were five different groups.

Three domains of classification

In 1977, Carl Woese, studying the sequences of 16S ribosomal RNA, realized that archaea had a separate line of evolutionary descent from bacteria. This determined the new classification with an extra taxonomic level above kingdoms called domains. The three domains are archaea, bacteria and eukarya.

Nature of science

Theories and other scientific knowledge claims may eventually be falsified. In this example, similarities in morphology due to convergent evolution rather than common ancestry suggested a classification that by cladistics has been shown to be false.

Practice problems

1. The FASTA sequences for cytochrome c proteins in humans (*Homo sapiens*) and mice (*Mus musculus*) were aligned using BLASTp.

Homo sapiens	1	MGDVEKGKKIF**IM**KC**S**QCHTVEKGGKHKTGPNLHGLFGRKTGQA**PGY**SYTA**A**NKNKGI**I**W	60
Mus musculus	1	MGDVEKGKKIF**VQ**KC**A**QCHTVEKGGKHKTGPNLHGLFGRKTGQA**AGF**SYTD**A**NKNKGI**T**W	60
Homo sapiens	61	GEDTLMEYLENPKKYIPGTKMIF**V**GIKKK**EE**RADLIAYKKATNE	105
Mus musculus	61	GEDTLMEYLENPKKYIPGTKMIF**A**GIKKK**G**ERADLIAYKKATNE	105

   a) State the number of amino acids in the studied protein that are different in these organisms.

   b) Suggest a reason cytochrome c is useful to construct cladograms.

   c) Yeast (*Saccharomyces cerevisiae*) cytochrome c protein has 64 differences from human cytochrome c. Identify whether mice or yeast are closer to humans in the cladogram.

2. What is a genome in an animal?

   A. Only DNA found in the nucleus of all cells

   B. Only mitochondrial DNA

   C. DNA that is expressed in the nucleus and mitochondria

   D. DNA that is expressed and DNA not expressed in all cells

- How can similarities between distantly related organisms be explained?
- What are some examples of ideas over which biologists disagree?

A4.1 Evolution and speciation

You should know:

✓ evolution occurs when heritable characteristics of a species change.

✓ evolution of homologous structures explains similarities in structure when there are differences in function.

✓ convergent evolution is the development of analogous structures in unrelated species, for example wings in birds and insects or the streamlined bodies of fish and cetaceans.

✓ analogous structures have similar function but a different ancestry.

✓ gradual evolutionary change in a species is not speciation.

✓ reproductive isolation and differential selection can lead to speciation.

You should be able to:

✓ distinguish Darwinian evolution from Lamarckism.

✓ explain the processes involved in evolution.

✓ describe evidence for evolution in terms of DNA, RNA and protein sequences; selective breeding of animal and crop plants; homologous structures, such as pentadactyl limbs.

✓ compare the pentadactyl limbs of mammals, birds, amphibians and reptiles with different methods of locomotion.

✓ explain speciation due to isolation.

✓ state that speciation increases the total number of species on Earth, whereas extinction decreases it.

Additional higher level:

✓ compare sympatric speciation, occurring without geographical isolation, with allopatric speciation, which requires isolation.

Additional higher level:
- ✔ adaptive radiation is a source of biodiversity.
- ✔ speciation can occur abruptly.
- ✔ plants can suffer abrupt speciation by hybridization and polyploidy.
- ✔ compare and contrast temporal, behavioural and geographical reproductive isolation.
- ✔ describe barriers to hybridization and sterility of interspecific hybrids as mechanisms for preventing the mixing of alleles between species.

The theory of evolution

In the early 19th century, Jean-Baptiste Lamarck proposed that organisms could pass on information acquired by use or disuse to their offspring. In the 20th century Charles Darwin and other scientists discredited his ideas by proposing **evolution** as change in the heritable characteristics of a population. Acquired changes that are not genetic in origin are not regarded as evolution.

There are three levels at which the evolutionary process acts. First is the origin of genetic novelties caused by variation. Second is the changing of frequencies within the population and third is the fixation of the diversity already attained on the preceding two levels. The first is attained by mutations and recombinations, the second by natural selection and the third level is obtained by isolating mechanisms.

Evidence for evolution

There is overwhelming evidence for the evolution of life on Earth in sequence analysis and selective breeding results. Data analysis of DNA, RNA and proteins gives evidence of the common ancestry between organisms through shared sequences. Artificial selection can also cause evolution, as can be seen in selective breeding of domesticated animals and varieties of crop plant. The difference between them and the original wild species shows how rapidly evolutionary changes can occur.

Evolution of **homologous structures** explains similarities in structure when there are differences in function. Comparative anatomy has shown that certain structural features are basically similar in different species. This homology may be regarded as evidence of evolution from a common ancestor. The forelimbs of tetrapod vertebrates have similar structures, as they all have a five-finger pattern (pentadactyl), but are adapted for different purposes. For example, bats use their wings to fly, while moles use their front limbs to dig. Homologous structures showing adaptations to different conditions or lifestyles are examples of **divergent evolution**.

Diversity of life has evolved and continues to evolve by selection. Natural selection increases the frequency of characteristics that make individuals better adapted and decreases the frequency of other characteristics leading to changes within the species. These changes in species can lead to evolution. **Analogous structures**, such as wings in birds and in butterflies, are examples of **convergent evolution**, where similar selection pressures produce similar structures in species that have no common ancestor.

- **Evolution** is the change in characteristics that can be inherited by different generations of a population. There is evidence of evolution from sequence analysis and selective breeding.

Nature of science

The theory of evolution by natural selection is based on observations and supporting evidence provided by many scientists, including Charles Darwin, and is unlikely to be falsified. However, as it is impossible to formally prove that it is true by correspondence, it is referred to as a theory.

- **Homologous structures** have similar form (common ancestor) but have different functions.
- **Divergent evolution** is the accumulation of differences between closely related populations within a species, which can lead to speciation.
- **Analogous structures** have similar form or function but are from a different ancestor.
- **Convergent evolution** of similar features in species of different populations over time creates analogous structures.

Unity and diversity

Example 1

The limbs of many vertebrates have a similar structure (the pentadactyl limb). In the diagram, all the organisms have humerus, radius and ulna bones.

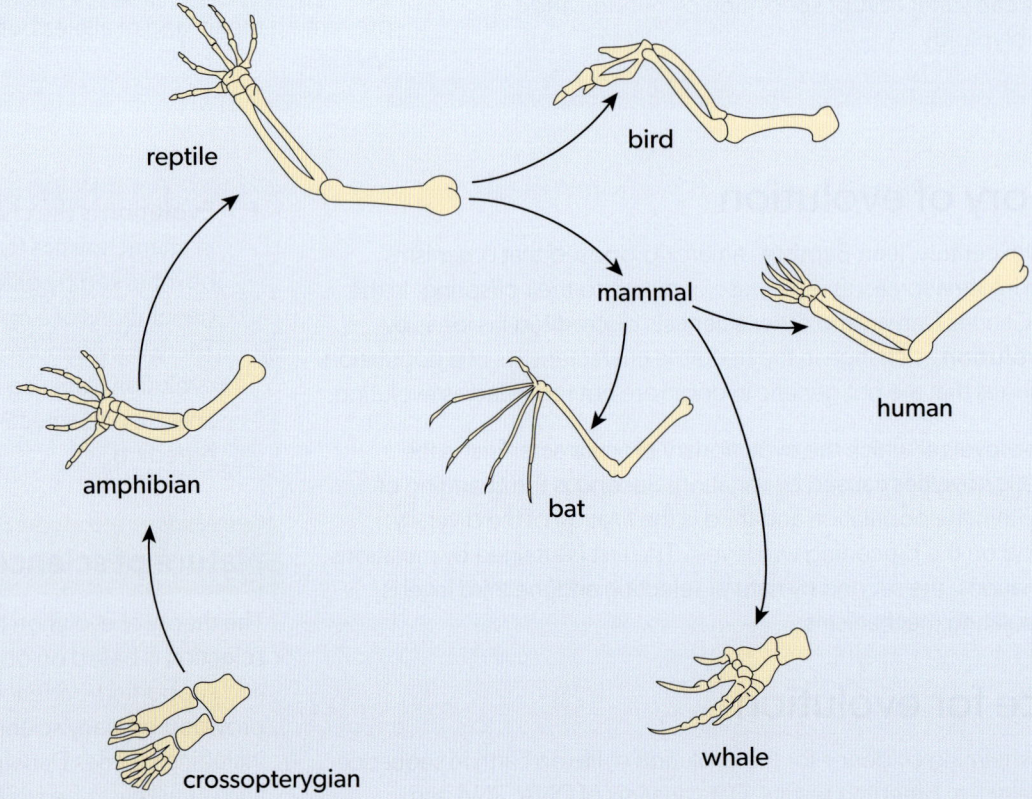

What hypothesis is currently accepted to account for the similarity in the limbs?

A. Organisms pass on characteristics that they acquire during their lifetime.

B. The pentadactyl limb is an ideal design for a broad range of purposes.

C. All the organisms have descended from a common ancestor.

D. Convergent evolution has resulted in each organism finding a similar solution to a mechanical problem.

Solution

The correct answer is **C**. These vertebrates descended from the same common ancestor. Although the limbs look very different, they all share the same five-fingered bone structure. This is an example of divergent evolution producing homologous structures.

Assessment tip

Many students confuse speciation with evolution. Evolution is change within a species, whereas speciation is the production of a new species.

Speciation

Speciation is when one or more species are formed by the splitting of an existing species. This means that the number of species will increase, whereas if a species becomes extinct, the number will decrease.

A4.1 Evolution and speciation

Sample student answer

The cladogram shows primate evolutionary changes in head-forelimb musculature in gibbons (*Hylobates*), orangutans (*Pongo*), gorillas (*Gorilla*), humans (*Homo*), chimpanzees (*Pan troglodytes*) and bonobos (*Pan paniscus*) using parsimony analysis. Each square represents one change (mutations, deletions or insertions) at the given time. Blue squares show changes that revert to the original ancestor state.

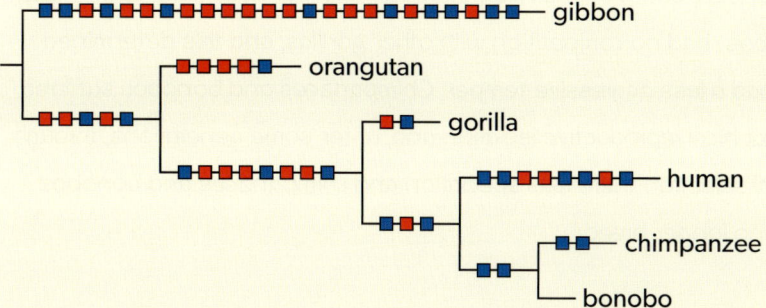

Source: Diogo, R., Molnar, J. L. & Wood, B. (2017). Bonobo anatomy reveals stasis and mosaicism in chimpanzee evolution and supports bonobos as the most appropriate extant model for the common ancestor of chimpanzees and humans. *Scientific Reports*, 7, 608. www.doi.org/10.1038/s41598-017-00548-3 (CC BY 4.0)

> **Examiner tip**
>
> Usually phylogenetic relationships are built using more than one gene sequence.

a) (i) Identify the number of changes occurring before orangutans diverged from the other primates. [1]

This answer could have achieved 1/1 mark:

> 5

(ii) Identify the number of changes in orangutans since they separated from other primates. [1]

This answer could have achieved 1/1 mark:

> 5

b) (i) Identify the primate that is most closely related to bonobos. [1]

This answer could have achieved 1/1 mark:

> Chimpanzees

(ii) Suggest one hypothesis on how bonobos diverged from other *Pan* populations. [4]

This answer could have achieved 4/4 marks:

> The ancestral bonobo populations diverged from other *Pan* populations (chimpanzees) in a period when the Congo River was formed. Because the two *Pan* species are poor swimmers, the formation of the Congo River was thought to have led to speciation of the bonobo. Shortly after the river crossing of bonobo ancestors, the banks of the Congo once again flooded with water, separating the river-crossing group of *Pan* in the south from their relatives north of the river forever after.

▲ This answer clearly explains one theory on how bonobos evolved through natural selection to eventually undergo speciation through reproductive isolation due to geographic isolation.

North of the river, the chimpanzees had to compete with other primates as well as gorillas for both food and territory. As a result, they evolved a tougher disposition to be able to defend themselves when threatened. Chimpanzees were evolutionarily selected for aggressive tendencies. By comparison, south of the river, the other chimpanzees (later evolving to bonobos) had no competition with other gorillas, and this determined they had a less aggressive temper. Chimpanzees and bonobos suffered geographical reproductive isolation and, after some generations, through natural selection there was speciation and chimpanzees and bonobos could no longer breed.

In speciation, populations are reproductively isolated and therefore do not interbreed. **Reproductive isolation** of populations can be sympatric or allopatric. **Sympatric isolation** occurs in the same place and includes temporal and behavioural isolation. **Allopatric isolation** is geographic isolation. In different environments there are different selection pressures as there are different habitats or niches to exploit. These changes cause natural selection. The allele frequencies change, causing the populations to diverge.

- **Reproductive isolation** is a mechanism that prevents production of offspring. It can be temporal, behavioural or geographical.
- **Sympatric isolation** occurs in the same geographical region.
- **Allopatric isolation** occurs in separated geographical regions.

Adaptive radiation as a source of biodiversity

Adaptive radiation is a process that describes the rapid speciation of one or more species to fill many ecological niches. It occurs in organisms of the same species that carry different characteristics and undergo different selective pressures. Natural selection increases the frequency of characteristics that make individuals better adapted and decreases the frequency of other characteristics, leading to changes within the species. The ancestral species will change so much and in so many ways that speciation causes different species to appear.

- **Adaptive radiation** is rapid speciation due to selective pressures leading to divergent evolution.

In subtopic B4.1, you will study the adaptations of organisms to an environment.

Darwin's finches, inhabiting the Galapagos archipelago, are a good example of adaptive radiation and speciation. Their common ancestor arrived on the Galapagos about 2 million years ago. Fifteen of the currently recognized species evolved from this common ancestor. Changes in the size and form of the beak have enabled different species to utilize different food resources, especially seeds of different sizes.

Example 2

The diagram shows the beaks of some finches on the Galapagos Islands. The medium ground finch, *Geospiza fortis*, subsists mainly on small seeds but members of the *G. fortis* population with larger beaks can tackle the bigger seeds. *G. magnirostris* is a larger species of finch with a large beak. *G. magnirostris* compete with the *G. fortis* for large seeds. During a drought, there were few large seeds available, so the *G. fortis* finches with smaller beaks had an advantage over the *G. magnirostris* finches with bigger beaks, as they could feed on smaller seeds. This trait was then passed down to the next generation, showing a strong shift towards smaller beak size among *G. fortis*.

(a) *G. fortis* (large beak)

Which process of evolution has occurred in the finches?

A. Convergent evolution of the beaks

B. Natural selection for the strongest beaks

C. Selection pressure caused by the exploitation of different food sources

D. Mutation of the beaks to adapt in different environments

(b) *G. fortis* (small beak)

Solution

The correct answer is **C**, as it shows selective pressure. It occurs to organisms of the same species that carry different characteristics and undergo different selective pressures (food sources). It is not convergent evolution, as they all come from the same ancestor. The strongest beak is not always selected for—it is the better adapted that is selected. The mutations occur before the process of selection. They are already in the genetic pool—they do not occur in response to a selection pressure.

(c) *G. magnirostris*

Barriers to hybridization and abrupt speciation in plants

Hybridization occurs when an animal or plant breeds with an individual of another species or variety. There are reproductive barriers that prevent mixing of alleles or gene flow between species and therefore maintain the genetic integrity of species. Reproductive barriers can occur before or after fertilization. Pre-fertilization barriers prevent mating and fertilization, for example, in plants it could involve the failure of the pollen to reach the stigma. Courtship behaviour such as songs of frogs (*Rana aurora*) often prevents hybridization in animal species. Males make calls to attract females during the breeding season in a good place to lay her eggs. Each species has its own call, so females are only attracted to calls from a male of the same species. Post-fertilization barriers result in the selection against hybrids. Examples include the lethality of interspecific hybrids and hybrid sterility. A mule is an example of a sterile hybrid between a female horse and a male donkey.

In plants, abrupt speciation is produced by either hybridization or **polyploidy**. Polyploid species are formed by multiplication of the genome. This can provide useful traits for agriculture, such as enhanced photosynthesis or enlargement of seed or fruit size by expansion of cell size. Hybridization occurs in the knotweed or smartweed (genus *Persicaria*) where scientists have found unusual specimens with various combinations of characters of different species such as *P. senegalensis*, *P. glabra*, *P. attenuata*, *P. setosula* and *P. decipiens*.

- **Polyploidy** is the multiplication of sets of chromosomes. It can lead to speciation as it produces hybrids.

- How does the theory of evolution by natural selection predict and explain the unity and diversity of life on Earth?
- What counts as strong evidence in biology?

A4.2 Conservation of biodiversity

You should know:
- biodiversity is the variety of life in all its forms, levels and combinations.
- evidence from fossils suggests that there are currently more species alive on Earth today than at any time in the past.
- there is a need for several approaches to conservation of biodiversity.

You should be able to:
- analyse data on number of species and the biodiversity crisis.
- explain causes of anthropogenic species extinction.
- explain causes of ecosystem loss and the current biodiversity crisis.
- understand the rationale behind focusing conservation efforts on evolutionarily distinct and globally endangered species (EDGE).

- **Biodiversity** is the variety of life in all its forms, levels and combinations.

Nature of science

Taxonomy usually focuses on pattern recognition, but there is also a taxonomic correction process that retests and updates existing species based on new evidence. "**Splitters**" recognize more species than "**lumpers**" in a taxonomic group.

For example, in 1923, the common gallinule (*Gallinula galeata*) of North America was lumped with the common moorhen (*G. chloropus*) of Europe because, according to their breeding ranges and morphologies, they were subspecies. Then, in 2011, those species were split again, recognizing the differences in their vocalizations, bill and shield morphologies, and mtDNA.

Biodiversity

Biodiversity is a property of life at all levels of biological organization, as it includes all ecosystems, species and alleles (genetic information) in a region. **Species richness** (number of species) and **species evenness** (relative abundance of each species) are the two main components of measuring the **species diversity** in an ecosystem. There is variability between living organisms (both inter and intraspecific) and the ecosystems of which they form part. **Genetic diversity** serves as a means for populations to adapt to changing environments. **Ecosystem diversity** addresses the variation in the complexity of a community, for example, resources and habitats differ between deserts and tropical rainforests. Diversity in the ecosystem is important for its integrity.

Several websites can provide an idea of the number of species identified, including the International Union for Conservation of Nature (IUCN) Red List and Our World in Data. There are a few million organisms described, but scientists believe there are many more to be discovered. Fossil evidence suggests that there are more species alive now than in the past.

- **Species richness** is the number of species in a community.
- **Species evenness** is the relative abundance of each species in a community.
- **Species diversity** measures the variation of species in a community.
- **Genetic diversity** is the variation in genetic information found in organisms.
- **Ecosystem diversity** is the biotic and abiotic variations within land and aquatic ecosystems.
- **Splitters** are taxonomists who classify organisms into many different smaller groups.
- **Lumpers** are taxonomists who classify organisms into one large group.

Causes of anthropogenic species extinction

In the history of Earth's biodiversity, there have been five mass extinction events, all caused by natural causes. Scientists now think that there is a sixth mass extinction underway, but this time caused entirely by humans. Considerable scientific evidence indicates that there is a biodiversity crisis of increasing extinctions by lowering species abundances.

Some people argue the sixth mass extinction is not taking place. They use the IUCN Red List to argue the rate of species loss does not differ from the background rate. However, the Red List is heavily biased as it is formed mainly by birds and mammals. Only a minute fraction of invertebrates has been evaluated against conservation criteria.

The pre-human ecosystem of the North Island in New Zealand was dominated by the giant moa (*Dinornis novaezealandiae*), which were endemic, flightless birds. Moa were the largest and most diverse terrestrial herbivores. There is strong evidence suggesting that hunting and habitat destruction following the arrival of humans resulted in the extinction of this bird species. Likewise, monk seals (*Neomonachus tropicalis*) used to be found in dense colonies throughout both mainland coasts and offshore islands and atolls in the Caribbean Ocean. Overfishing and hunting reduced animal biomass so that the monk seals could no longer survive on the fish resources that remained in the Caribbean reefs and became extinct.

- **Anthropogenic species extinction** is caused by human actions.

▲ Figure 1 Giant moa bird

Sample student answer

Scientists conducted animal surveys in protected forests in the Taman Negara National Park (TNM), Malaysia, in the years 2016 and 2000. Using camera-traps to detect all mammal species larger than 1 kg, they calculated the detection frequencies as the number of independent photographs of each species divided by total camera trap days per 100 trap days. The table shows IUCN Red List categories as CR = critically endangered, EN = endangered, VU = vulnerable, NT = near threatened and LC = least concern.

Species—IUCN classification	TNM 2016	TNM 2000
Canidae		
Dhole (*Cuon alpinus*)—EN	0.17	0.14
Felidae		
Malayan tiger (*Panthera tigris jacksoni*)—CR	0.42	0.51
Indochinese leopard (*Panthera pardus delacouri*)—CR	1.58	2.61
Clouded leopard (*Neofelis nebulosa*)—VU	0.42	0.12
Asian golden cat (*Catopuma temminckii*)—NT	0.57	0.48
Marbled cat (*Pardofelis marmorata*)—NT	0.89	0.02
Leopard cat (*Prionailurus bengalensis*)—LC	2.98	1.36
Ursidae		
Sun bear (*Ursus malayanus*)—VU	4.42	2.21
Mustelidae		
Yellow-throated marten (*Martes flavigula*)—LC	2.24	0.12
Malayan weasel (*Mustela nudipes*)—LC	0.07	–
Otter spp. (*Lutra* spp.)	–	–
Herpestidae		
Crab-eating mongoose (*Herpestes urva*) – LC	–	–

Unity and diversity

Viverridae		
Malay civet (*Viverra tangalunga*)—LC	0.81	1.41
Large Indian civet (*Viverra zibetha*)—LC	0.02	0.05
Masked palm civet (*Paguma larvata*)—LC	0.22	–
Common palm civet (*Paradoxurus hermaphroditus*)—LC	0.17	0.02
Banded civet (*Hemigalus derbyanus*)—NT	0.27	0.02
Binturong (*Arctictis binturong*)—VU	0.15	0.02
Prionodontidae		
Banded linsang (*Prionodon linsang*)—LC	0.20	–
Suidae		
Eurasian wild pig (*Sus scrofa*)—LC	8.78	3.04
Cervidae		
Sambar (*Rusa unicolor*)—VU	0.44	0.46
Red muntjac (*Muntiacus muntjac*)—LC	7.15	3.18
Mouse deer (*Tragulus* spp.)—LC	6.02	0.65
Bovidae		
Gaur (*Bos gaurus*)—VU	0.07	0.12
Sumatran serow (*Capricornis sumatraensis*)—VU	0.10	0.02
Elephantidae		
Asian elephant (*Elephas maximus*)—EN	1.46	3.85
Tapiridae		
Malayan tapir (*Tapirus indicus*)—EN	2.57	7.31
Cercopithecidae		
Southern pig-tailed macaque (*Macaca nemestrina*)—VU	1.21	0.25
Long-tailed macaque (*Macaca fascicularis*)—LC	0.02	0.02
White-thighed surili (*Presbytis siamensis*)—NT	0.44	–
Hystricidae		
Malayan porcupine (*Hystrix brachyuran*)—LC	4.34	1.29
Asiatic brush-tailed porcupine (*Atherurus macrourus*)—LC	0.44	0.14
Manidae		
Malayan pangolin (*Manis javanica*)—CR	0.02	0.07

Source: adapted from Clements, G. R., et al. (2021). Conservation status of large mammals in protected logged forests of the greater Taman Negara Landscape, Peninsular Malaysia. *Biodiversitas*, **22**(1), 272–277.

Examiner tip

When referring to organisms in an examination, either the common name or the scientific name is acceptable.

a) List species that are considered endangered in the IUCN Red List category. [3]

This answer could have achieved 3/3 marks:

> Dhole, Asian elephant, Malayan tapir

b) (i) Identify a species that, although not threatened, has not been detected in 2000. [1]

This answer could have achieved 1/1 mark:

> White-thighed surili

(ii) Suggest one reason this species was not detected in 2000. [1]

This answer could have achieved 0/1 mark:

> They are extinct.

▼ This answer is not necessarily true, as the species could still be present in other ecosystems. There are many reasons this species was not detected. One reason could be the species has emigrated to another area or has been killed in this one. Another reason could be the animals are still in the park, but are not detected, for example, the camaras did not activate when the animal entered the trap, or the animals did not go close to the traps. Any reasonable answer would be accepted by examiners.

A4.2 Conservation of biodiversity

c) Poaching of large species might be increasing in Malaysia. Using the data provided, discuss whether the TNP is serving as a means of conservation of the *Felidae* family. [3]

This answer could have achieved 0/3 marks:

The TNP is a very good ex situ protector.

▼ This answer did not score marks as the TNP is an *in situ* reserve for conservation of species. A discussion must include a range of arguments. In all cases, the *Felidae* richness (number of species) is the same, as there are six species present. The evenness does change, as the number of organisms in each species is much lower in 2000 compared with 2016. Two of the species that are critically endangered have increased in numbers while four have decreased. The marbled cat has decreased a great deal. Overall, the findings suggest that the TNP is an important site for the conservation of endangered species, but more work needs to be done.

Causes of ecosystem loss

As a family of plants, *Dipterocarpaceae* are the best-known trees in the tropics. The dipterocarp forests of Southeast Asia are a dominant and valuable part of the world's tropical rainforest, yet are some of the most threatened. As for most ecosystems, these forests are threatened by an array of drivers, each of which increases the probability of extinction of species. Tree plantations and deforestation, hunting and trade for food, medicine or ornamentation, mining, development of roads, urbanization, reservoir construction, wetland drainage, fires, pollution, invasive species due to global transport, disease and climate change are some of the drivers that threaten the ecosystems.

Example 3

In Australia, studies were performed on wild grassland sites containing shrubs to determine the effect that fires have on the biodiversity of birds living there. The birds occupy different habitats in the ecosystem, and each is affected differently by the fires. Counts were made of the numbers of birds of several species immediately before the fire, and then at intervals in the following years.

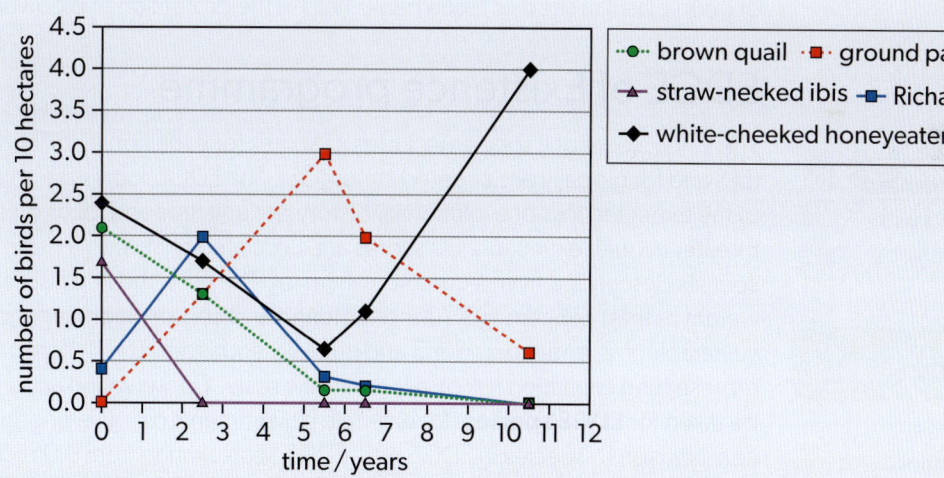

Source: Gill, A. M. et al. (1999). *Australia's Biodiversity – Responses to Fire, Plant, Birds and Invertebrates.* www.rest.neptune-prod.its.unimelb.edu.au/server/api/core/bitstreams/c9a40e0a-a0ba-5bf9-a761-d6c909f110a6/content

a) State the time after the fire before the greatest number of ground parrots was found.
b) The white-cheeked honeyeater eats nectar from flowering plants. Using the data, predict the effect the fire had on these flowering plants.
c) Immediately before the fire (0 years), the Simpson diversity index for 10 hectares of the ecosystem was 5.4. Predict, giving a reason, whether you would expect this value to increase, decrease or remain unchanged 10.5 years after the fire.
d) Suggest **two** reasons why the results varied for the different bird species.

Solution

a) 5.5 years

b) The flowering plants were initially destroyed by the fire but after several years the number of flowering plants started to increase. The numbers of flowering plants increased to higher levels than before the fire after 9 years.

c) The index would decrease because there are fewer species after 10.5 years, which determines a lower species diversity or richness.

d) Different habitats or niches, different food sources, different breeding sites or different responses to predators could change the number of birds in the different species.

Nature of science

Surveys need to be repeated to provide evidence of change in species richness and evenness and be published in peer-reviewed sources. The IPBES Global Assessment Report on Biodiversity and Ecosystem Services is an intergovernmental report that gives evidence for a biodiversity crisis as it includes results from reliable surveys of biodiversity in a wide range of habitats around the world.

- **EDGE species** have an evolutionary distinctness and are threatened with extinction.

Nature of science

Which species should be prioritized for conservation efforts has complex ethical, environmental, political, social, cultural and economic implications and therefore needs to be debated.

Assessment tip

You will not be asked to calculate the EDGE score in an exam.

Biodiversity crisis and conservation approaches

Biodiversity provides humans with invaluable resources. Expansion of human population through increase in population growth has increased resource exploitation and altered the use of land, causing a biodiversity crisis. Alteration of land use by deforestation and clearance of land for agriculture or urbanization has resulted in the fragmentation or loss of habitats, leading to biodiversity loss. Hunting and other forms of over-exploitation have directly and indirectly affected food chains and webs. Global transport of goods to supply humans has increased pollution and the spread of pests, diseases and invasive species.

An overall prognosis of regional biodiversity and priority actions to protect biodiversity in the future needs to be discussed. No single approach by itself is sufficient, and different species require different measures. *In situ* conservation of species in natural habitats, management of nature reserves, rewilding and reclamation of degraded ecosystems are approaches followed in many countries. Also, *ex situ* conservation in zoos and botanic gardens and storage of germ plasm in seed or tissue banks help in the protection of biodiversity.

EDGE of Existence programme

The EDGE score of a species includes its scores for evolutionary distinctness (ED) and for globally endangered status (GE). The EDGE scores are an estimate of the expected loss of evolutionary history per unit time and focus on species threatened with extinction. ED scores are calculated relative to a clade of species descended from a common ancestor. The GE is the Red List category weight as least concern = 0, near threatened and conservation dependent = 1, vulnerable = 2, endangered = 3 and critically endangered = 4, here representing extinction risk on a logarithmic scale. Conservation actions can be taken for **EDGE species**, for example governments can give priority for conservation to species in EDGE lists. EDGE scores range from 0.0565 to 6.48 and are approximately normally distributed around a mean of 2.63.

Example 4

What is likely to happen to a species classification when the ED in the EDGE score is tripled?

A. Increases in threat level

B. Changes threat level from endangered to vulnerable

C. Does not affect the threat level

D. Only affects the threat level if it is classified as vulnerable

Solution

The correct answer is **A**, as if the ED increases, the EDGE score also increases, regardless of the GE score.

Practice problems

1. Studies from biochemistry provide evidence of the evolutionary relationships between primates. Scientists found that the sequence of haemoglobin in gorillas differs by one amino acid from haemoglobin in chimpanzees and humans, whose sequence is the same. They made models showing the possible relationships between humans, chimpanzees and gorillas.

   Which model best represents the relationship according to haemoglobin structure?

   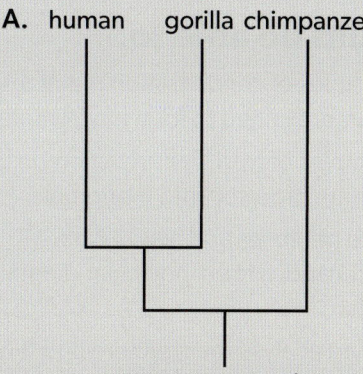
   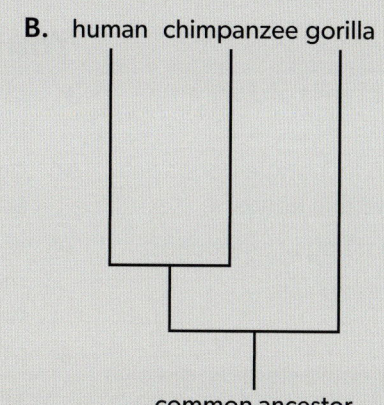
   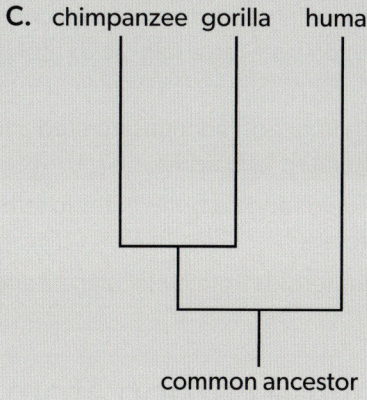
   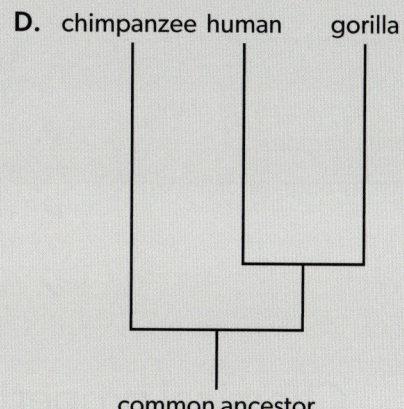

2. Outline factors that increase variation in a population.
3. Explain how analogous and homologous structures exemplify commonality and diversity.
4. Suggest recommendations for a sustainable future in the mixed dipterocarp forest in Southeast Asia that is suffering a biodiversity crisis.

- In what ways is diversity a property of life at all levels of biological organization?
- How does variation contribute to the stability of ecological communities?

B Form and function

B1.1 Carbohydrates and lipids

You should know:
- covalent bonds are sharing of pairs of electrons between atoms.
- carbon atoms can form up to four single bonds or a combination of single and double bonds.
- polysaccharides are energy storage compounds.
- cellulose has a structural role in plants.
- glycoproteins have a role in cell–cell recognition.
- lipids, such as fats, oils, waxes and steroids, dissolve in non-polar solvents but are only sparingly soluble in aqueous solvents.
- triglycerides have a role in energy storage and insulation.
- phospholipid bilayers are a consequence of the hydrophobic and hydrophilic regions.
- non-polar steroids can pass through phospholipid bilayers.

You should be able to:
- give examples of condensation reactions in which monomers are joined to form macromolecules.
- give examples of hydrolysis reactions in which polymers are digested into monomers.
- recognize pentoses and hexoses as monosaccharides from molecular diagrams in ring forms.
- identify and analyse the structure and function of cellulose, starch and glycogen.
- describe the hydrophobic properties of lipids.
- describe how triglycerides are formed by condensation from three fatty acids and one glycerol.
- compare saturated, monounsaturated and polyunsaturated fatty acids.
- compare lipids and carbohydrates in terms of energy storage.
- recognize steroids from molecular diagrams.

Chemical properties of the carbon atom

All living organisms are created from carbon compounds, including carbohydrates, proteins, lipids and nucleic acids. These compounds are used by living organisms in a complex web of chemical reactions called metabolism.

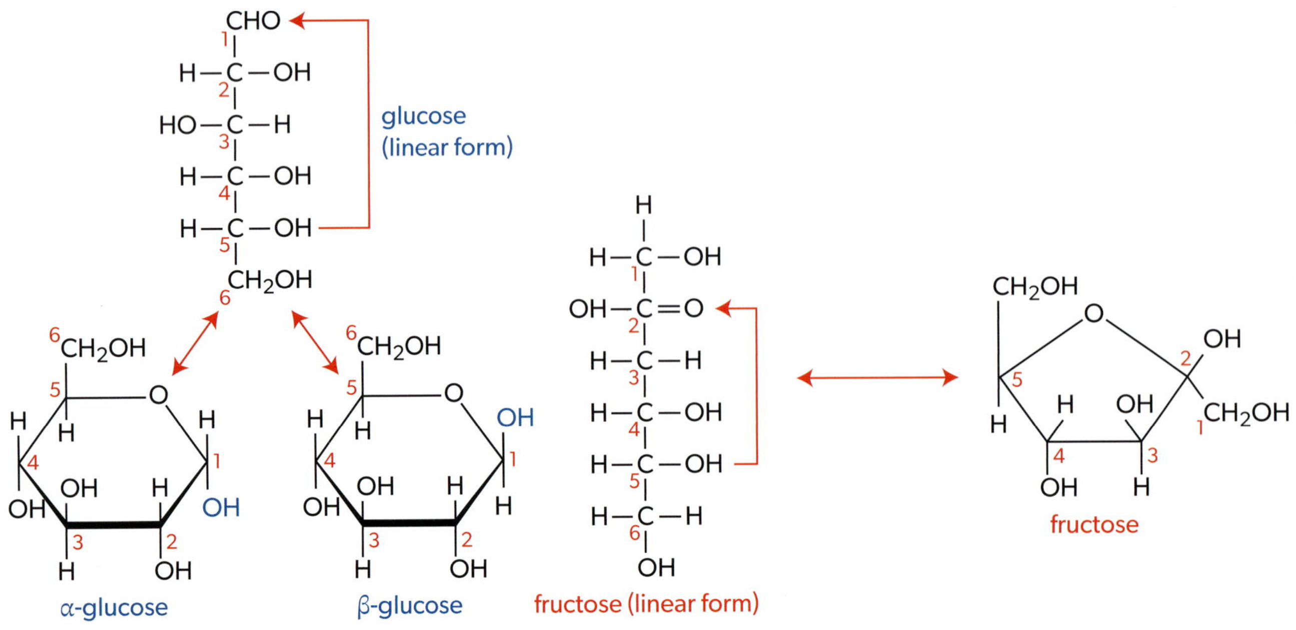

▲ Figure 1 Linear and ring forms of glucose and fructose

Each carbon atom can form up to four single bonds or a combination of single and double bonds with other carbon atoms or atoms of other non-metallic elements. In **carbohydrates**, carbon atoms can join with other carbon atoms or other atoms such as oxygen or hydrogen. The molecule of glucose is shown in a linear manner (**Figure 1**), but experimental evidence indicates that for monosaccharides containing five or more carbon atoms, such open-chain structures are in equilibrium with two cyclic structures. According to which plane the hydroxyl group of carbon 1 is found, glucose can be alpha (found in starch and glycogen) or beta (found in cellulose).

- **Carbohydrates** are carbon compounds containing hydrogen and oxygen. They serve as energy storage and structural components of organisms.

Example 1

Five test tubes with different glucose solutions were tested for the presence of glucose. A positive test results in an orange precipitate, which is seen in varying densities in the tubes.

In an experiment, a test tube containing only lipids and another containing a disaccharide were subjected to carbohydrate digestion where the disaccharide was broken down into two monomers. The results after a glucose test are shown in the photograph.

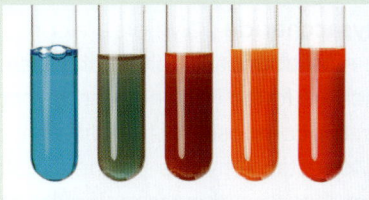

a) Identify the test tube containing the dissacharide.
b) Outline the reaction taking place.
c) Identify the type of bond that has been broken in the disaccharide.

Solution

a) Tube 2—the disaccharide is digested giving a glucose solution that reacts positively for the presence of glucose.
b) **Hydrolysis**. Water molecules are split to provide the –H and –OH groups that are incorporated to produce monomers.
c) Glycosidic bond (a **covalent bond** formed by condensation).

Nature of science

Scientific conventions are based on international agreement. For example, the International System of Units (SI) uses the unit prefixes "kilo", "centi", "milli", "micro" and "nano" for values ×1,000, ×100, ×0.001, ×10^{-6} and ×10^{-9} respectively.

- In **hydrolysis**, water molecules are split to provide the –H and –OH groups that are incorporated to produce monomers from polymers.
- **Covalent bonds** are formed when pairs of electrons are shared between atoms.

Polysaccharides store energy

Glucose is a nearly universal and accessible source of energy, as it is easily oxidized to form adenosine triphosphate (ATP). Because it is very soluble in water it can be easily transported and is chemically stable.

Monosaccharide monomers, such as glucose and fructose, can be linked together with glycosidic bonds to form **disaccharides**, such as sucrose. Because this reaction liberates water it is called a **condensation** reaction. Sucrose (which is found in plants), lactose (the sugar found in milk) and maltose (the sugar found in seeds) are also disaccharides.

- **Monosaccharides** are simple sugars and are the basic building blocks of carbohydrates.
- **Disaccharides** are two monosaccharides joined together.
- **Condensation** is the union of monomers into polymers.

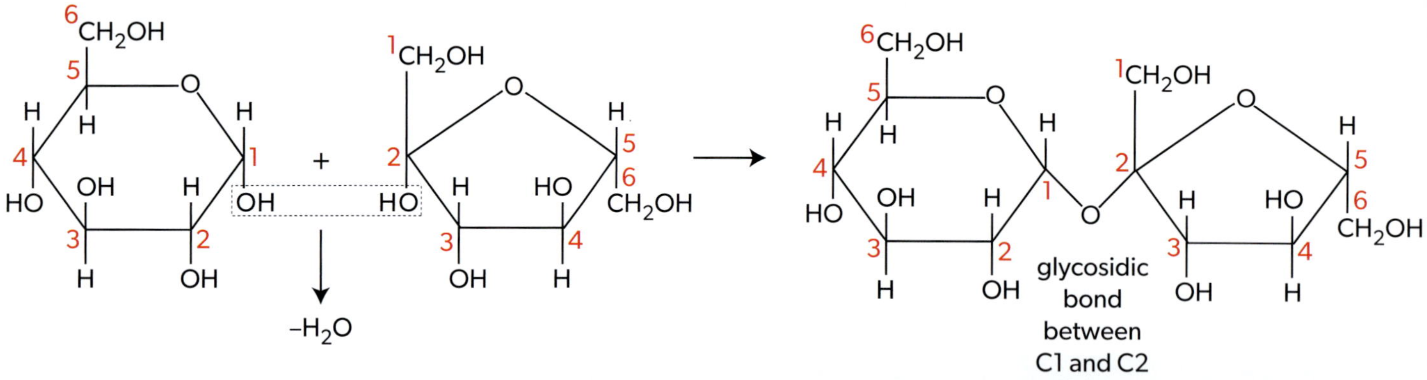

Form and function

▲ **Figure 2** The formation of sucrose (a disaccharide) from the condensation of glucose and fructose (two monosaccharides)

- **Polysaccharides** are polymers of monosaccharides.

Polysaccharide polymers are composed of long chains of monosaccharides. The most abundant polysaccharides in nature are starch and cellulose, which are found in plants, and glycogen, which is mainly found in animals. Starch is composed of two different polymers of α-glucose, amylose and amylopectin. Amylose is a linear molecule, whereas amylopectin is branched. In amylose, the glucose molecules are joined through α-1,4-glycosidic bonds. In amylopectin, these same bonds are formed in the main branch, but other α-glucose molecules are joined at carbon 6 forming α-1,6-glycosidic bonds. The compact nature of starch in plants and glycogen in animals due to coiling and branching during polymerization, the relative insolubility of these compounds due to large molecular size, and the relative ease of adding or removing α-glucose monomers by condensation and hydrolysis makes it an ideal molecule to build or mobilize energy stores.

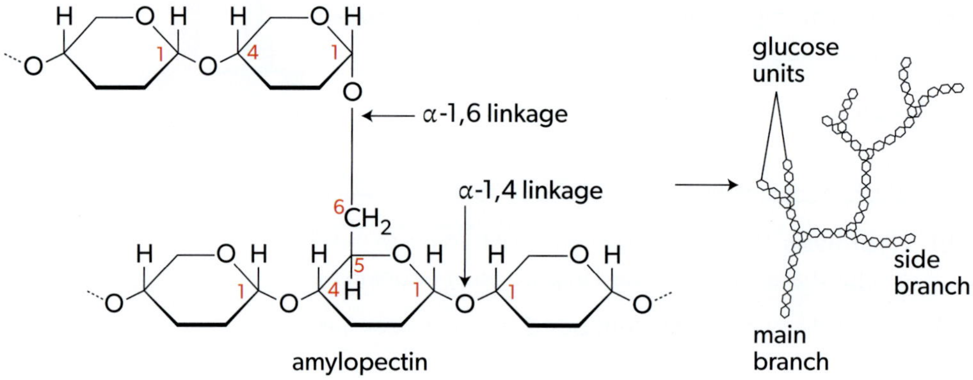

▲ **Figure 3** Structure of starch

Cellulose is a polymer of β-glucose molecules joined by 1,4-glycosidic bonds. Because the hydroxyl group on carbon 1 is on the opposite side of the chain to the hydroxyl group on carbon 4, every second glucose is upside-down to position these two hydroxyl groups together. This makes cellulose a strong molecule, difficult to digest in the human body. Many cellulose molecules join to form microfibres. The fibres found in the cell walls of plants are made from these cellulose microfibres.

Example 2

Explain how the bonding in starch and cellulose molecules affects their structure.

Solution

Starch is formed by α-glucose, whereas cellulose is formed by β-glucose. In α-glucose the hydroxyl group on carbon 1 is on the same side of the chain as the one on carbon 4, allowing the formation of the 1,4-glycosidic bond. In β-glucose, the hydroxyl groups are found on different sides of the chain. For the β-1,4-glycosidic bonds to form, every second β-glucose needs to be upside-down to position the hydroxyl groups on the same side of the chain.

Role of glycoproteins in cell–cell recognition

Conjugated proteins are proteins that are linked to molecules of other types. For example, glycoproteins have carbohydrate components. Many glycoproteins are found at the surface of cells. They carry short carbohydrate chains consisting of several sugar molecules, which usually project outward from the cell. Recognition by the immune system often depends on the precise structure of the carbohydrate chains of glycoproteins. An example is the ABO antigens in blood group recognition.

> In subtopic C3.2 "Antigens trigger antibody production" you will study antigens as recognition molecules that trigger antibody production.

Lipids

Lipids are carbon compounds such as glycerides and steroids that are insoluble in water (hydrophobic). They contain double the energy of carbohydrates. **Fatty acids** are the main components of simple lipids. They are formed by a long chain of carbons and hydrogens with an acid group at carbon 1. The carbons can be joined by single or double bonds.

There are many types of lipids. Some have a simple structure, such as monoglycerides, formed from glycerol and one fatty acid, while others have very complex molecules, such as oils, waxes and steroids. Triglycerides are formed by condensation from three fatty acids and one glycerol molecule. These fatty acids do not need to be the same. In some cases, one carbon joins to a phosphate group instead of a fatty acid, forming a phospholipid.

- **Lipids** are carbon compounds containing mainly hydrogen and some oxygen. They form a diverse group comprising oils, fats, membrane components and hormones.
- **Fatty acids** are formed by long chains of carbons (4 to 34) with hydrogens attached and an acid group at one end.

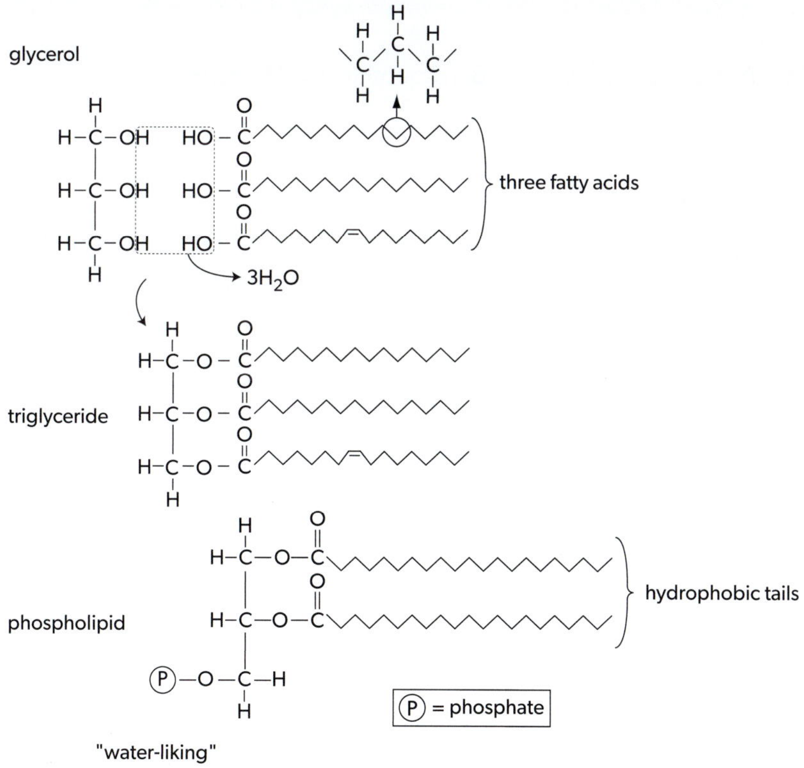

◀ Figure 4 Formation of triglycerides and structure of a phospholipid

Saturated, monounsaturated and polyunsaturated fatty acids

A fatty acid with no double bonds is said to be **saturated**. If it has one or more double bonds it is **unsaturated**. Monounsaturated fatty acids have only one double bond, whereas polyunsaturated fatty acids have two or more double bonds. The greater the number of double bonds in a fatty acid, the lower the melting point. This is the case for oils that are liquid at room temperature due to the high proportion of unsaturated fatty acids in their composition.

Fats are used as energy storage in plants and in endotherms.

- **Saturated** fatty acids do not have any double or triple bonds.
- **Unsaturated** fatty acids have double or triple bonds between two or more carbons.

Form and function

Example 3

Which molecule represents a lipid?

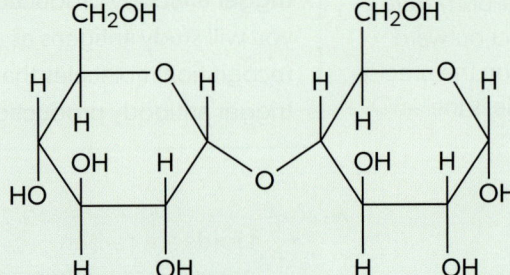

C.

$$H_2CO-\overset{O}{\underset{\|}{C}}-CH_2(CH_2)_{15}CH_3$$
$$HCO-\overset{O}{\underset{\|}{C}}-CH_2(CH_2)_{15}CH_3$$
$$H_2CO-\overset{O}{\underset{\|}{C}}-CH_2(CH_2)_{15}CH_3$$

D. adenine structure

Solution

The correct answer is **C**, as it shows a triglyceride. **A** shows a disaccharide, **B** a dipeptide and **D** adenine, a nitrogenous base.

Sample student answer

Draw the structure of a saturated fatty acid. [2]

This answer could have achieved 1/2 marks:

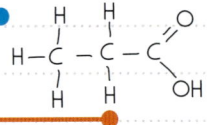

▲ One mark was given for the acid group and hydrogens shown.

▼ The structure only has three carbons and the shortest fatty acid known is butyric acid, with four carbons.

Properties of triglycerides

When the availability of food is high, organisms store energy in the form of triglycerides. These can be mobilized as free fatty acids to be used as energy when there is energy deprivation. Triglycerides are good storage molecules because they are formed from long chains of carbons with many carbon–carbon and carbon–hydrogen bonds that, when oxidized, release a lot of energy. As triglycerides are poor conductors of heat, they are ideal for thermal insulation. In organisms, especially those living in cold climates, fat storage tissue is deposited under the skin.

Formation of phospholipid bilayers

Phospholipids form bilayers because they are **amphipathic**. This means they have a hydrophilic part and a hydrophobic part. The hydrophilic heads face both the outside and the inside of the cell while the hydrophobic part is in the middle of the bilayer. The low melting point of phospholipids in the bilayer is determined by the kinking of the long chain of fatty acids occurring at unsaturated bonds. This determines that some phospholipids are found in the liquid state while others are in the solid state, making the membrane fluid.

- **Amphipathic** molecules contain a hydrophilic (water-loving) and a hydrophobic (water-repelling) part.

You will study the structure of cell membranes in subtopic B2.1.

Example 4

Which features of phospholipids give them their amphipathic properties?

A. Basic phosphate groups and acidic lipids

B. Acidic phosphate groups and basic lipids

C. Hydrophobic phosphate groups and hydrophilic fatty acids

D. Hydrophilic phosphate groups and hydrophobic fatty acids

Solution

The correct answer is **D**, as the phosphate groups are charged and therefore soluble in water (hydrophilic) and the fatty acids are not soluble in water (hydrophobic).

Non-polar steroids pass through the phospholipid bilayer

Steroids are another type of lipid. Non-polar steroids pass through the phospholipid bilayer because they are hydrophobic. This characteristic is important to steroid hormones, which pass directly through the cell and nuclear membranes to act on gene regulation.

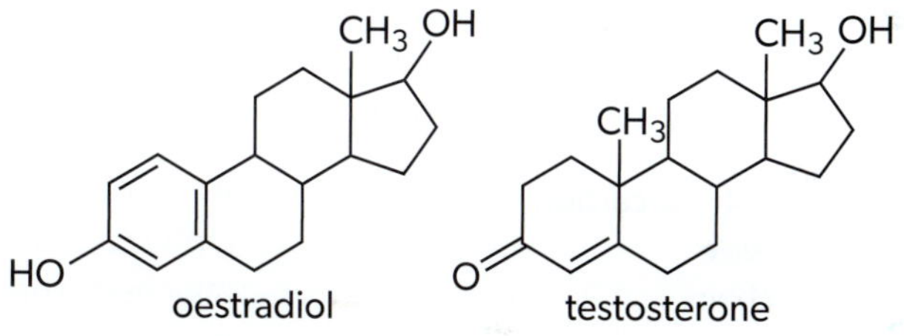

▲ Figure 5 Structure of the steroid hormones oestradiol and testosterone

Example 5

What is a difference between carbohydrates and lipids in energy storage?

A. Carbohydrates are used for long-term storage and lipids for short-term storage.

B. Carbohydrates contain more energy per 100 g than lipids.

C. Carbohydrates are more easily transported to where energy is required than lipids.

D. Carbohydrates store food only in plants, whereas lipids store food in plants and animals.

Solution

The correct answer is **C**, as carbohydrates can be hydrolysed to smaller molecules such as glucose or sucrose that are easily transported because they are soluble. Lipids contain more energy than carbohydrates, but as they are not soluble they serve as a long-term energy store. Both carbohydrates and lipids serve as stores in animals and plants.

- How can compounds synthesized by living organisms accumulate and become carbon sinks?
- What are the roles of oxidation and reduction in biological systems?

Form and function

B1.2 Proteins

You should know:
- ✓ essential amino acids must be obtained from food and non-essential amino acids can be made from other amino acids.
- ✓ a huge range of polypeptides is possible because the amino acids can be linked together in any sequence.

Additional higher level:
- ✓ the chemical diversity of R-groups of amino acids determines the properties of assembled polypeptides.
- ✓ the three-dimensional conformation of a protein is determined by the amino acid sequence (primary structure).
- ✓ the secondary structure is the formation of alpha helices and beta-pleated sheets stabilized by hydrogen bonding.
- ✓ the tertiary structure is the further folding of the polypeptide stabilized by interactions between R-groups.
- ✓ the quaternary structure of non-conjugated (insulin and collagen) and conjugated proteins (haemoglobin).

You should be able to:
- ✓ draw the generalized structure of an amino acid.
- ✓ draw molecular diagrams to show the formation of a peptide bond by the joining of two amino acids in a condensation reaction.
- ✓ explain how heat and changes to pH can denature a protein.

Additional higher level:
- ✓ recognize beta sheets and alpha helices from protein models.
- ✓ describe the effects of polar and non-polar amino acids on the tertiary structure of proteins.
- ✓ describe how the structure of insulin and collagen suit their function.
- ✓ compare the relationship of form and function between globular and fibrous proteins.

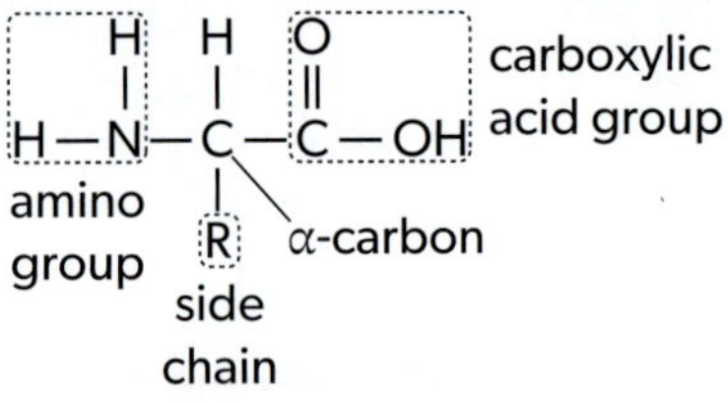

▲ Figure 6 Structure of an amino acid

Amino acid structure

Amino acids are the building blocks of proteins. Each amino acid has an alpha carbon atom with amine group, carboxyl group, R-group and hydrogen attached. These amino acids are joined by condensation in the ribosomes to form polypeptides. The bond between amino acids is called a **peptide bond**.

> **Sample student answer**
>
> Draw molecular diagrams to show the condensation reaction between two amino acids to form a dipeptide. [4]
>
> *This answer could have achieved 4/4 marks:*

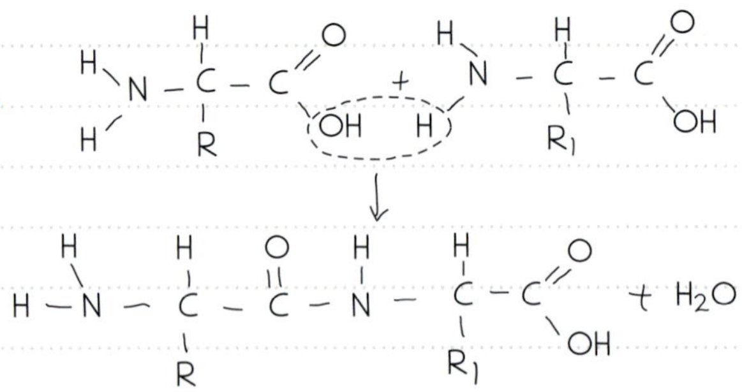

▲ The answer scored full marks for showing each amino acid with an acid group (COOH) with the double bond between the C and O at one end and an amine group (NH_2) at the other end. It also shows the α-carbon in the middle with H and R-group attached and the peptide bond correctly drawn between N and C showing the loss of water.

Essential and non-essential amino acids

Human health requires a balanced diet. **Essential amino acids** are those that cannot be synthesized by the body and therefore they must be included in the diet. People following a vegan diet need to make sure they take in essential amino acids. Non-essential amino acids can be made from other amino acids.

- A **peptide bond** is the bond between amino acids.
- **Essential amino acids** cannot be synthesized by the body, therefore they must be included in the diet.

Sample student answer

The table summarizes the relative content of essential amino acids in different foods. Cysteine and tyrosine are classified as being "conditionally essential". The quantity of each amino acid in a hen egg is set as 1.0 and all other values are relative to the hen egg standard.

	Hen egg	Human milk	Cow milk
isoleucine	1.0	1.1	1.1
leucine	1.0	1.4	1.3
valine	1.0	1.0	1.0
threonine	1.0	1.0	0.9
methionine and cysteine	1.0	1.1	0.7
tryptophan	1.0	1.6	1.3
lysine	1.0	1.0	1.3
phenylalanine and tyrosine	1.0	1.0	0.9
histidine	1.0	0.9	1.1

a) Outline what is meant by the term essential amino acid. [2]

This answer could have achieved 0/2 marks:

> An essential amino acid is required by the human body to remain at homeostasis state and to carry out functions and processes. The human body requires the essential amino acids along with other substances to conduct metabolic processes.

▼ This answer does not score marks because it never mentions that essential amino acids must be obtained from the diet as they cannot be synthesized by the body from other amino acids.

b) Evaluate human milk as an overall source of essential amino acids. [2]

This answer could have achieved 0/2 marks:

> Human milk provides an infant with the first essential amino acids so that infants who consume human milk do not require an additional source, such as formula.

▼ This answer does not score marks as it does not mention that human milk contains all the essential amino acids. In five out of nine essential amino acids the levels are higher than cow milk and in four out of nine are higher than hen egg. Only histidine is slightly lower.

Variety of amino acid sequences

There are 20 different amino acids. A DNA molecule is used as a blueprint to produce RNA that will carry the information to produce a polypeptide in a process called translation. The three-dimensional shape of a protein is determined by the amino acid sequence. Peptide chains can have any number of amino acids, from a few to thousands, in any order. Examples of polypeptides are some antibiotics and peptide hormones such as insulin.

Form and function

Effect of pH and temperature on protein structure

Changes in temperature and pH can alter the spatial disposition of a protein. If the change is extreme, the protein can **denature**. When temperature is increased, the increased vibrations within the molecule can cause the interactions between the R-groups of different amino acids to be broken, changing the structure of the protein. When the pH is changed, the amino acids can either receive hydrogen ions (in acid pH) or lose them (in alkaline pH), changing the protein structure.

- **Denaturation** is a change in the spatial disposition of the protein due to extreme temperature or pH.

Sample student answer

Keratin is a protein found in hair, nails, wool, horns and feathers. Keratinase is a protein that functions as an enzyme that digests keratin. The graphs show the relative keratinase activity obtained in experiments of keratin digestion at different pH values and at different temperatures.

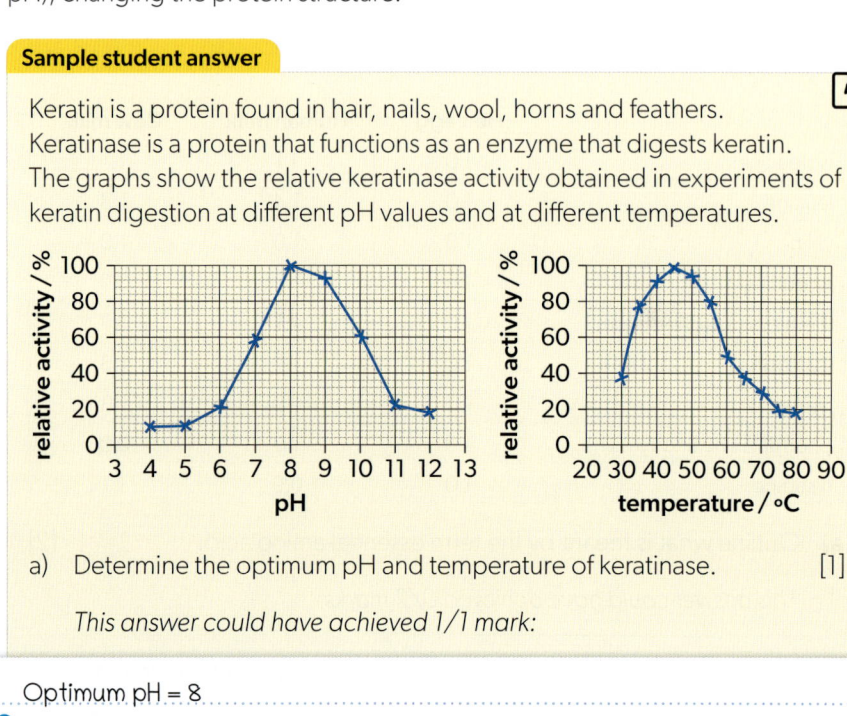

a) Determine the optimum pH and temperature of keratinase. [1]

This answer could have achieved 1/1 mark:

Optimum pH = 8
Optimum temperature = 45 °C

▲ The student identified the optimum temperature in the range 44 to 48 °C and the optimum pH between 7.8 and 8.5.

b) Suggest changes occurring in the reaction vessel that could be used to indicate keratinase activity. [2]

This answer could have achieved 2/2 marks:

The change in mass of keratin.
The change in mass of the newly produced amino acids.

▲ Marks could be scored for mentioning changes in colour or absorbance or any chemical changes indicating the presence of amino acids.

c) State two conditions that should be kept constant in both experiments. [2]

This answer could have achieved 2/2 marks:

The mass of keratinase used.
The time of keratinase–keratin reaction.

▲ The answers are correct. The amount of buffer would have also scored a mark, but not temperature or pH, as these were the variables of the experiment.

d) Suggest a reason for the results observed at 80 °C. [1]

This answer could have achieved 1/1 mark:

Denaturing of keratinase protein.

▲ The answer is correct, as at high temperatures the structure is lost and therefore the protein is denatured.

R-groups of amino acids

Amino acids are classified according to their side chain (**R-group**). These side chains interact in such a way that they determine the form and function of the peptide. R-groups are hydrophobic or hydrophilic. Hydrophobic amino acids have non-polar side chains, whereas hydrophilic R-groups are polar or charged, acidic or basic.

- **R-groups** of amino acids are side chains that are the basis for the immense diversity in proteins.

> **Assessment tip**
>
> You are not required to give specific examples of R-groups.

▲ Figure 7 Classification of some amino acids

Primary and secondary structure of proteins

The **primary structure**—the sequence of amino acids and the exact position of each amino acid—determines the three-dimensional shape of proteins. Proteins therefore have precise, predictable and repeatable structures, despite their complexity. The **secondary structure** of proteins is produced by hydrogen bonding in regular positions that causes pleating (beta sheets) and coiling (alpha helices).

Tertiary structure of proteins

Tertiary structure is the further folding of the polypeptide into globular or fibrous structures, stabilized by interactions between R-groups. The bonds maintaining tertiary structure are disulfide covalent bonds, hydrophobic interactions, ionic bonds and hydrogen bonds. Disulfide bonds are formed between two cysteine amino acids (**Figure 8**). Amine and carboxyl groups in R-groups can become positively charged by binding of hydrogen ions or negatively charged by dissociating from hydrogen ions. Through electrostatic attractions they can then participate in ionic bonding. The polarity of amino acids also affects tertiary structure of a protein. Globular proteins are soluble in water because the polar hydrophilic amino acids are on the outside of the protein, in the part in contact with water, and the non-polar hydrophobic amino acids are clustered in the centre. Integral proteins have their hydrophobic parts embedded in the lipid bilayer.

- **Primary structure** of proteins or polypeptides is the sequence and number of amino acids. Covalent bonds are involved.
- **Secondary structure** is the formation of alpha helices and beta-pleated sheets stabilized by hydrogen bonding.
- **Tertiary structure** is the further folding of the polypeptide into globular or fibrous structures. Disulfide covalent bonds, hydrophobic interactions, ionic bonds and hydrogen bonds are involved.

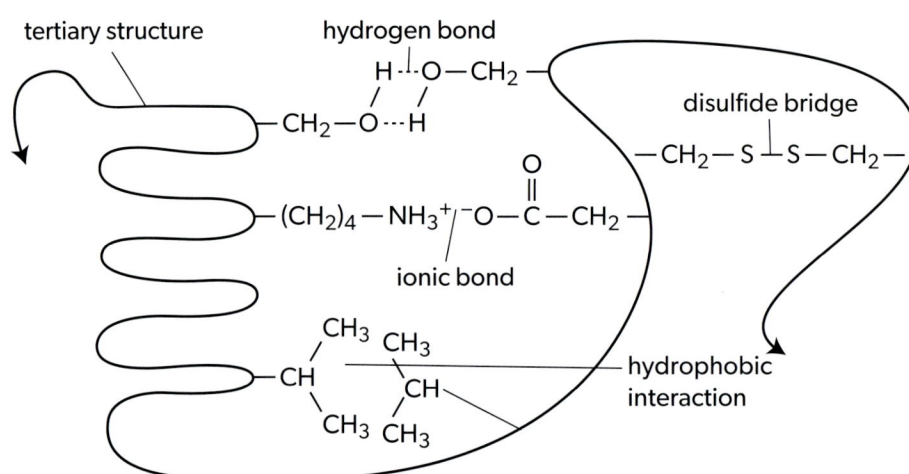

▲ Figure 8 Bonds that stabilize tertiary structure of proteins

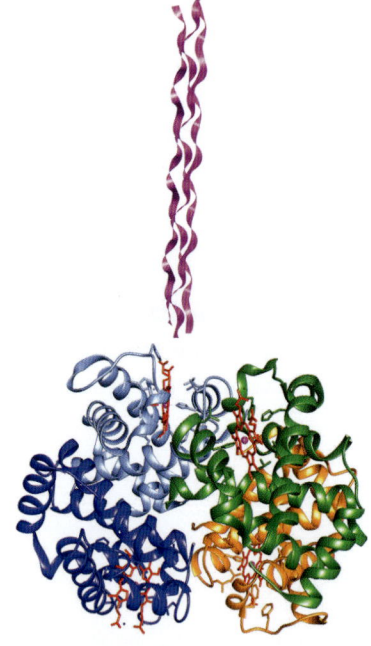

▲ Figure 9 Quaternary structure of collagen (top) showing the three fibrous peptides; and haemoglobin (bottom) showing the four polypeptides and four haem groups (red)

Quaternary structure of non-conjugated and conjugated proteins

Quaternary structure exists in proteins with more than one polypeptide chain. Proteins can also be associated with other molecules such as a prosthetic group—such proteins are said to be "**conjugated**". Haemoglobin is a conjugated protein formed by four peptides and containing four haem prosthetic groups. Insulin and collagen are examples of non-conjugated proteins.

Globular and fibrous proteins

Globular proteins are spherical in shape and with the hydrophilic R-groups usually facing towards the outer part of the protein. This makes them water-soluble, allowing their transport in blood plasma or phloem. Examples of globular proteins are haemoglobin and most enzymes. Fibrous proteins, however, are usually linear and insoluble in water. Collagen is required for structure and is used to make connective tissue, therefore it is essential that it is not soluble in water.

- **Quaternary structure** exists in proteins with more than one polypeptide chain.
- **Conjugated proteins** are associated with prosthetic groups.

Nature of science

Advances in technology, such as cryogenic electron microscopy, have allowed imaging of proteins and their interaction with other proteins. Protein structures can be observed using molecular visualization software such as Protein Data Bank.

- How do abiotic factors influence the form of molecules?
- What is the relationship between the genome and the proteome of an organism?

Practice problems

1. Compare and contrast the structure of glycogen and starch.

2. The image shows a model of human insulin hormone obtained using molecular visualization software.

   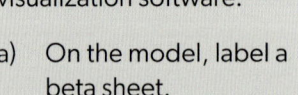

   a) On the model, label a beta sheet.

   b) Describe the structure of insulin.

   c) Outline the function of insulin.

   d) The strong homology seen in the insulin sequence of diverse species suggests that it has been conserved across much of animal evolutionary history. Comment on this statement.

B2.1 Membranes and membrane transport

You should know:
- phospholipids form bilayers due to their amphipathic properties.
- lipid bilayers act as effective barriers between aqueous solutions.
- channel proteins assist in facilitated diffusion.
- pump proteins transfer particles against a concentration gradient using ATP.
- glycoproteins and glycolipids are involved in cell adhesion and recognition.

Additional higher level:
- composition of fatty acids and cholesterol influence the fluidity of cell membranes.
- neurotransmitter-gated and voltage-gated channels allow ion transport in neurons.
- the sodium-dependent glucose cotransporter allows indirect active transport of glucose.
- cell-adhesion molecules (CAMs) are involved in cell–cell junctions.

You should be able to:
- describe simple diffusion of oxygen and carbon dioxide across membranes.
- compare the structure and function of integral and peripheral membrane proteins.
- explain the selective permeability in membranes.
- explain how water moves across membranes by osmosis with the aid of aquaporins—water channels.
- draw the fluid mosaic model in two dimensions to include peripheral and integral proteins, glycoproteins, phospholipids and cholesterol.

Additional higher level:
- explain how cholesterol acts as a modulator in membrane fluidity.
- describe membrane fluidity and the fusion and formation of vesicles for transport by exocytosis and endocytosis (bulk transport).
- explain the importance of sodium–potassium pumps in generating membrane potentials.

Lipid bilayers as cell membranes

The cell membrane is formed by a double layer (**bilayer**) of phospholipids. Phospholipids are amphipathic, which means they have a hydrophilic part and a hydrophobic part. The hydrophilic heads face both the outside and the inside of the cell while the hydrophobic part is in the middle of the bilayer. The cell membrane controls the entrance and exit of substances to and from the cell. Large or charged molecules do not pass through the hydrophobic hydrocarbon chains. Ions with positive or negative charges cannot easily diffuse through, whereas polar molecules, which have partial positive and negative charges over their surface, diffuse at very low rates. Small molecules such as oxygen and carbon dioxide pass across the membrane between the phospholipids by simple diffusion.

Membrane proteins

Proteins are embedded in the phospholipid bilayer. Some proteins are found crossing from side to side (integral **transmembrane proteins**), some partly inside (**integral**), whereas others are only on the outside (**peripheral**). Lipoproteins and glycoproteins can also be found on the outside of the cell membrane. Integral transmembrane proteins contain alpha helices with mostly hydrophobic amino acids, but also contain polar amino acids in the interior that allow the diffusion of molecules in channels.

- Membranes are **bilayers** of phospholipids with scattered proteins.
- **Transmembrane proteins** are integral membrane proteins that span across the membrane.
- **Integral proteins** are embedded in the phospholipid bilayer and protrude on only one side of the membrane.
- **Peripheral proteins** are temporarily attached either to the surface of the phospholipid bilayer or to integral proteins.

Form and function

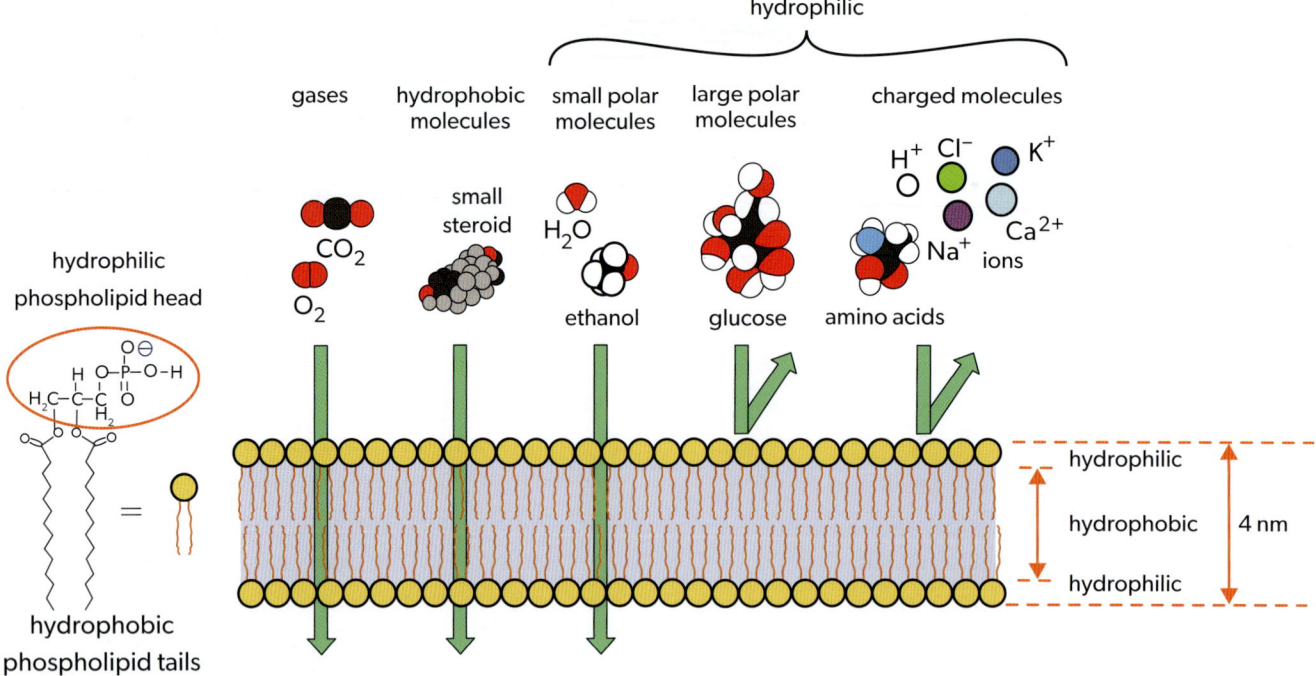

▲ Figure 1 Permeability of the phospholipid bilayer

Source: Cooper, G. M. (2000). *The cell: a molecular approach.* 2nd edition. Sinauer Associates. (CC BY-NC)

- **Osmosis** is the diffusion of water through a selectively permeable membrane, from a higher water potential (lower solute concentration) to a lower water potential (higher solute concentration).
- **Aquaporins** are integral membrane proteins that serve as channels in the transfer of water across the membrane.

Movement of water across membranes

Water can pass by **osmosis** through biological membranes by two pathways: simple diffusion through the lipid bilayer, or water-selective facilitated diffusion through integral membrane channel proteins called **aquaporins**. Water molecules with random motion collide with the lipid bilayer before diffusing to the other side. The cell membrane is impermeable to solutes, resulting in a difference in solute concentration on either side of the membrane. Water moves from a lower to a higher solute concentration solution. Aquaporins increase the water permeability under osmotic gradients. They are present in highest concentrations in tissues where rapid transmembrane water movement is important (for example in renal tubules).

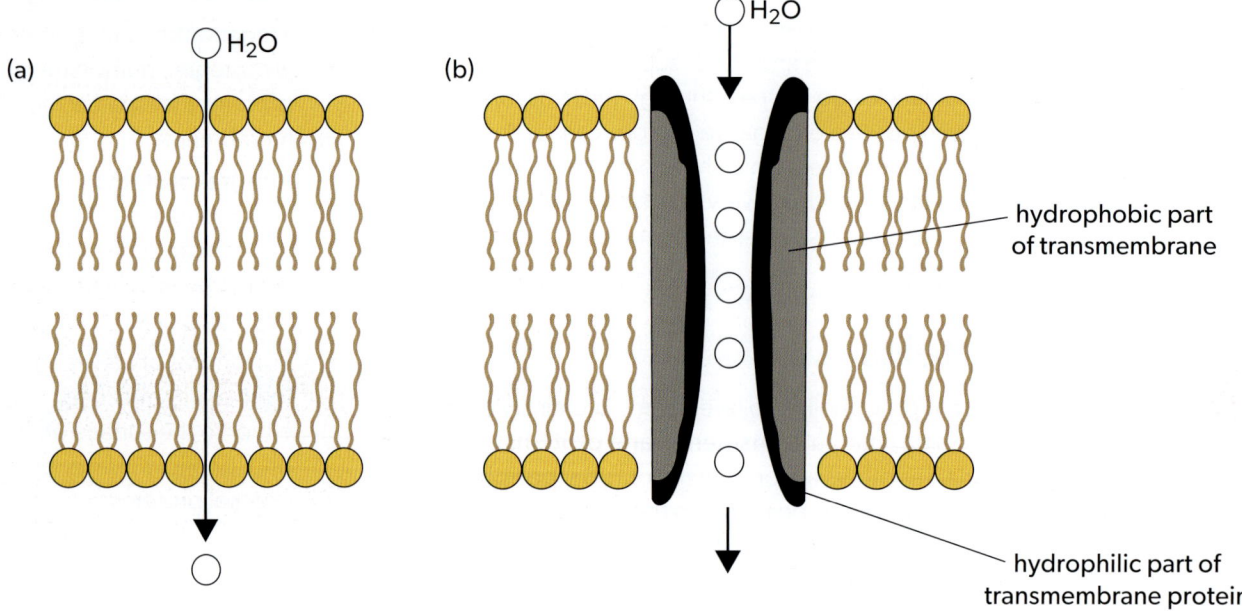

▲ Figure 2 Water transport by (a) simple diffusion and (b) facilitated diffusion through aquaporins allowing transport through the core of the transmembrane protein driven by osmotic gradients

Source: Reuss, L. (2012). Water transport across cell membranes. eLS. www.doi.org/10.1002/9780470015902.a0020621.pub2

Transport across cell membranes

Channel proteins are transmembrane proteins that can open highly selective pores through the membrane, allowing **facilitated diffusion** of any molecule of the appropriate size and charge. Ion channels are present in the membranes of all cells, especially in nerve and muscle, where their regulated opening and closing is responsible for the transmission of electric signals. Substances can pass in or out through these channels by **active** or **passive transport**. Active transport usually occurs against a concentration gradient and therefore requires pump proteins that use energy from ATP.

Permeability by simple diffusion is not selective and depends only on the size and hydrophilic or hydrophobic properties of particles. However, facilitated diffusion and active transport allow selective permeability in membranes.

You will study water potential in more detail in subtopic D2.3.

- **Facilitated diffusion** is the passive transport of molecules or ions across the cell membrane through specific transmembrane proteins (channel proteins).
- **Passive transport** is the movement across the membrane without the use of energy.
- **Active transport** is the movement across the membrane requiring energy in the form of ATP.

Structure and function of glycoproteins and glycolipids

Glycoproteins and glycolipids are molecules with carbohydrates attached to them. Carbohydrate residues of membrane-bound glycolipids and glycoproteins are located on the outside of the membrane. These molecules play an important part in cell adhesion and in cell recognition.

Cell adhesion occurs through anchoring junctions that attach cells to neighbouring cells. An example is cadherins, which act as transmembrane adhesion glycoproteins in cell-to-cell adhesion. Cell recognition occurs when molecules from different cells bind together. An example of glycoproteins involved in cell-to-cell recognition is the antigens on red blood cell membranes. These are recognized by antibodies and therefore determine A, B, AB and O blood groups.

Fluid mosaic model of membrane structure

Singer and Nicolson proposed the structure of the cell membrane as a medley or mosaic of different molecules that have movement in two dimensions, therefore, it is fluid.

Example 1

The drawing shows the structure of the cell membrane.

What is a characteristic of the cell membrane?

A. It has extrinsic proteins only.

B. The phospholipid bilayer is arranged with the hydrophilic parts in the middle.

C. It is a fluid structure of phospholipids with embedded proteins and cholesterol.

D. It has a continuous layer of proteins.

Singer–Nicolson

Solution

The correct answer is **C**, as although there are extrinsic proteins, there are also transmembrane and intrinsic proteins. The phospholipids are arranged as a bilayer with the hydrophobic part in the middle. A continuous layer of proteins would not explain its fluidity.

Form and function

Sample student answer

Draw a labelled diagram that shows the positions of proteins within the cell membrane. [3]

This answer could have achieved 2/3 marks:

▲ This answer scored one mark for the labelled integral protein shown crossing the membrane (transmembrane). The second mark was for the glycoprotein.

▼ The phospholipid bilayer is drawn but not labelled. Also the protein labelled as peripheral protein is shown embedded in the bilayer, not on the membrane surface, so it is really an integral protein.

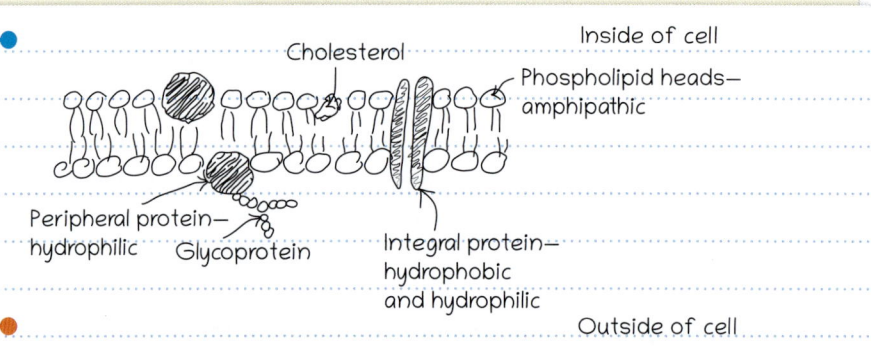

- **Cholesterol** is a lipid that acts as a modulator in membrane fluidity.

Fatty acid content and membrane fluidity

The fluidity of membranes depends on the temperature and on the amount of **cholesterol** and types of fatty acids forming the membrane. The low melting point of phospholipids in the bilayer is determined by the kinking of the long chain of fatty acids occurring at unsaturated bonds. This determines that some phospholipids are found in the liquid state while others are in the solid state, making the membrane fluid. Saturated fatty acids do not have double bonds and therefore are tightly packed, so they have higher melting points and make membranes stronger at higher temperatures. Cholesterol embedded in the membrane layer controls membrane fluidity and permeability to some solutes. Cholesterol acts as a modulator (adjustor) of membrane fluidity, stabilizing membranes at higher temperatures and preventing stiffening at lower temperatures. Bilayers consisting of high-fluidity phospholipids are more permeable to water. The presence of cholesterol decreases the fluidity and decreases permeability of the bilayer to water.

Organisms can adapt to different environments by change in shape or movement of cells given by the fluidity of membranes. For example, fish species normally living in the Arctic have a different membrane composition than fish living in hot-springs water. Differences in fluidity are obtained by differences in the saturation of membrane phospholipids. At increased cell temperatures, the proportion of saturated fatty acids increases and the proportion of unsaturated fatty acids decreases.

Example 2

Which plasma membrane is the least fluid at high temperatures?

A. 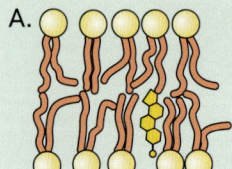 B. 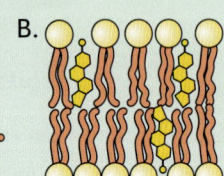 C. 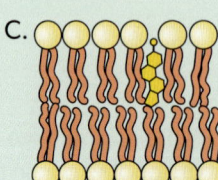 D.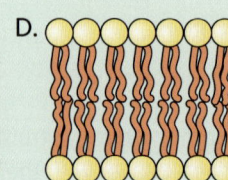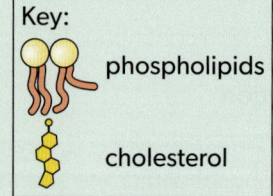

Solution

The correct answer is **B**, as it has more cholesterol and is formed only by unsaturated fatty acids (this can be seen by the lack of "kinking" in the tails). The most fluid would be **A**, as it has mostly unsaturated fatty acids and some cholesterol. **D** would be more fluid than **C** as cholesterol stabilizes membranes at high temperatures.

Fusion and formation of vesicles in exocytosis and endocytosis

Substances that cannot enter through channel proteins because they are too large require bulk transport – this is transport in membrane-bound vesicles. Bulk transport into the cell is called **endocytosis** and bulk transport exiting the cell is called **exocytosis**. In endocytosis, the fluidity of the cell membrane allows the membrane to surround the particle to be ingested. In exocytosis, vesicles formed in the Golgi apparatus fuse with the membrane to transport the substances out of the cell.

- Bulk transport through vesicles includes **exocytosis** (leaving the cell) and **endocytosis** (entering the cell).

You will study the structure of vesicles in subtopic B2.2.

Example 3

Use the diagram of part of a cell to answer the questions.

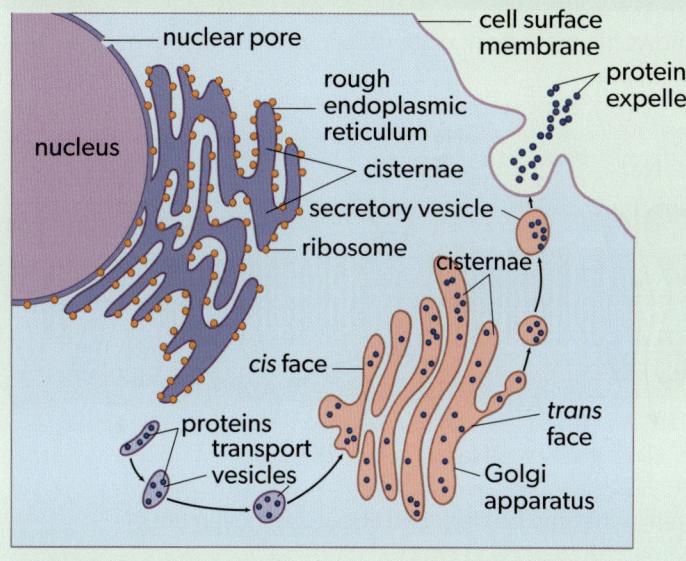

a) Which organelle(s) other than the nucleus show(s) that this is a eukaryotic cell?

I. Golgi apparatus
II. Ribosomes
III. Rough endoplasmic reticulum

A. I only
B. III only
C. I and III only
D. I, II and III

b) What process is shown in the diagram?

A. Translation
B. Transcription
C. Exocytosis
D. Diffusion

Solution

a) The correct answer is **C**, as ribosomes are also present in prokaryotes (although these are 70S instead of 80S as in eukaryotes).

b) The correct answer is **C**, as the exocytic vesicles can be seen secreting proteins. Of course, transcription and translation and even diffusion must have occurred, but these process are not shown in the diagram.

Gated ion channels in neurons

Some channels open when a **ligand** binds to them, for example a neurotransmitter, and others when there is a change in the electric potential across the membrane (voltage-gated channels). Nicotinic acetylcholine receptors (for example, the GABA receptor) are an example of neurotransmitter-gated ion channels, found in skeletal muscle cells and in neurones. These receptors are protein subunits arranged around a central pore. The binding of a ligand in the extracellular part causes allosteric changes in the channel and opens the pore.

Voltage-gated sodium or potassium channels are transmembrane channels specific for each of these ions and are sensitive to changes in membrane potentials. In action potentials, they play a crucial role in returning the depolarized cell to a resting state. These channels are inactivated by a "ball and chain" model, where the pore is blocked by a globular protein subunit (ball) attached by a flexible chain of amino acids (**Figure 3**). The voltage sensor is a region of the protein bearing charged amino acids that relocate upon changes

- A **ligand** is a molecule that binds to another molecule that has a biological purpose.

Form and function

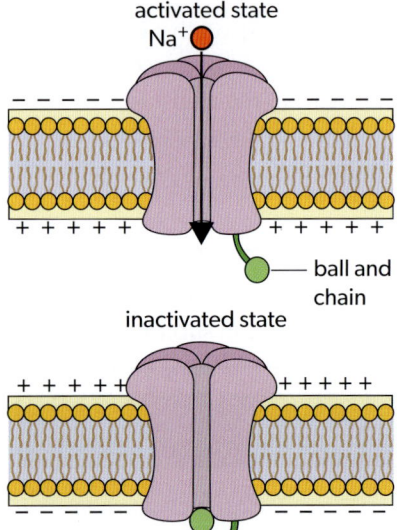

▲ Figure 3 Voltage-gated sodium channel

> You will study membrane potential in subtopic C2.2.

in the membrane electric field. Changes in membrane potential cause the transmembrane part of the channel to move in the lipid bilayer. This movement results in a conformational change that causes the pore to open or close.

Sodium–potassium pumps are exchange transporters

The sodium–potassium pump is an example of an exchange transporter found on cell membranes that requires energy. It transports three sodium ions ($3Na^+$) out of the cell and two potassium ions ($2K^+$) into the cell against a concentration gradient using one ATP molecule. It helps to maintain the osmotic equilibrium and the membrane potential in cells.

Example 4

The diagram shows the movement of ions that can occur across a membrane of a neuron.

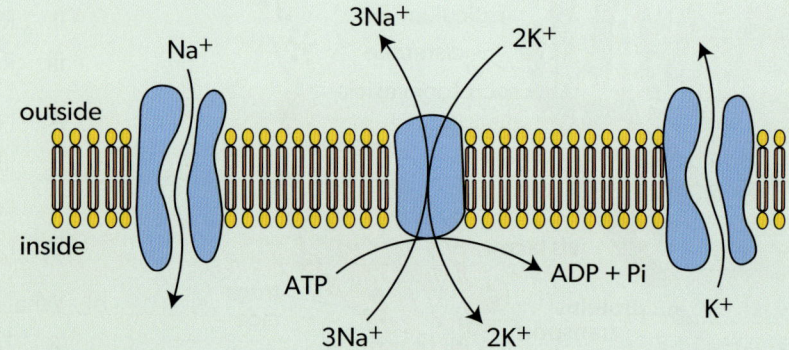

From the diagram, what can be deduced about the movement of sodium ions?

A. They are actively pumped out and some re-enter by facilitated diffusion.

B. They are actively pumped out and some re-enter by simple diffusion.

C. They diffuse out of the cell along with potassium ions.

D. There is net movement of sodium into the cell.

Solution

The correct answer is **A**. The channel on the left shows sodium ions entering by facilitated diffusion (without the use of energy) through a voltage-gated channel. In the centre the sodium ions leave and potassium ions enter by active transport through a sodium–potassium pump using ATP. The channel on the right shows potassium ions leaving by facilitated diffusion. Sodium ions do not re-enter by simple diffusion as a channel protein is required. Sodium ions do not diffuse out of the cell—they are pumped out by the sodium–potassium pump. There is net movement of sodium out of the cell, as more sodium ions leave the cell than enter.

- **Sodium cotransport** is the process by which sodium is transported outside the cell by active transport to allow the entrance of other substances, such as glucose, into the cell by cotransport with the sodium ions.

Sodium-dependent glucose cotransporters

Membrane transporters that use energy stored in sodium gradients to drive nutrients into cells are an example of indirect active transport. These **cotransporters** are important in glucose absorption by cells in the small intestine and glucose reabsorption by cells in the nephron. Sodium-dependent glucose cotransporters in the intestine are channel transmembrane proteins embedded in the membrane with a sodium binding site that transports glucose into the epithelial cells. Glucose then leaves the epithelium through facilitated diffusion protein channels into blood vessels (**Figure 4**). Sodium–potassium ion pumps serve to stabilize the membrane potential.

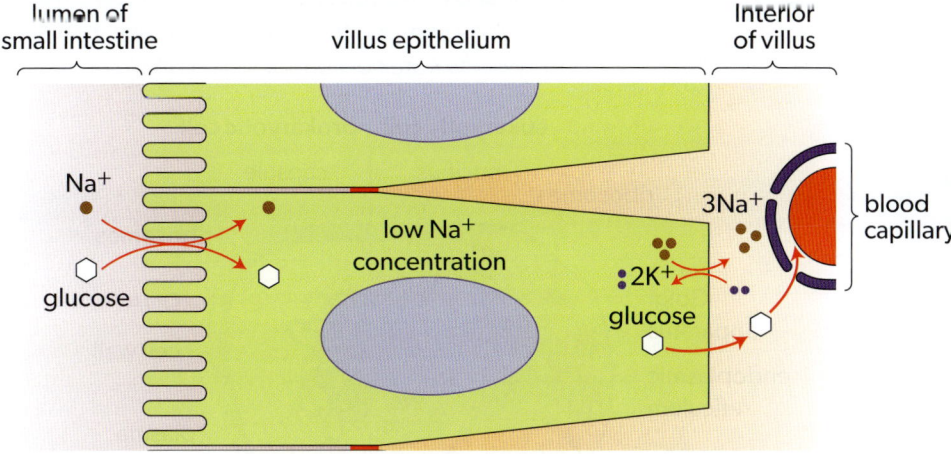

◀ Figure 4 Absorption of glucose in the villus of the small intestine

The glucose reaching the kidneys is filtered through the glomerulus and passes to the proximal convoluted tubule, where most of it is reabsorbed. This is done using sodium-dependent glucose cotransporters. Sodium–potassium ion active transport pumps remove sodium from the tubule wall and the sodium is put back into the blood.

Cell-adhesion molecules

Cell adhesion is fundamental in the development and maintenance of tissues. It allows communication between cells and their regulation. Cell adhesion regulates the cell cycle, its differentiation, migration and survival through signals. Cell adhesion occurs through interaction with cell-adhesion molecules (CAMs). Integrins, cadherins, selectins and immunoglobulins are examples of CAMs. Different forms of CAMs are used for different types of cell–cell junctions.

- What processes depend on active transport in biological systems?
- What are the roles of cell membranes in the interaction of a cell with its environment?

B2.2 Organelles and compartmentalization

You should know:

✔ organelles are discrete subunits of cells with specific functions.

✔ the nucleus and cytoplasm are in separate compartments.

Additional higher level:

✔ mitochondria are adapted to produce ATP by aerobic cell respiration.

✔ chloroplasts are adapted for photosynthesis.

✔ the nucleus has a double membrane with pores.

✔ free ribosomes synthesize proteins for retention in the cell.

✔ ribosomes on the rough endoplasmic reticulum synthesize proteins for transport within the cell and secretion.

You should be able to:

✔ give examples of cell parts that are organelles and recognize those that are not.

✔ discuss the advantages of compartmentalization in the nucleus and in cytoplasm.

Additional higher level:

✔ explain adaptation to function in mitochondria and chloroplasts.

✔ describe the structures and functions of the endoplasmic reticulum, Golgi apparatus and vesicles.

✔ explain the function of clathrin in vesicle formation.

Form and function

Nature of science

The study of the function of individual organelles became possible with cell ultracentrifugation and cell fractionation.

- **Organelles** are subunits of cells that perform a specific function.

Organelles

Organelles are discrete subunits in cells. Each has adaptations which allow it to perform its specific functions. **Figure 5** shows some organelles and other cell components found in a eukaryotic and prokaryotic cell. Organelles shown in the diagram include nuclei, vesicles, ribosomes and the plasma membrane; however, the cell wall, cytoskeleton and cytoplasm are not considered organelles.

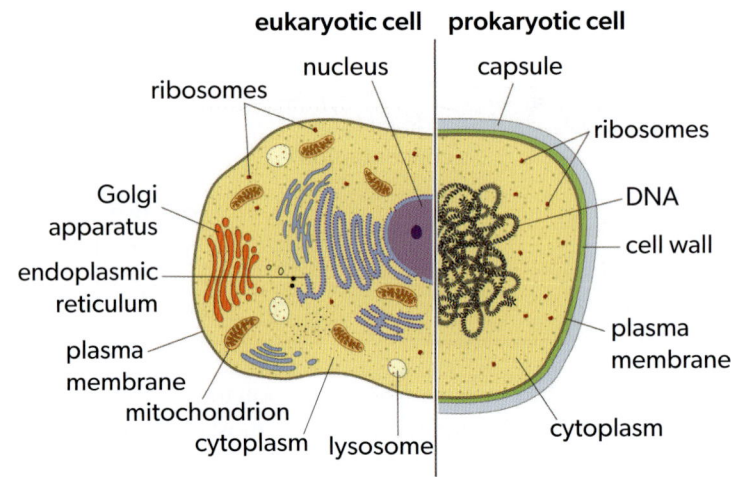

▶ Figure 5 Components of a eukaryotic and a prokaryotic cell

Assessment tip

You can be asked in paper 1B about experimental processes. In cell fractionation, cell contents are separated by lysis and centrifugation. Nuclear content is found in the pellet and cytoplasmic content in the supernatant. These contents can be further separated in a sucrose gradient by density centrifugation.

Advantage of separation of the nucleus and cytoplasm

The nucleus is surrounded by a double membrane nuclear envelope, which establishes a barrier between the nuclear contents and the cytoplasm. This separation into compartments allows for protection of genetic material, regulated exchange of materials through nuclear pores and a distinctive environment for the processes of replication, transcription and translation. The separation of the nucleus allows gene transcription and translation to be separated. The post-transcriptional modification of mRNA can happen before the mRNA meets ribosomes in the cytoplasm. These modifications protect the mRNA from enzymatic degradation and allow for alternative splicing, which is not possible in prokaryotes, where mRNA may immediately meet ribosomes.

Advantages of compartmentalization in the cytoplasm of cells

Cell compartmentalization in the cytoplasm allows the separation of biochemical processes and concentration of metabolites and enzymes in one compartment. Lysosomes are spherical membrane-bound organelles that are rich in digestive enzymes. If they were not contained in a membrane-bound system, these enzymes would degrade proteins, nucleic acids and certain polysaccharides. White blood cells that encounter foreign matter engulf it into a phagocytic vacuole of invaginated plasma membrane, into which cytoplasmic granules release their contents of lethal enzymes. It is important, therefore, to compartmentalize cells to protect from such destructive enzymes.

Adaptations of the mitochondrion for aerobic cell respiration

The structure of a mitochondrion is adapted to the functions it performs, for production of ATP by aerobic cell respiration. The mitochondrial matrix contains enzymes and substrates for the link reaction and Krebs cycle. It has

B2.2 Organelles and compartmentalization

many ribosomes and DNA for protein synthesis. It has a double membrane with a small intermembrane gap between the inner and outer membranes where a gradient of protons can develop. The cristae are folds in the inner membrane that give it a large surface area for ATP synthesis by chemiosmosis, as this is where the proton pumping and electron transport chains occur.

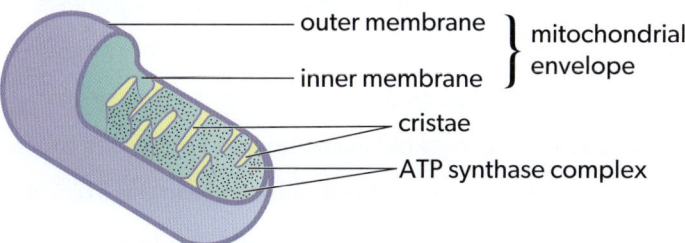

▲ Figure 6 Stereogram of a mitochondrion

Adaptations of the chloroplast for photosynthesis

Sample student answer

a) Mitochondria are thought to have evolved from prokaryotic cells. Describe **two** adaptations of the mitochondria, each related to their function. [2]

This answer could have achieved 1/2 marks:

> Mitochondria have two membranes which are needed for oxidative phosphorylation and the electron transport chain. The intermembrane space is vital to the production of ATP in these steps.
> Also the folded cristae increase surface area for the processes of cellular respiration and ATP production.

▼ Although the answer correctly mentions that mitochondria have two membranes, it does not mention how this helps in the oxidative phosphorylation or electron transfer chain. It also mentions the intermembrane gap, but does not mention the building up of the proton gradient.

▲ The mark is scored for the cristae that increase surface area for ATP synthesis.

b) The diagram shows the structure of a chloroplast.

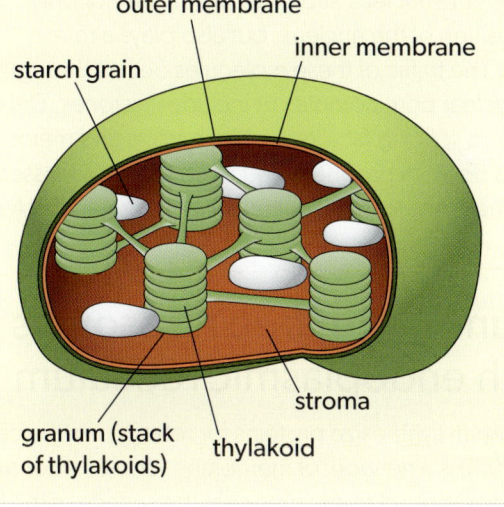

Examiner tip

Remember that only the first two characteristics will be marked in a two-mark question. In this case, because the answer is shown in two paragraphs, each paragraph was considered independently.

> Explain how the chloroplast is adapted to perform the light-independent process of photosynthesis. [8]
>
> *This answer could have achieved 7/8 marks:*

> The chloroplasts are adapted to perform the process of photosynthesis. They have the inner membrane called a thylakoid membrane that stacks into grana to provide a large surface area. Absorption of light occurs in photosystems embedded in the thylakoid membrane and generates excited electrons which are carried by carriers such as cytochrome in the thylakoid membranes. The membrane allows these proteins to be one next to the other. Light-dependent reactions take place in the intermembrane space of the thylakoids. The electrons lost from Photosystem II get replaced by the oxidation of water, which is lysed into protons and oxygen. The electron transfer in the thylakoid membrane causes these protons to be actively pumped across the thylakoid membrane into the thylakoid space. The protons then flow down their electrochemical potential gradient through ATP synthase, creating ATP by phosphorylation of ADP in the process of chemiosmosis. Oxygen diffuses out of the chloroplasts. Excited electrons from Photosystem I are used to reduce NADP forming NADPH to be used in the light-independent reactions in the stroma of the chloroplast.

▲ This answer explains how the chloroplast is adapted to perform its function. It could have added that it contains ribosomes and DNA to synthesise proteins involved in photosynthesis, for example the proteins of the electron transfer chain. Also, that there are small volumes of fluid inside thylakoids, and compartmentalization of enzymes and substrates of the Calvin cycle in the stroma.

▼ Although the answer was good enough to score many marks, it has some faults. For example, water is not oxidized and it is not lysed into just protons and oxygen.

Functional benefits of the double membrane of the nucleus

The nucleus is surrounded by a double membrane with nuclear pore complexes. The outer membrane is continuous with the endoplasmic reticulum. The nuclear pores are responsible for the selective transport of proteins and RNA between the nucleus and cytoplasm. This not only changes the internal composition of the nucleus, but also plays a role in regulating gene expression. The traffic of these molecules occurs through regulated channels in the nuclear pore complex. It includes histones, DNA polymerases, transcription and splicing factors. During mitosis and meiosis, the nuclear envelope breaks down in prophase and once these processes end it reassembles around the chromosomes, where the contained DNA is replicated and transcribed.

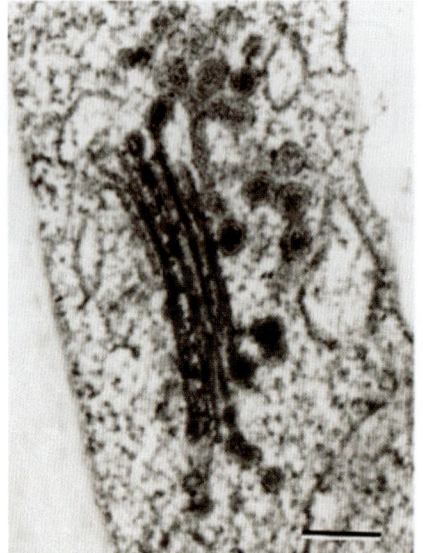

▲ Figure 7 Cryogenic electron micrograph of the Golgi apparatus

Source: Andreeva, A. V., et al. (1998). The structure and function of the Golgi apparatus: a hundred years of questions. *Journal of Experimental Botany*, **49**(325), 1281–1291. www.doi.org/10.1093/jxb/49.325.1281

Structure and function of free ribosomes and of the rough endoplasmic reticulum

Free ribosomes in the cytoplasm synthesize proteins for retention in the cell. The endoplasmic reticulum (ER) is a network of membrane tubules extending throughout the cell. Membrane-bound ribosomes coat the surface of the ER, creating the rough ER (RER). These ribosomes synthesize proteins for transport within the cell and for secretion. The Golgi apparatus consists of a series of flattened, tubular membrane-stacked structures, surrounded by vesicles.

The Golgi apparatus has two main functions: post-translational protein modification (for example, glycosylation) and sorting, packing, routing and recycling of proteins to the appropriate cellular destinations.

Structure and function of vesicles in cells

During **endocytosis**, the membrane that will form the vesicle is coated with the protein **clathrin**. Clathrin provides a mechanical support and the energy necessary to deform the membrane, pulling it inwards, causing the plasma membrane to start curving to form the vesicle.

- **Endocytosis** is the uptake of material into the cell from the surface.
- **Clathrin** is a protein that covers endocytic vesicles formed from the plasma membrane.

Example 5

Using the diagram, explain the function of vesicles in neurons.

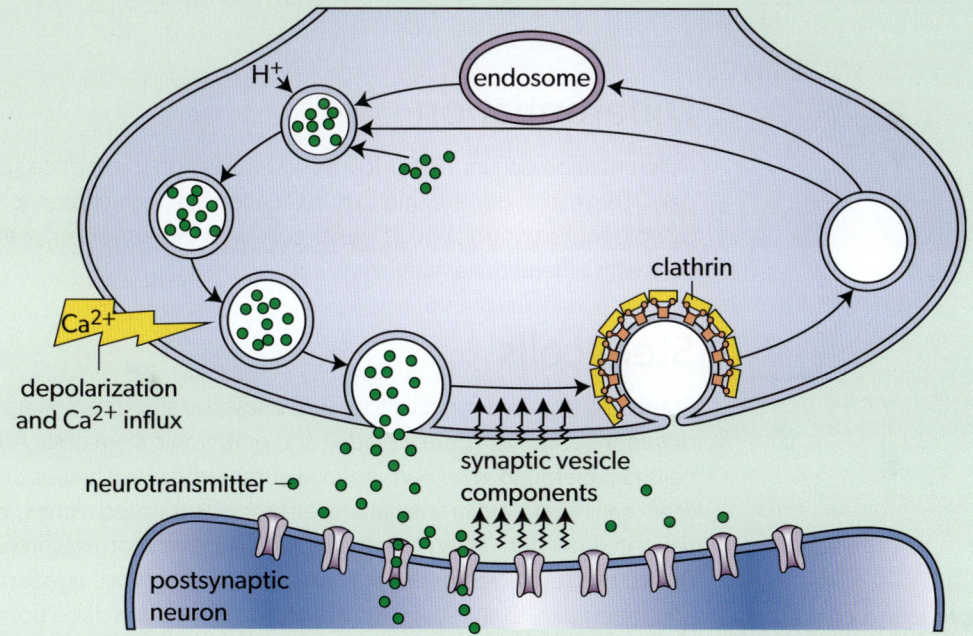

Source: McMahon, H. T. & Boucrot, E. (2011). Molecular mechanism and physiological functions of clathrin-mediated endocytosis. *Nature Reviews Molecular Cell Biology*, 12(8), 517–533. www.doi.org/10.1038/nrm3151

Solution

Action potentials and depolarization-induced calcium ion influx trigger synaptic vesicle fusion and neurotransmitter release from the presynaptic neuron. Synaptic vesicle components are sorted into clathrin-coated pits. Endocytosis of neurotransmitters occurs in clathrin-coated vesicles. Recycled synaptic vesicles are translocated to endosomes, where they are reloaded with neurotransmitters for another round of release.

- What are examples of structure–function correlations at each level of biological organization?
- What separation techniques are used by biologists?

Form and function

B2.3 Cell specialization

You should know:
- unspecialized cells develop into specialized cells from the early-stage embryo by differentiation.
- stem cells divide endlessly and differentiate along different pathways.
- surface area-to-volume ratios and constraints on cell size.

Additional higher level:
- the structure of cardiac muscle cells allows propagation of stimuli through the heart wall.
- the structure of striated muscle cells allows contraction.
- gametes are adapted to their function.

You should be able to:
- describe cell adaptations.
- describe examples of location and function of stem cell niches in adult humans.
- distinguish between totipotent, pluripotent and multipotent stem cells.
- explain specialization according to size.
- model surface area-to-volume relationships.

Additional higher level:
- describe adaptations of cells to increase surface area-to-volume ratios.
- compare type 1 and type 2 pneumocytes in terms of structure and function.

Differentiation

Differentiation occurs in cells following fertilization to produce specialized cells. These changes are due to modifications in gene expression determined by epigenetic control, which makes cells with the same genome transform into cells with different characteristics.

Stem cells

Stem cells have the capacity to divide endlessly and differentiate along different pathways. There are adult and embryonic stem cells. Adult stem cells can be found in humans in small numbers in many tissues of the body. Stem cells are found in specialized environments called niches, which regulate how stem cells participate in tissue generation, maintenance and repair. This regulation is humoral, neuronal, through cell signalling, physical and metabolic. Examples of **multipotent** stem cells include bone marrow, intestinal epithelium and hair follicles. In humans, the zygote that is formed after fertilization is extremely undifferentiated (**totipotent**). Embryonic stem cells are **pluripotent** as they can generate any cell type.

- **Multipotent** stem cells can differentiate into all cell types within one particular lineage.
- **Totipotent** cells can develop into any type of cell and a new organism.
- **Pluripotent** stem cells (for example, human embryonic stem cells) can develop into all cell types.

Cell size as an aspect of specialization

In multicellular organisms, the degree of specialization depends on the size of the cell. The lack of specialization in small multicellular organisms might be due to a minimum threshold size at which specialization becomes advantageous.

Adult human cells vary in sizes:
- Striated muscle fibres: 2–3 cm long and 100 µm in diameter.
- Female gamete (egg cell): a diameter of about 100 µm.
- Male gamete (sperm): a length of about 50 µm and a diameter of about 5 µm.
- White blood cell: 12–17 µm in diameter.
- Red blood cell: 6–8 µm in diameter.
- Neuron: 10–25 µm in diameter and 4–100 µm long.

Example 6

A study simulated digital organisms to determine the transition from undifferentiated organisms, which contain only a single cell type, to multicellular organisms with specialized cells. The bar graphs show the number of simulations that evolved organisms with (blue bars) and without (red bars) specialized cells after 200, 500, 5,000 and 10,000 generations, grouped according to the size of the organism. A chi-squared test (χ^2) was performed and the p-values obtained are shown.

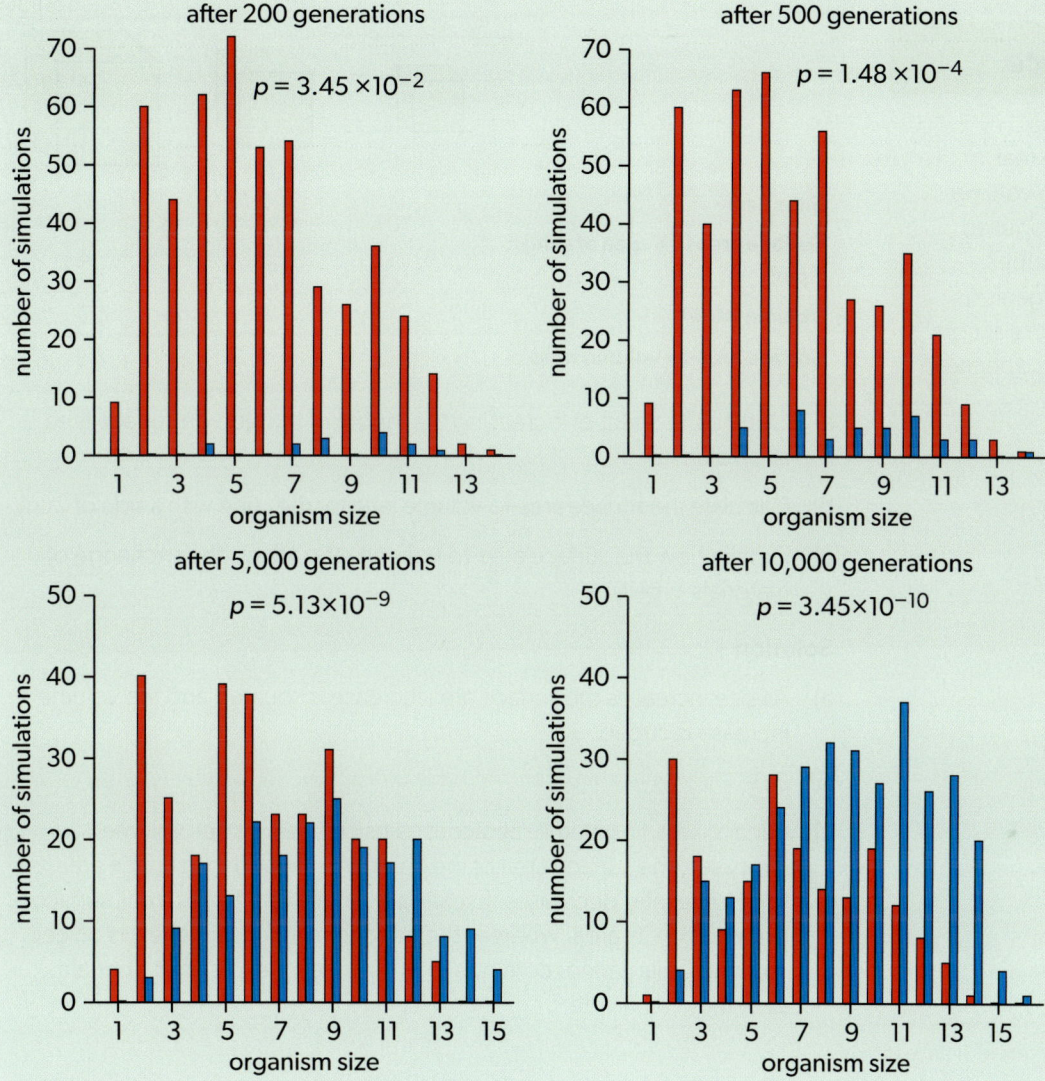

Source: Willensdorfer, M. (2008). Organism size promotes the evolution of specialized cells in multicellular digital organisms. *Journal of Evolutionary Biology*, 21(1), 104–110. www.doi.org/10.1111/j.1420-9101.2007.01466.x

a) State the independent variable in this simulation.

b) Explain the use of the chi-squared test in this simulation.

c) Scientists concluded that the evolution of simple multicellular organisms, which are composed of many functionally identical cells, accelerates the evolution of more complex organisms with specialized cells. Justify this conclusion using the data of the simulations.

Solution

a) Organism size

b) The chi-squared (χ^2) test can be used to determine whether there is a significant difference between the expected frequencies and the observed frequencies in one or more categories. The alternative hypothesis is accepted (and the null hypothesis is rejected) because the difference between the observed results and expected results is statistically significant (with a $p < 0.05$, where p is the probability).

c) Larger organisms are composed of more cells than smaller organisms. Large organisms show a significant production of specialized cells earlier than small organisms.

Surface area-to-volume ratios and constraints on cell size

Assessment tip

This is a paper 1B question.

Nature of science

Models can be used as representations of the real world. Surface area-to-volume relationship can be modelled using different-sized cubes. Although the living organisms are more complex than a simple cube, the scale factors operate in a similar manner.

Example 7

The table shows the properties of three cubes according to the length of the side.

Side / cm	1	2	3
Surface area (6 × area of side) / cm²	6	24	54
Volume (side³)	1	8	27
Surface area-to-volume ratio	6:1		2:1

a) Outline the effect of increasing the length of the side on surface area and volume.

b) Calculate the surface area-to-volume ratio for the cube with a side of 2 cm.

c) Explain how the surface area-to-volume ratio affects the exchange of materials in cells.

Solution

a) As size increases the surface area increase is squared and the volume increase is cubed.

b) 3:1

c) Surface area-to-volume ratio constrains cell size. As cell volume increases, the surface area of the cell increases and the surface area-to-volume ratio decreases. Exchange of materials across a cell surface depends on its area, whereas the need for exchange depends on cell volume. Cells cannot continue to grow indefinitely as after a certain size, they will divide.

Adaptations to increase surface area-to-volume ratios of cells

- **Adaptations** to increase surface area-to-volume ratios include flattening of cells, microvilli and invaginations.

Some cells have **adaptations** to increase surface area-to-volume ratios, such as the flattening of cells, microvilli and invaginations. Erythrocytes (red blood cells) are shaped as a semicircular disc, increasing the surface area for absorption of oxygen. The proximal convoluted tubule cells in the nephron are simple tall cuboidal epithelium, with many microvilli (brush border) which increase surface area for reabsorption of substances.

Adaptations of type I and type II pneumocytes in alveoli

The epithelium in alveoli of the lungs has more than one cell type because different adaptations are required for the overall function of the tissue. **Type I pneumocytes**, which are numerous, very large and thin to reduce distance for diffusion, are where gaseous exchange occurs. **Type II pneumocytes**, which are cube-shaped and occur among the type I cells, produce surfactant. They have many secretory vesicles (lamellar bodies) in the cytoplasm that discharge the surfactant to the alveolar lumen.

- **Type I pneumocytes** are cells in the alveoli involved in the process of gas exchange.
- **Type II pneumocytes** secrete pulmonary surfactant, a fluid that decreases the surface tension within the alveoli.

Adaptations of cardiac muscle cells and striated muscle fibres

Cardiac and skeletal are the only striated muscles in the body—they are made up of sarcomeres with contractile myofibrils with filaments of actin and myosin that allow for contraction. The structure of cardiac muscle cells (myocytes) allows propagation of stimuli through the wall of the heart. Cardiac muscle contraction is involuntary—not under conscious control. The cardiac muscle has numerous short, cylindrical cells arranged end to end, resulting in long, branched fibres. The nucleus of cardiac muscle cells is found in the centre of the cell.

The skeletal muscles are attached to bones and are responsible for movement and posture. They are voluntary, and therefore are under conscious control. They are formed by unbranched striated muscle cells with a cylindrical shape and blunt ends. The striated muscle cells are much larger than any average cell and with many nuclei per cell (multinucleate). They also contain a specialized endoplasmic reticulum, which stores and releases calcium ions to regulate muscle contraction. Striated muscle fibres contain many myofibrils, each made up of contractile sarcomeres.

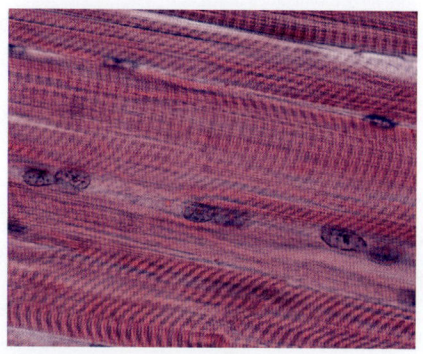

▲ Figure 8 Light microscope image of a myofibril

You will learn about the involvement of striated muscle cells in movement in subtopic B3.3.

Adaptations of sperm and egg cells

The human gametes are adapted to their function. The female egg or ovum is much larger than the sperm. The cytoplasm contains nutrients for the developing embryo. The outer glycoprotein layer (zona pellucida) changes after fertilization so that no further sperm can penetrate. Follicle cells outside the zona pellucida (corona radiata) protect the egg cell. The male gamete is the sperm and is formed by a head, neck and tail. The head contains the acrosome with enzymes that assist in the acrosome reaction to allow the sperm nucleus to enter the female egg in fertilization. The neck is full of mitochondria to provide energy for movement performed by the tail.

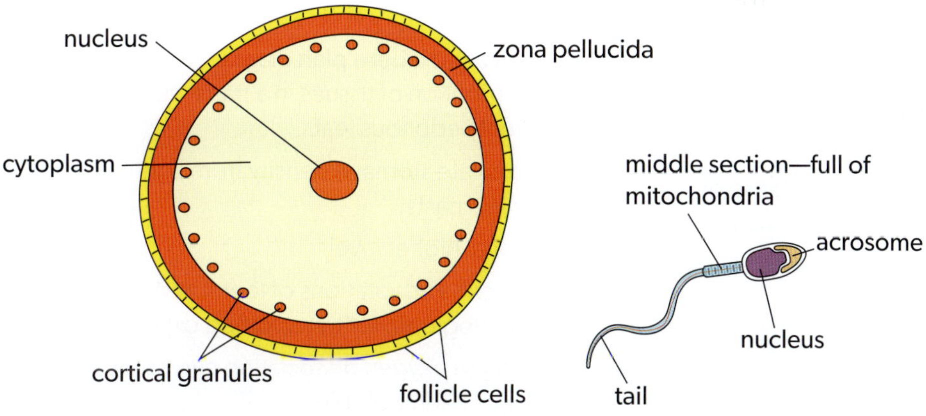

▲ Figure 9 An egg cell (left) and a sperm (right)

Form and function

AHL

Practice problems

1. The diagram shows the closed and open conformations of a voltage-gated potassium channel.

   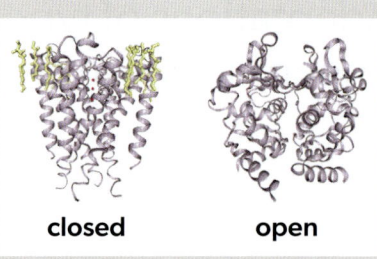
   closed open

   a) Explain how the disposition of the proteins of the potassium channel in the membrane assists in the movement of ions.

   b) Suggest the mode of transport of potassium through these channels.

2. Cholesterol is a steroid lipid found in the cell plasma membrane. The diagram shows a molecule of cholesterol.

   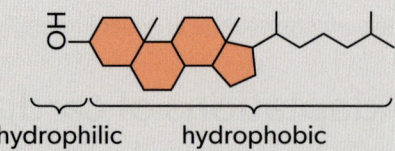
   hydrophilic hydrophobic

   a) State the group of lipids cholesterol belongs to.

   b) Deduce the position of cholesterol in the phospholipid bilayer according to the hydrophobic and hydrophilic parts.

3. Discuss whether a striated muscle fibre is a cell.

- What are the advantages of small size and large size in biological systems?
- How do cells become differentiated?

B3.1 Gas exchange

You should know:

- the challenges of gas exchange become greater with increasing size of organism as surface area-to-volume ratio decreases.
- concentration gradients are maintained at exchange surfaces in animals.
- adaptations for gas exchange in leaves include: the waxy cuticle, epidermis, air spaces, spongy mesophyll, stomatal guard cells and veins.
- transpiration is the inevitable consequence of gas exchange in the leaf.
- factors affecting the rate of transpiration in a leaf.

Additional higher level:

- there is cooperative binding of oxygen to haem groups and allosteric binding of carbon dioxide in haemoglobin.
- the Bohr shift explains the increased release of oxygen by haemoglobin in respiring tissues.
- foetal haemoglobin has more affinity for oxygen and less for carbon dioxide.

You should be able to:

- describe the properties of gas-exchange surfaces: permeability, thin tissue layer, moisture, a large surface area.
- explain adaptations of mammalian lungs for gas exchange: the presence of surfactant, a branched network of bronchioles, extensive capillary beds and a large surface area.
- explain lung ventilation, including the role of the diaphragm, intercostal muscles, abdominal muscles and ribs.
- measure lung volume—tidal volume, vital capacity, and inspiratory and expiratory reserves.
- draw and label a plan diagram to show the distribution of tissues in a transverse section of a dicotyledonous leaf.
- calculate stomatal density from light micrographs or leaf casts.

Additional higher level:

- describe adaptations of foetal and adult haemoglobin for transporting oxygen.
- analyse oxygen dissociation curves for adult and foetal haemoglobin.

Gas-exchange surfaces

All organisms require **gas exchange**, as oxygen is required for respiration. As organisms increase in size, the surface area-to-volume ratio decreases and the distance from the centre of an organism to its exterior increases. This determines the need for adaptations or special organs for gas exchange. Maintaining this concentration gradient at exchange surfaces in animals includes a dense network of blood vessels and continuous blood flow. It also includes ventilation of air in lungs or water in gills.

Gas exchange occurs in type I pneumocytes of the alveoli. Alveoli are found at the end of a branched network of bronchioles in the mammalian lungs. The alveoli are adapted for gas exchange. They are permeable, moist, are one cell thick and have capillaries close by.

- **Gas exchange** is the process by which gases move passively by diffusion across a surface. For example, the intake of oxygen and release of carbon dioxide between the alveoli and blood capillaries of the lungs.

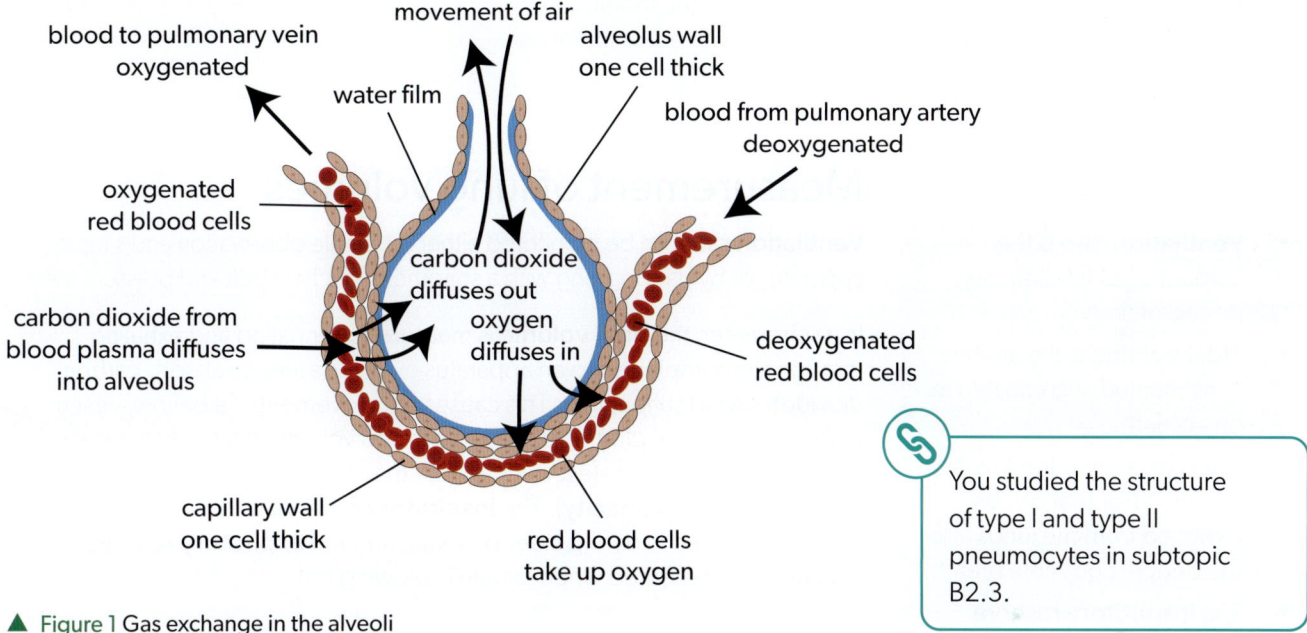

▲ Figure 1 Gas exchange in the alveoli

> You studied the structure of type I and type II pneumocytes in subtopic B2.3.

Ventilation of the lungs

Ventilation is the movement of air into and out of the lungs. The lungs are actively ventilated to ensure that gas exchange can occur passively. Ventilation maintains concentration gradients of oxygen and carbon dioxide between air in alveoli and blood flowing in adjacent capillaries. It is performed by four different sets of muscles. The external and internal intercostal muscles, and the diaphragm and abdominal muscles are examples of antagonistic muscle action. In inhalation, the external intercostal muscles and the diaphragm contract, expanding the thoracic cavity, allowing air to flow into the lungs. In exhalation, these two sets of muscles relax. In contrast, during inhalation, the internal intercostal muscles and the abdominal muscles are relaxed. They contract only during exhalation.

Form and function

Example 1

A model was used to explain the action of muscles on the ribcage in lung ventilation.

a) State the muscles represented by the ribbons.

b) Explain, in terms of ventilation, what happens when the ribbons are pulled up.

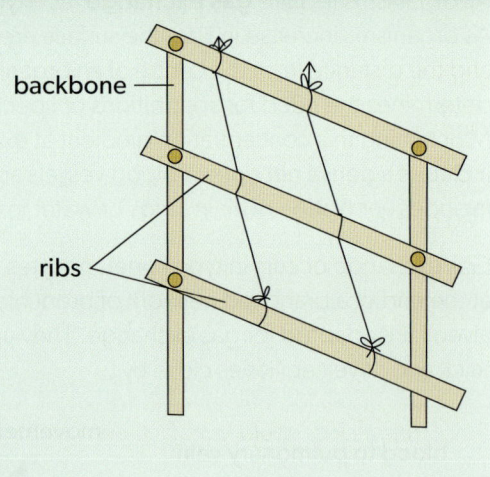

Solution

a) The ribbons represent the (external) intercostal muscles.

b) Pulling the ribbons up represents the contraction of the external intercostal muscles. In this case, the ribs pivot over the backbone and the ribcage rises. This increases the volume of the thoracic cavity. Consequently, air pressure in the cavity decreases causing air to come into the lung.

Measurement of lung volumes

- **Ventilation rate** is the amount of air inhaled in a period of time.
- **Tidal volume** is the amount of air inhaled and exhaled in one breath.
- **Vital capacity** is the greatest volume of air that can be removed from the lungs after the deepest possible breath.
- The **inspiratory reserve** is the volume of air that can be inspired forcefully; **expiratory reserve** is the volume of air that can be expired forcefully.

Ventilation rate can be monitored either by simple observation and simple apparatus or by data logging with a spirometer or chest belt and pressure meter.

In a spirometer, the **tidal volume** is measured by inhaling and exhaling through a mouthpiece into an apparatus over soda lime to absorb carbon dioxide to avoid suffocation. This causes the movement of a bellow, which makes a pen move, plotting a chart of exhaled volume of air. When a deep breath is taken, the greatest volume of air that can be removed from the lungs is measured (**vital capacity**). The **inspiratory reserve** is the volume measured when air is forced in during a deep breath, and the **expiratory reserve** is the volume of air measured when forcefully blowing out.

Sample student answer

The lung volume was recorded as a student breathed into a spirometer.

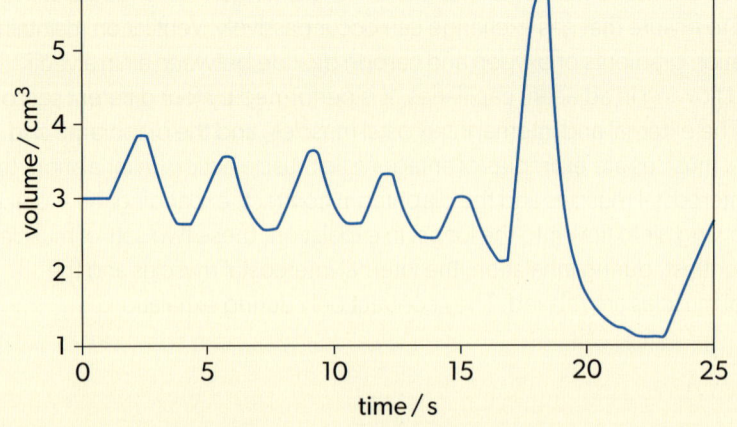

a) Describe the ventilation pattern of this student. [2]

This answer could have achieved 2/2 marks:

> The student's ventilation pattern is normal for 5 breaths, each time the tidal volume is a bit less, and suddenly at 17 seconds makes a deep inspiration and then slowly expires.

▲ The student described the normal breaths at the beginning of the experiment and the deep breath after 5 breaths (or at 16/17 seconds).

b) Use the data to estimate the ventilation rate during normal breathing. Show your working. [2]

This answer could have achieved 1/2 marks:

ventilation rate = breaths per minute
15 seconds: 4.5 breaths
60 seconds × breaths = 60 × 4.5 / 15 = 18 breaths
Answer = 18 breaths

▲ One mark for the correct working.

▼ No marks for the answer as the units are incorrect—it should be breaths per minute.

c) Sketch on the diagram the results expected for the student after mild exercise starting from time 0 seconds. [2]

This answer could have achieved 2/2 marks:

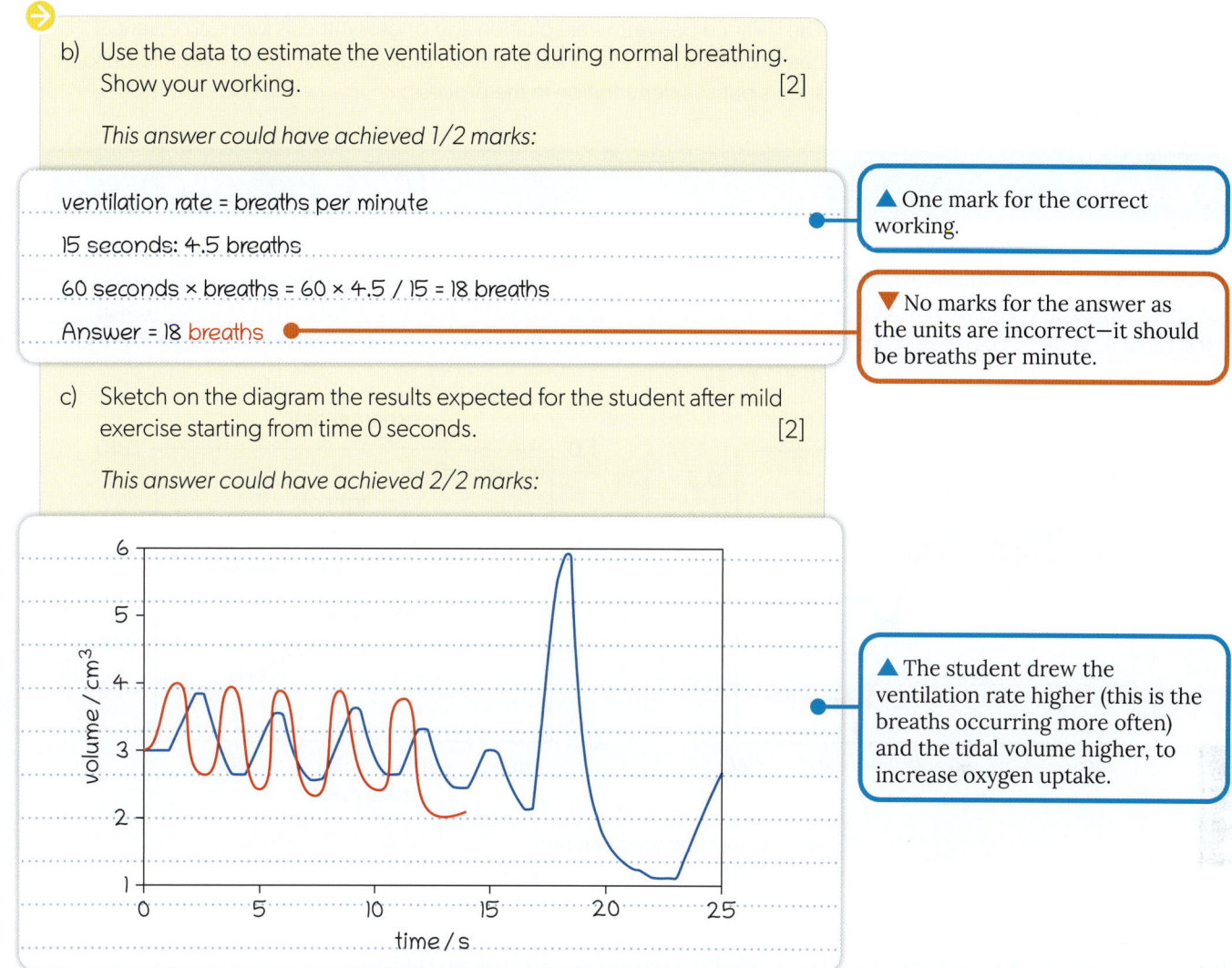

▲ The student drew the ventilation rate higher (this is the breaths occurring more often) and the tidal volume higher, to increase oxygen uptake.

Adaptations for gas exchange in leaves

Leaves have a large surface area but are flat, with a thin layer of epidermis, making diffusion distances short. Cells in the palisade mesophyll, where most photosynthesis takes place, are stacked close together and full of chloroplasts to capture as much light energy as possible. Cells in the spongy mesophyll are separated with air gaps between them, allowing for a faster diffusion of gases. The veins act as support for the blade of the leaf and carry the vascular tissue with xylem and phloem. In the epidermis, there are **stomata**. These are formed by two guard cells that open and close a gap between them through which gas can enter or leave.

Gas exchange is important to plants, as carbon dioxide is required for the process of photosynthesis, in which oxygen is produced. Water leaves the plants mainly in a gaseous state. Gases enter the leaves from an area of high concentration to an area of low concentration by diffusion, mainly through the stomata. Leaves carry out gas exchange through the upper and lower surfaces at different rates, depending on the stomatal density, stomatal opening, waxy cuticle thickness and permeability. Gas exchange increases when the stomata are open and decreases when stomata are closed (especially during the night). There are usually more stomata in the lower surface of the leaf and the cuticle is thicker on the upper surface, therefore there is more exchange on the lower surface.

- **Stomata** are two guard cells with a pore in the centre. Opening of stomatal pores aids in gas exchange.

Form and function

The balance between transpiration and photosynthesis forms an essential compromise in the existence of plants. Stomata must remain open to build sugars but risk dehydration in the process.

Example 2

Scientists measured movement of carbon dioxide and water vapour in grape (*Vitis vinifera*) leaves. They measured the carbon dioxide assimilation rate per unit of leaf area (A) and the water transpiration rate per unit of leaf area (E) with the lights on (negative time on the graph) and the lights off (positive time on the graph). In the control leaves, the surfaces were not covered (control). In the experimental leaves, the lower leaf surface was sealed to force all gas exchange through the upper surface (experimental).

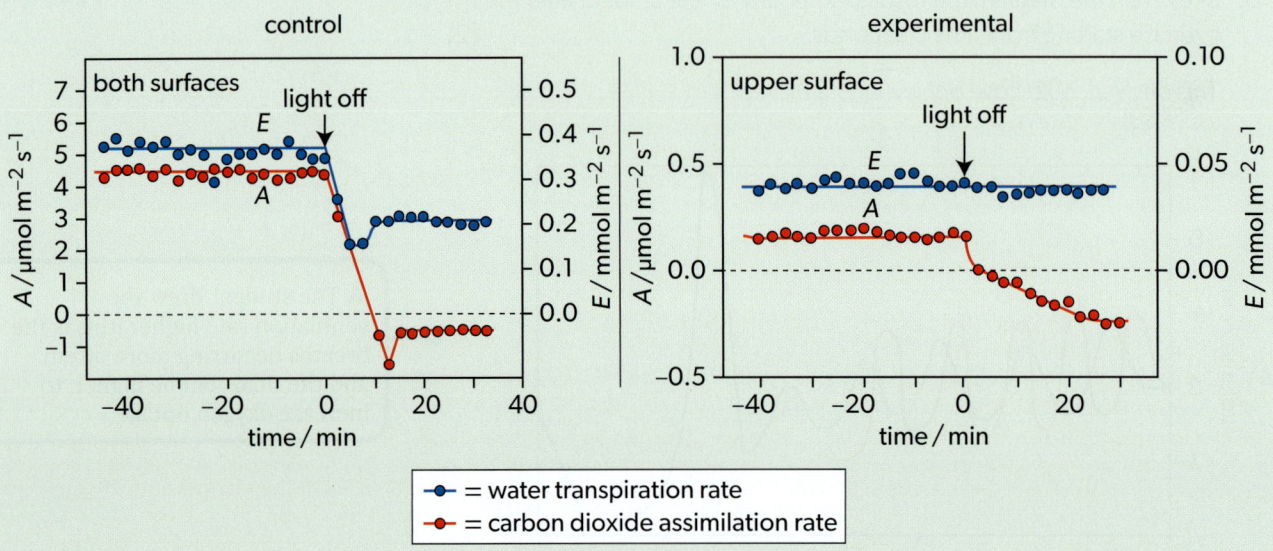

Source: Boyer, J. S., Wong, S. C. & Farquhar, C. D. (1997). CO_2 and water vapor exchange across leaf cuticle (epidermis) at various water potentials. *Plant Physiology*, 114(1), 185–191. www.doi.org/10.1104/pp.114.1.185

a) Compare and contrast the trend in transpiration rate with and without light for the control and for the experimental leaves.

b) Explain the reason for negative assimilation and transpiration decrease when the leaf was in the dark.

c) The cuticle allows some gas exchange but not as much as the stomata. Justify this statement using the results of the experiment.

Solution

a) In the control leaves, the transpiration rates are always higher than in the experimental leaves. In both, the transpiration rates remain constant when the light is on. In control leaves, when the light is turned off, the transpiration rate decreases, while in the experimental leaves there is no change.

b) The opening of the stomatal pore is determined by the osmotic potential of guard cells and their turgor pressure. In most plants the stomata open when there is an increase in light, and they close when there is a decrease in light. Blue light induces ions such as protons (H^+) and potassium (K^+) to enter guard cells. This makes water enter the cells, increasing their turgor pressure, causing the pore to open. When darkened, the stomata close, decreasing gas exchange.

c) The cuticle is on the upper surface. When the lower surface was sealed, the stomata were also sealed and the upper surface became the only one exchanging gas, therefore transpiration and assimilation were much less than in the unsealed leaf.

Distribution of tissues in a leaf

Sample student answer

The micrograph shows a transverse section of a leaf from a flowering plant.

a) Draw a plan diagram in the box with labels to show the distribution of tissues in this transverse section of a leaf. [3]

This answer could have achieved 1/3 marks:

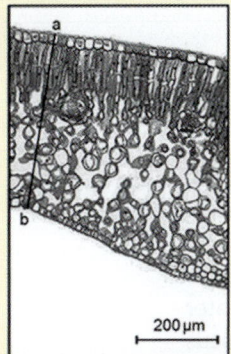
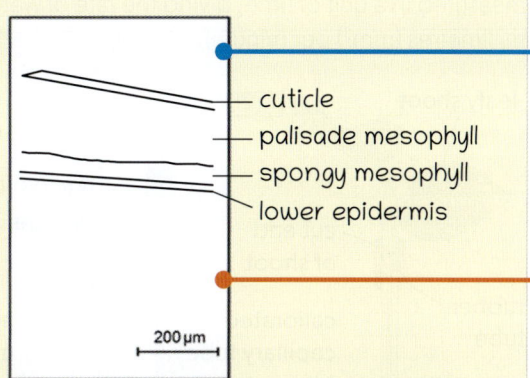

▲ One mark could be achieved for the lower epidermis.

▼ No further marks are achieved, although the labels are correct, because the proportions are not kept—the palisade is too wide compared with the spongy mesophyll. The cuticle is too thick. The upper epidermis is not shown under the cuticle.

Source: Houseman, J. (2014, September 30). English: Photomicrograph of a dicot leaf. A-Lower epidermis, B-Lower palisade mesophyll, C-Upper epidermis, D-Upper palisade mesophyll, E- Spongy mesophyll, F-Leaf vein. Scale=0.2mm. Wikimedia Commons. https://commons.wikimedia.org/wiki/File:Dicot_leaf_L.jpg (CC BY-SA 4.0)

b) Calculate the actual thickness of the leaf, from upper to lower surface along the line a–b. Show your working. [2]

This answer could have achieved 2/2 marks:

a–b = 3 cm

Scale bar: 1.5 cm = 200 μm

Therefore a–b = 2 × 200 μm = 400 μm

Answer = 400 μm

▲ Correct working and correct answer.

c) Suggest a reason for using a lower power objective lens when first focusing on a slide under the microscope. [1]

This answer could have achieved 1/1 mark:

The leaf is easier to see.

▲ Although rather vague, the answer is correct. It is easier to find the specimen using a low power objective lens as there is a larger field of view of focus. It is easier to locate the most interesting part of the specimen with a low power objective lens. At the same time, most microscopes are calibrated so when they are in focus using the low power objective lens, they will be close to focus using the higher power, so this helps focusing. Also, it reduces the risk of cracking the slide.

d) Identify **one** adaptation of the leaf for the absorption of light, visible in this micrograph. [1]

This answer could have achieved 0/1 mark:

Many chloroplasts.

▼ This answer is incomplete, as the cells containing the chloroplasts are not mentioned. A double layer of large palisade cells with lots of chloroplasts on the upper surface maximizes light absorption. Also, a thin, transparent epidermis and a transparent cuticle allow the light to pass through.

Factors affecting transpiration

Transpiration is the inevitable consequence of gas exchange in the leaf. During photosynthesis in the leaves, carbon dioxide is required and oxygen is produced. As the stomata open for gas exchange, water is also lost by transpiration.

Different factors affecting transpiration can be tested using a potometer. The capillary tube represents the xylem vessels. The water reservoir is used to top up the capillary tube to start a new measurement. The distance moved can be measured in a unit of time, giving the rate of water uptake (measured in cubic millimetres (mm^3) per minute).

▲ Figure 2 A potometer

Some factors increase transpiration when they are increased. These include water availability, temperature (if too high it decreases), carbon dioxide concentration (if too high it decreases), air movement or wind, light intensity, leaf area and stomatal density. Humidity, atmospheric pressure, sunken stomata and a thick cuticle all decrease the rate of transpiration if they are increased. An experimental way of increasing the atmospheric humidity is to place the potometer under a bell jar or in a plastic bag. This will maintain the water vapour eliminated by transpiration in the environment. A method of decreasing the environmental humidity is to use a fan. The fan blows away the transpired water vapour, drying the environment.

Stomatal density

Example 3

The light micrograph shows part of a leaf surface measuring 1 mm by 1 mm.

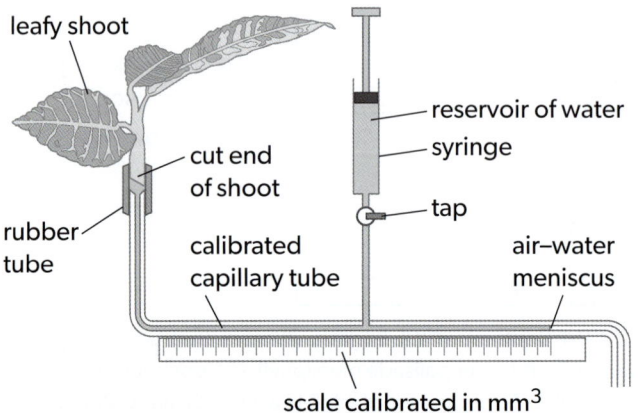

a) Calculate the stomatal density.

b) Identify the types of cells present in this micrograph.

c) Evaluate the need for investigating both leaf surfaces to calculate the stomatal density.

Solution

a) The stomatal density is 4 per mm^2.

b) The stomata are formed by two guard cells. The other cells shown are epithelial cells of the epidermis.

c) Both upper and lower epidermis must be investigated to compare the stomatal distribution, but most stomata are in the lower surface of the leaf. A larger area of the leaf and more leaves should be tested for the results to be reliable.

Nature of science

When working out stomatal density in an investigation you should take lots of repeats under different high-power fields of view. This is because there is a lot of variability in biological material, so the number of stomata in each field of view will be slightly different. Taking repeats will make your results more reliable.

Oxygen dissociation curves

Haemoglobin is a protein in the erythrocytes (red blood cells) formed by four peptides with four haem groups, each with an iron ion. This molecule binds oxygen in the alveoli and releases it in tissues. At low partial pressure of oxygen, few haem groups are bound to oxygen, so haemoglobin does not carry much oxygen. At higher oxygen pressure more haem groups are bound to oxygen, making it easier for more oxygen to be picked up. This occurs because the haemoglobin molecules suffer a slight change in the quaternary structure allowing more oxygen molecules through cooperative binding. The haemoglobin becomes saturated at very high oxygen pressure as all the haem groups become bound. This produces a sigmoid shaped curve (S-curve). Hydrogen ions (for example in acids) and carbon dioxide bind to different sites of haemoglobin. This **allosteric** binding modifies the quaternary structure of haemoglobin and lowers its affinity for oxygen.

- **Haemoglobin** is a protein in the erythrocytes formed by four peptides with four haem groups, each with an iron ion.
- An **allosteric** effect is when a change to one part of a molecule structurally changes another part of the molecule.

Example 4

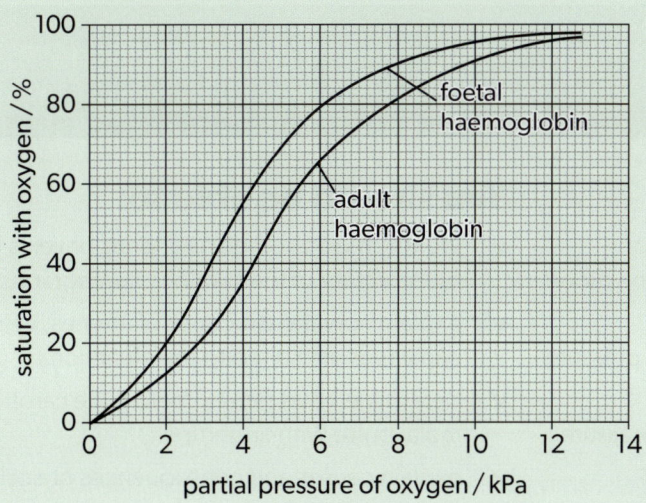

What explains the difference between the adult and foetal haemoglobin saturation curves?

A. Foetal haemoglobin has a higher affinity for oxygen.
B. Foetal haemoglobin is adult haemoglobin that has been purified.
C. Foetal haemoglobin has two peptides while adult haemoglobin has four.
D. Foetal haemoglobin has more haem groups than adult haemoglobin.

Solution

The correct answer is **A**. Foetal haemoglobin is different from adult haemoglobin. It has a greater affinity for oxygen, becoming saturated at a lower partial pressure of oxygen. Conversely, the binding of carbon dioxide is lower in foetal haemoglobin, therefore having a lower allosteric effect than in adult haemoglobin. This leads to higher affinity of oxygen of foetal haemoglobin and allows the transfer of oxygen in the placenta.

Bohr shift

The **Bohr shift** explains the increased release of oxygen by haemoglobin in respiring tissues. The oxygen dissociation curve is shifted to the right as the oxygen binding affinity is inversely related to acidity and carbon dioxide concentration. A shift to the right in the sigmoid dissociation curve means a decrease in oxygen affinity, therefore oxygen is released.

- The **Bohr shift** is the shift of the oxygen dissociation curve to the right with increased acidity or carbon dioxide in blood.

Form and function

AHL

- How do multicellular organisms solve the problem of access to materials for all their cells?
- What is the relationship between gas exchange and metabolic processes in cells?

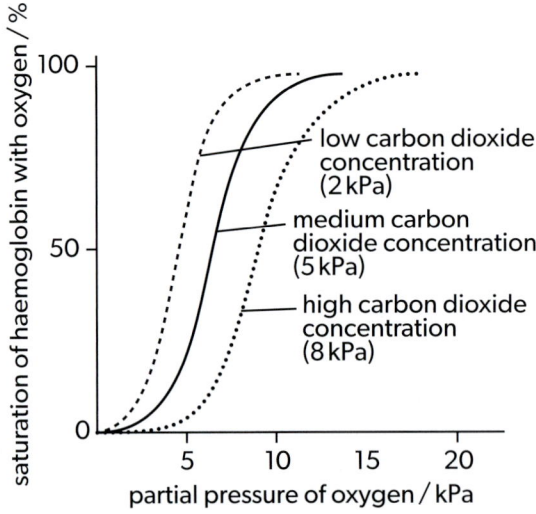

▲ Figure 3 The Bohr shift

B3.2 Transport

You should know:

- ✔ capillaries are adapted to allow exchange of materials between cells in the tissues and the blood in the capillary.
- ✔ arteries are adapted to carry blood at high pressure away from the heart.
- ✔ veins are adapted to collect blood at low pressure to return it to the heart.
- ✔ plants transport water from the roots to the leaves to replace losses from transpiration.

Additional higher level:

- ✔ tissue fluid is formed by pressure filtration of plasma in capillaries.
- ✔ substances are exchanged between tissue fluid and cells in tissue.
- ✔ excess tissue fluid is drained into lymph ducts.
- ✔ difference between the single circulation of bony fish and the double circulation of mammals.
- ✔ how the mammalian heart is adapted for delivering pressurized blood to the arteries.
- ✔ stages in the cardiac cycle.
- ✔ root pressure in xylem vessels is generated by active transport of mineral ions.
- ✔ adaptations of phloem sieve tubes and companion cells for the translocation of sap.

You should be able to:

- ✔ identify blood vessels as arteries or veins from the structure of their walls in micrographs.
- ✔ explain how the structures of arteries, veins and capillaries are suited to their function.
- ✔ determine heart rate by feeling the carotid or radial pulse with fingertips.
- ✔ explain causes and consequences of occlusion of the coronary arteries.
- ✔ evaluate epidemiological data relating to the incidence of coronary heart disease.
- ✔ explain how structure and function are correlated in the xylem of plants.
- ✔ draw plan diagrams to show the distribution of tissues in stems and roots in dicotyledonous plants.

Additional higher level:

- ✔ identify these structures on a diagram of the heart: cardiac muscle, pacemaker, atria, ventricles, atrioventricular and semilunar valves, septum and coronary vessels.
- ✔ trace the unidirectional flow of blood from named veins to arteries.
- ✔ interpret systolic and diastolic blood pressure measurements from data and graphs.

Adaptations of blood vessels

The heart and blood vessels make up the circulatory system. There are three main types of blood vessels: arteries that carry blood away from the heart, veins that carry blood towards the heart and capillaries that connect both these vessels.

Capillaries are adapted for exchange of materials between blood and the internal or external environment. They have a large surface area due to branching, and narrow diameters and thin walls. Some capillaries have small pores, called fenestrations, where exchange needs to be particularly rapid (for example, in glomeruli of nephrons in kidneys). Sinusoids are discontinuous capillaries with multiple fenestrations (for example, in the liver).

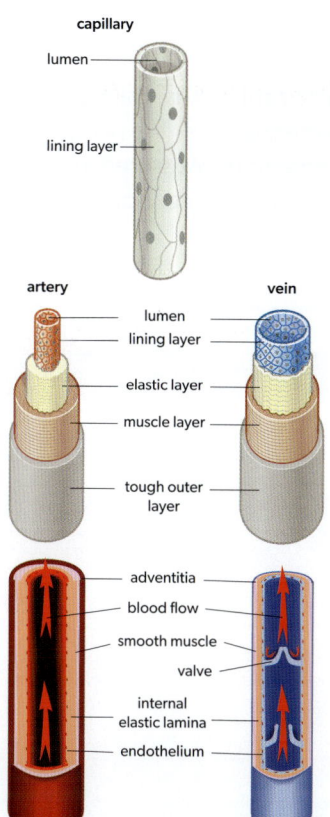

▲ Figure 4 Blood vessels

Example 5

Compare and contrast the structure of arteries, veins and capillaries.

Solution

	Arteries	Veins	Capillaries
Walls	thick walls to withstand high pressure	thin walls to allow muscles to exert pressure on them	walls one layer of cells thick to allow diffusion of substances
Fibres in wall	collagen and elastic fibres in outer layer to give wall strength and flexibility	thin outer layer of collagen and elastic fibres for protection	no fibres
Lumen	narrow lumen to maintain high pressure	wide lumen so greater volume of blood can pass	narrow lumen to fit in small spaces
Valves	no valves	valves to avoid backflow	no valves
Direction of blood flow	away from heart	towards heart	connect arteries to veins
Pores	no pores	no pores	pores to allow lymphocytes to leave

Assessment tip

You do not need to answer the question in the form of a table, but it helps to ensure you are really comparing and contrasting.

Example 6

This light micrograph shows an artery and a vein.

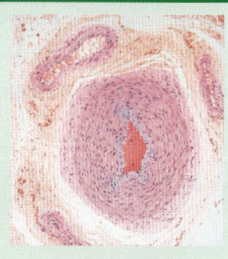

a) Label the artery and the vein in this micrograph.

b) List **two** characteristics of arteries visible in the micrograph.

Solution

a) Artery (central/right), vein (top left)

b) A thicker wall and a smaller relative diameter of the lumen.

Measuring pulse rate

- **Pulse rate** or heart rate is the frequency of the heart contractions measured in beats per minute (bpm).

The pulse gives an idea of the heart rate. To measure your **pulse rate**, press an artery with your fingertips and count the beats for 1 minute (or a fraction of a minute and multiply to obtain a minute). Usually, you would measure your pulse in your wrist (radial pulse) or on the side of your neck (carotid pulse). While this traditional method requires contact with skin, digital methods such as smartphone apps, although less accurate, enable contactless monitoring by capturing subtle light changes of skin through a video camera.

Example 7

A sample of healthy students (160 males and 158 females) were selected based on their smoking habits. The students carried out a maximal treadmill test and their pulse was recorded in beats per minute (bpm) during and after stopping exercise.

Smoking status	Pulse rate at rest / bpm		Pulse rate after exercise / bpm	
	Females	Males	Females	Males
non-smokers	88.1 ± 20	70.4 ± 22	148.0 ± 16	149.3 ± 22
smokers	94.6 ± 18	83.0 ± 19	141.0 ± 20	143.2 ± 14

a) Compare and contrast resting pulse rate in males and females.
b) From the data, explain the effect of smoking on pulse rate.
c) Comment on the reliability of the data obtained.

Solution

a) Both males and females have a higher resting pulse rate in smokers. Females have a higher resting pulse rate than males (in both smokers and non-smokers).

b) Smokers have higher resting pulse rates than non-smokers. Both female and male smokers show a slower pulse rate increase during exercise. Nicotine in tobacco causes vasoconstriction and raises blood pressure.

c) The standard deviation is high, so the results are not that reliable.

Causes and consequences of occlusion of the coronary arteries

- **Coronary artery occlusion** occurs when a coronary artery is blocked by a plaque.

The main risk factors of coronary heart disease are raised blood pressure (hypertension), smoking tobacco, high blood glucose levels (diabetes), physical inactivity, being overweight or obese, and high cholesterol in blood.

Example 8

High cholesterol in blood is a risk factor for coronary heart disease. Blood cholesterol levels were measured in a group of teenage boys once a month for 1 year. Years later, when they were adults, the blood cholesterol levels were measured again in the same way. The results were recorded in the scatter graph. Each dot represents the cholesterol level as a young teenager and as an adult for each male in the study.

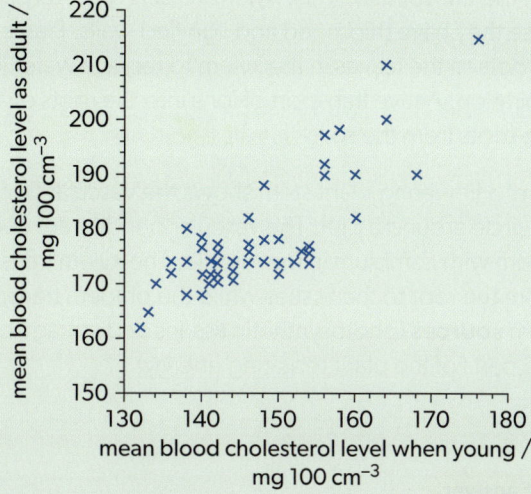

What does this graph show about blood cholesterol levels when young and as an adult?

A. There is a positive correlation.
B. There is a negative correlation.
C. There is no relationship.
D. They remain constant throughout the lifetime of a male.

Solution

The correct answer is **A**, there is a positive correlation as the trend line is increasing. The males that had a low cholesterol level when young, also had a low level (although a bit higher) in adulthood. In the same way, if it was high in youth, it was also high in adulthood.

Nature of science

A correlation coefficient shows if there is a statistical relationship between two variables. Values are between −1 and +1, where +1 is a perfect positive correlation. The fact that there is a correlation does not necessarily establish a causal link—this means that they are related but that one does not necessarily cause the other.

High levels of saturated or trans fats may contribute to hardening of the arteries (arteriosclerosis) or the formation of plaques or atheroma on the artery walls (atherosclerosis). An atheroma consists mainly of macrophages, lipids and connective tissue. It forms a swelling in the artery wall which slows down or even blocks the flow of blood. The loss of blood flow to the heart can result in coronary heart disease, a heart attack or myocardial infarction, or other cardiovascular diseases.

Transport of water in plants

Xylem vessels are continuous tubes of piled cells arranged one after the other. These cells have thickened walls that become lignified. This strengthens the walls, so that they can withstand low pressures without collapsing. The cells in xylem vessels die after some time, leaving only the walls of the non-living cells. This provides a larger space for water transport, and determines that the water movement in these vessels must be passive, as there are no mitochondria

- **Xylem** is the tissue that carries water from root to leaf.

Form and function

to supply energy. The divisions between cells also disappear, making a continuous tube. Small holes or pits allow substances to pass into and out of the vessels.

Evaporation of water in the mesophyll of the leaf creates a tension (negative pressure potential), which creates a pulling force called the transpiration pull. Water is drawn through cell walls out of the xylem in the vascular bundle of the leaf by capillary action due to **adhesion** to cellulose. Low pressure causes a suction force in the xylem. **Cohesion** allows water to be pulled up under tension in the transpiration pull. Xylem vessels resist tension and do not collapse because they have thickened and lignified walls. Plants transport water from the roots to the leaves in the xylem to replace water lost from the leaves by transpiration. Active transport of ions into the roots enables osmosis of water into the roots from the soil.

A cross-section of a flowering plant stem shows the vascular bundles distributed in a circle around a pith. The vascular bundles are composed of xylem and phloem with cambium in the middle. The xylem transports water and minerals from the root to the tissues while the phloem transports organic compounds from **sources** (photosynthetic tissues and storage organs) to **sinks** (roots and parts of the plant requiring energy).

- **Adhesion** is the attraction of water to other substances, for example the xylem walls.
- **Cohesion** is the attraction between water molecules due to their hydrogen bonds.
- A **source** is a part of the plant where carbohydrates and amino acids are loaded into the phloem.
- A **sink** is where the carbohydrates and amino acids are unloaded to be used.

Sample student answer

The light micrograph shows a transverse section of a flowering plant stem at ×400. The xylem has already been labelled.

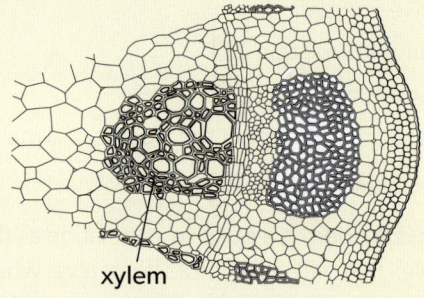

Draw a plan diagram of the micrograph with labels to show the distribution of tissues. [3]

This answer could have achieved 1/3 marks:

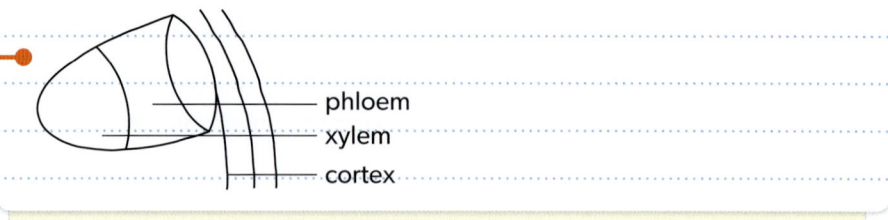

▼ The quality of the diagram is not good as the proportions need to be kept. One mark is scored for cortex. The vascular bundle and epidermis (that should have been narrower) have not been identified. The phloem does not score a mark because the label line points to the cambium (although too wide) not the phloem, which is to the right.

Distribution of tissues in a transverse section of a root

> **Example 9**
>
> Compare and contrast the transverse section of a dicotyledonous stem and root using a diagram.
>
> **Solution**
>
> cross-section of stem — epidermis, cortex, xylem and phloem (vascular bundle)
>
> cross-section of root — cortex, epidermis, xylem and phloem (vascular bundle)

Tissue fluid

Tissue fluid surrounds cells within tissues and transports nutrients and waste products between cells and blood capillaries. Higher hydrostatic and osmotic pressure exerted from arterioles or protein leakage into the tissues (which occurs during inflammation) force plasma from the capillaries into the tissues. Lower pressure in venules allows tissue fluid to drain back into capillaries.

Blood plasma contains the components of blood except for the blood cells. The plasma proteins, mainly clotting factors, albumin and globulin, and some enzymes, are too large to leave the capillaries, so are not found in tissue fluid.

- **Tissue fluid** surrounds cells within tissues and transports nutrients and waste products between cells and blood capillaries.

> **Example 10**
>
> Compare and contrast plasma and tissue fluid composition.
>
> **Solution**
>
> | | Plasma | Tissue fluid |
> | --- | --- | --- |
> | plasma proteins | present | absent |
> | proteins secreted by cells | absent | present (few) |
> | clotting factors | present | absent |
> | glucose | more | less |
> | oxygen | more | less |
> | carbon dioxide | less | more |
> | salts | present | present |

The lymphatic system is a linear system that originates from a network of capillaries that collect macromolecules, cells and tissue fluid from the tissue bed in each organ. Lymph fluid, which is similar to tissue fluid, flows through the many lymphatic ducts with valves to prevent backflow. These ducts consist of thin endothelial cells with discontinuous junctions that allow the entry of lymph. The lymph flows through sporadically distributed lymph nodes, and eventually joins the blood system circulation through the subclavian vein into the vena cava.

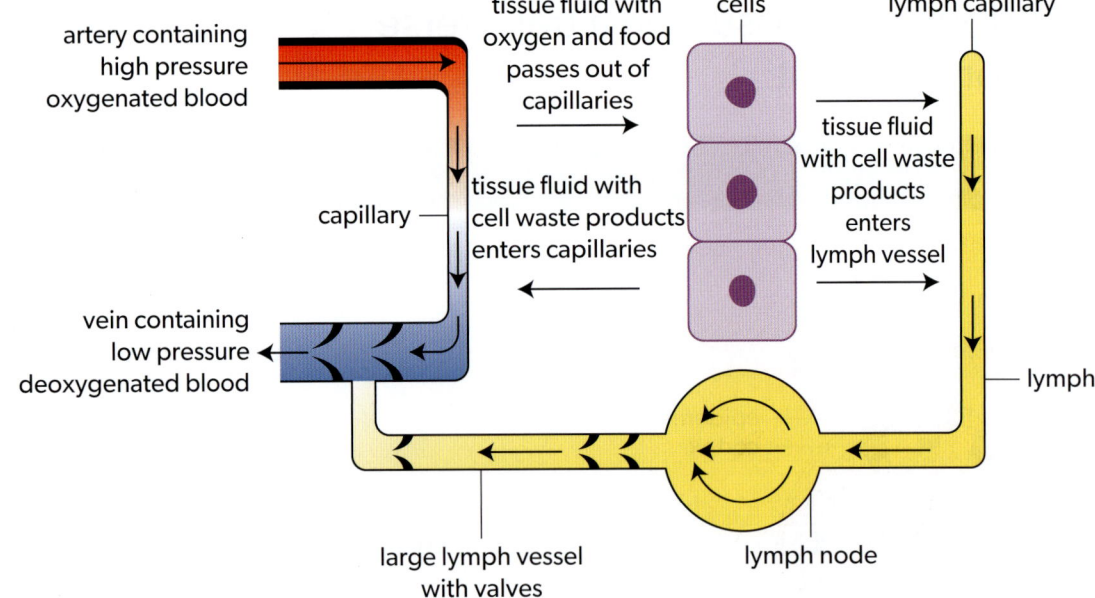

▲ Figure 5 The lymphatic system

Comparing single and double circulations

In mammals, blood flows twice through the heart, once to be oxygenated in the lungs and a second time for circulation to the rest of the body. In contrast, in bony fish, the blood has a single circulation—this means the blood only passes once through the heart.

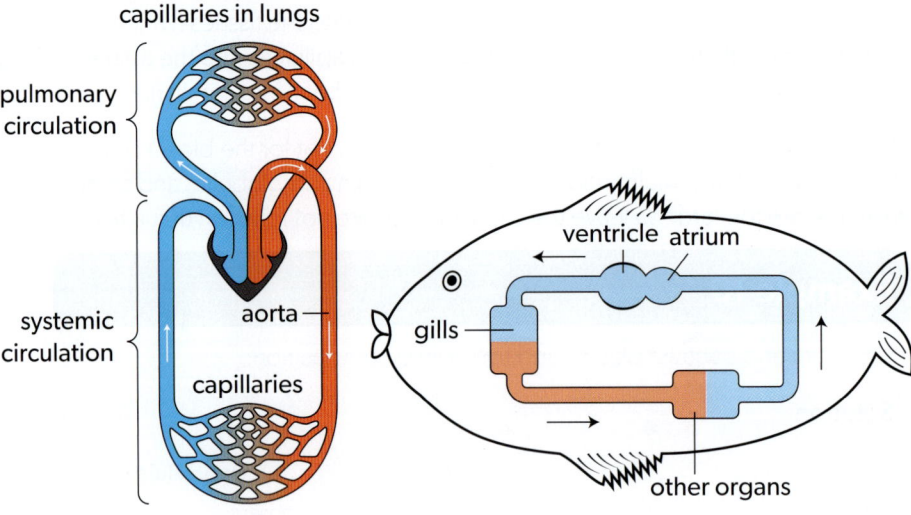

▲ Figure 6 Double circulation in humans (left) compared with single circulation in fish (right)

Adaptations of the mammalian heart and the cardiac cycle

The pressure in the different chambers of the heart changes according to whether the ventricles are contracting (systole) or relaxing after a contraction (diastole). A whole heartbeat cycle can be seen on a heart pressure graph. The first part of a heartbeat can be seen as the first parabolic curve, where the atria are contracting (atrial systole) and the second part of a heartbeat can be seen as the second, larger curve, where the ventricles are contracting (ventricular systole). In healthy adults you can hear two normal heart sounds in each heartbeat. These are often described as a "lub", the closing of the atrioventricular (AV) valves, and a "dub", the closing of the semilunar valves.

Example 11

a) The diagram shows a human heart.

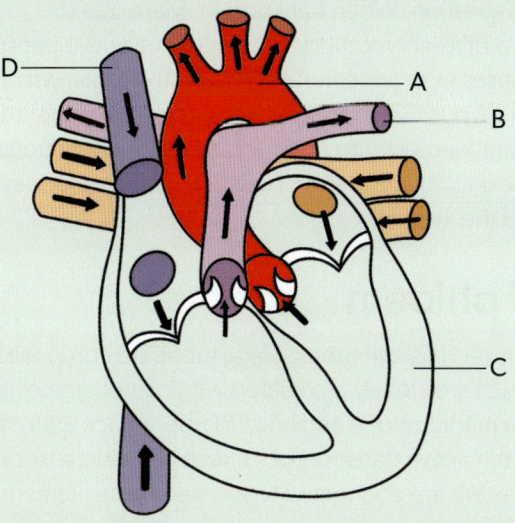

Identify labels A to D.

b) The graph shows pressure changes during the cardiac cycle.

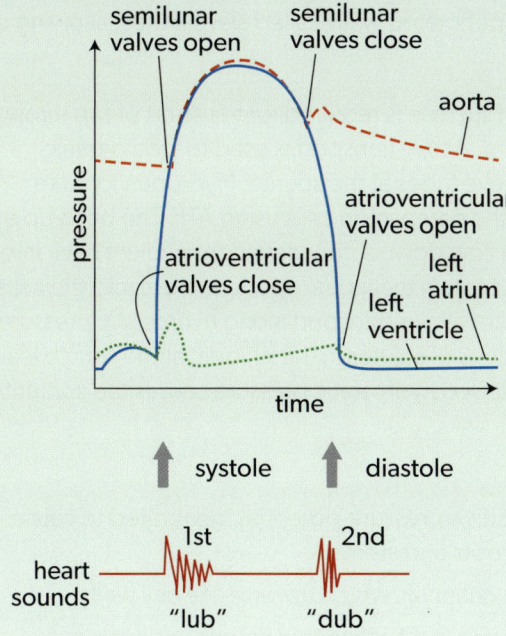

(i) Explain why the pressure in the ventricle increases during atrial contraction.

(ii) The pressure in the aorta is around 80 mmHg just before the semilunar valves open. Predict the pressure in the ventricle at the time these valves open.

(iii) Describe the change in pressure in the left ventricle after the semilunar valves close.

Solution

a) A—aorta, B—pulmonary artery, C—left ventricle wall, D—vena cava.

b) (i) During atrial contraction, the atrioventricular (AV) valve is open. Therefore, the blood is flowing from the left atrium into the left ventricle, increasing the pressure.

(ii) The pressure must be just above 80 mmHg to allow the blood to flow from the ventricles into the aorta. The ventricular pressure must be high enough to open the valves, letting blood out of the heart.

(iii) The pressure in the left ventricle decreases, as blood has left the ventricle and entered the aorta to be carried to the rest of the body. The atria are filling with blood, but because the AV valve is closed, no blood is flowing into the ventricle.

The sinoatrial (SA) node consists of involuntary myogenic muscle cells that contract rhythmically at around 70 beats per minute. Following each contraction, an impulse spreads through the atria to the atrioventricular (AV) node. Contraction of the atria causes blood to move into the ventricles. The AV node transmits the impulse to the ventricles, via the Purkinje fibres in the bundle of His. Signals from the SA node that cause contraction cannot pass directly from atria to ventricles. There is a delay between the arrival and passing on of a stimulus at the AV node. This delay allows time for atrial systole before the AV valves (bicuspid and tricuspid valves) close. Once the AV valves close, the ventricles depolarize and contract sending blood to the lungs and to the rest of the body. The atria repolarize when the ventricular systole is initiated (when the ventricles contract). Conducting fibres ensure coordinated contraction of the entire ventricle wall. Artificial **pacemakers** can be used in cases of SA problems to regulate the heart rate.

Assessment tip

Part (b) in example 11 is referring to the left ventricle, but as the same happens in both ventricles simultaneously, no marks will be taken off if you just refer to the ventricles instead of the left ventricle.

- The **pacemaker** controls heartbeat.

Root pressure in xylem

Transpiration requires leaves and special conditions. The lack of leaves in deciduous plants during winter and early spring, or high humidity environments, prevent transpiration. When transport in xylem due to transpiration is insufficient, a different method is needed. The active transport of minerals into the root causes high concentration inside the roots, which increases **root pressure**. This positive root pressure is a positive pressure potential that causes water movement into roots and stems. A continuous column of water is formed due to the cohesion between water molecules and the adhesion to the walls of the xylem.

Adaptations of phloem

> See subtopic A2.2 "Atypical cell structures in eukaryotes" for a drawing of a phloem sieve tube element.

The phloem tissue is composed of sieve tubes. Sieve tubes are composed of sieve tube elements, which are closely associated with companion cells. Companion cells are rich in mitochondria and have a large surface area of cell membrane to support the active transport of sucrose and aid transport in the phloem. Companion cells are connected to the sieve tube elements by **plasmodesmata**, which facilitate exchange of molecules. Sieve tube elements maintain the sucrose and organic molecule concentration that has been established by active transport. Sieve tube elements have a reduced cytoplasm and organelles, no nucleus, and their cell walls are rigid to establish the pressure necessary for sap to flow in the phloem. Individual sieve tube elements are separated by perforated walls called sieve plates, allowing quick flow from cell to cell.

A concentration gradient of sucrose is required for transport of sap through **translocation** in the phloem. Active transport is used to load organic compounds into phloem sieve tubes at the source. Hydrogen ions are actively transported out of the companion cells using ATP. The build-up of hydrogen ions makes them flow down a concentration gradient back into the companion cells through a protein that cotransports sucrose into the cells. The incompressibility of water allows transport along hydrostatic pressure gradients. High concentrations of solutes in the phloem at the source lead to water uptake by osmosis. Raised hydrostatic pressure causes the contents of the phloem to flow towards sinks.

- **Root pressure** is positive pressure potential, generated to cause water movement in roots and stems.
- **Plasmodesmata** are channels which traverse the cell walls.
- **Translocation** is the transport of organic compounds in plant phloem. It requires energy.

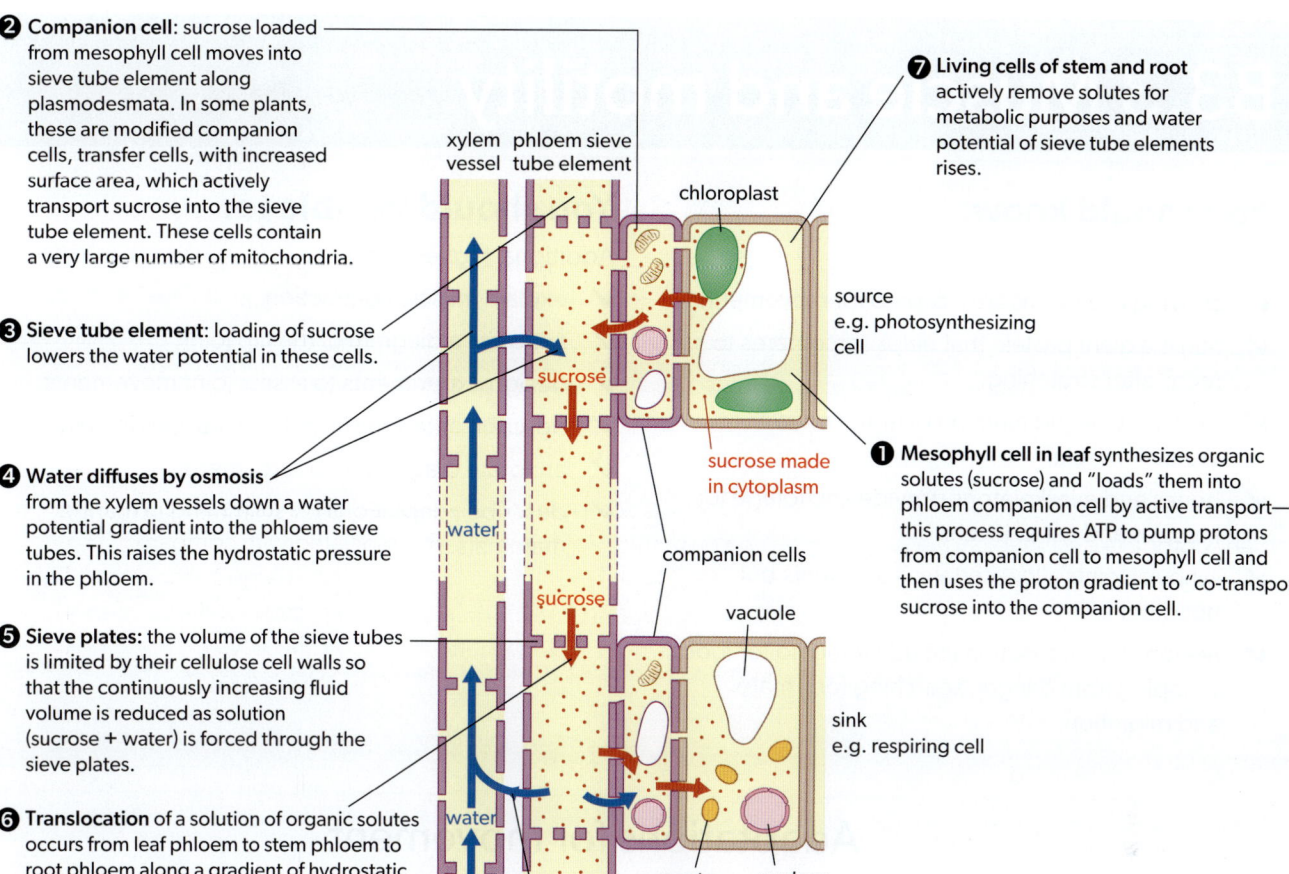

❷ **Companion cell**: sucrose loaded from mesophyll cell moves into sieve tube element along plasmodesmata. In some plants, these are modified companion cells, transfer cells, with increased surface area, which actively transport sucrose into the sieve tube element. These cells contain a very large number of mitochondria.

❸ **Sieve tube element**: loading of sucrose lowers the water potential in these cells.

❹ **Water diffuses by osmosis** from the xylem vessels down a water potential gradient into the phloem sieve tubes. This raises the hydrostatic pressure in the phloem.

❺ **Sieve plates**: the volume of the sieve tubes is limited by their cellulose cell walls so that the continuously increasing fluid volume is reduced as solution (sucrose + water) is forced through the sieve plates.

❻ **Translocation** of a solution of organic solutes occurs from leaf phloem to stem phloem to root phloem along a gradient of hydrostatic pressure: mass flow.

❼ **Living cells of stem and root** actively remove solutes for metabolic purposes and water potential of sieve tube elements rises.

❶ **Mesophyll cell in leaf** synthesizes organic solutes (sucrose) and "loads" them into phloem companion cell by active transport—this process requires ATP to pump protons from companion cell to mesophyll cell and then uses the proton gradient to "co-transport" sucrose into the companion cell.

❽ **Water diffuses by osmosis** (i.e. down the water potential gradient) from the sieve tube elements. This water joins the water absorbed by root hairs and diffuses into the xylem vessels.

▲ Figure 7 Transport in plants

- How do pressure differences contribute to the movement of materials in an organism?
- What processes happen in cycles at each level of biological organization?

B3.3 Muscle and motility

You should know:
Additional higher level:
- ✔ all living organisms are adapted for movement.
- ✔ titin is a giant protein that helps sarcomeres to recoil after stretching.
- ✔ the structure and function of motor units in skeletal muscle.
- ✔ bones and exoskeletons provide anchorage for muscles and act as levers.
- ✔ synovial joints allow certain movements but not others.
- ✔ reasons for locomotion include foraging for food, escaping from danger, searching for a mate and migration.

You should be able to:
Additional higher level:
- ✔ explain muscle contraction.
- ✔ annotate a diagram of the hip joint.
- ✔ design experiments to assess joint movements.
- ✔ describe antagonistic pairs of muscles in lungs.
- ✔ annotate diagrams of a sarcomere.
- ✔ describe adaptations for swimming in marine mammals.

Adaptations for movement

Movement is a characteristic of all living organisms. This includes locomotion from one place to another or slight movements like the one performed by sunflowers (*Helianthus annuus*) searching for light. Adult sponges (*Lissodendoryx colombiensis*), although sessile, have slight amoeba-like movement and can open and close their mouth-like opening.

Mechanism of muscle contraction

In subtopic B2.3, there is a description of striated muscle cells.

Muscles are suited for their function. They are formed by single muscle fibres that form bundles of fibres, each with many cells, and capillaries. Muscle cell fibres contain many myofibrils, each made up of contractile sarcomeres, with several repeating units that alternate between light and dark visible bands. The sarcomere contracts by the sliding of actin and myosin filaments. Titin is a giant protein (around 34,000 amino acids) that connects the Z-line, which organizes the thin actin filaments, and the M-line, which organizes the thick myosin filaments. Titin holds the whole contractile apparatus in shape as the muscle flexes, helps sarcomeres to recoil after stretching and prevents overstretching.

The propagation of a nerve action potential releases calcium ions into the muscle cell. The calcium ion exposes the myosin binding site on the actin filament. The binding of actin and myosin causes the muscle to contract. The contraction of skeletal muscle is achieved by the sliding of actin and myosin filaments. The myosin then binds ATP, which allows it to release the actin and return to the starting position. This movement requires the hydrolysis of ATP for the filaments to slide.

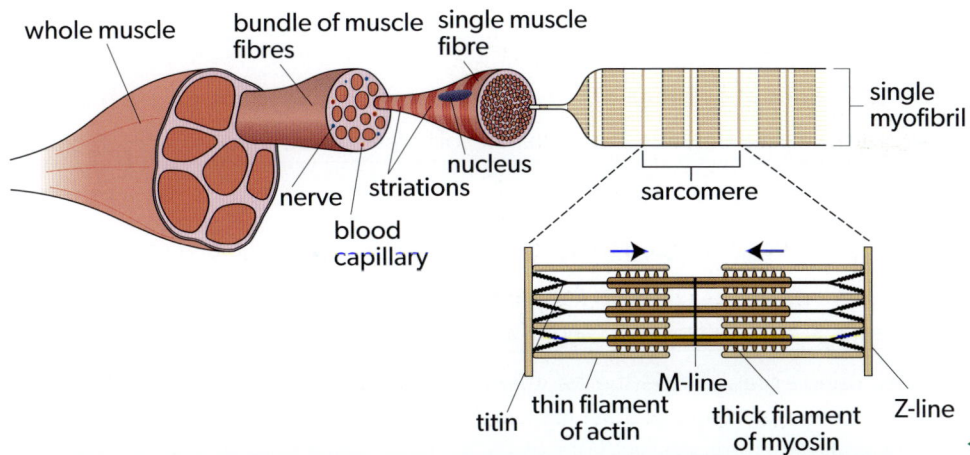

◀ Figure 8 Structure of muscles

Example 12

What is the function of titin in muscle contraction?

A. To connect the thin myosin M-line with the thick Z-line

B. To connect the actin bands together

C. To bind calcium ions

D. To slide on actin to shorten the sarcomere

Solution

The correct answer is **A**. Titin contributes to force transmission at the Z disc and resting tension in the I band region by connecting the M-line to the Z-line.

Structure and function of motor units in skeletal muscle

The **motor unit** in a skeletal muscle is formed by the motor neuron, the skeletal muscle fibres and the neuromuscular junctions that connect them. Groups of motor units coordinate the contractions of a muscle. Muscles with more motor units can control force output more finely.

- The **motor unit** in a skeletal muscle is the motor neuron, the skeletal muscle fibres and the neuromuscular junctions.

Example 13

Fast (F) motor units are important for short exercises that require large forces. In contrast, slow (S) motor units exert low force and have a slower contraction speed.

The distribution of muscle fibres in a motor unit was studied in the soleus (SOL) and tibialis anterior (TA) muscles of adult cats. The position and size of muscle fibres were mapped in cross-sections taken along the length of the muscle. The size of the area occupied by the motor unit (territory size) was calculated and expressed as a percentage of the entire muscle cross-section.

Motor unit	Territory size / %	Number of fibres	Fibre diameter	
			Mean (µm)	Range (µm)
SOL1	41	1090	55	41–71
SOL2	76	1060	59	41–87
SOL3	43	749	50	36–73
TA1	8	815	60	19–93
TA2	18	900	69	28–120
TA3	22	982	53	22–87
TA4	12	830	53	21–79

Source: Bodine, S. C., et al. (1968). Spatial distribution of motor unit fibers in the cat soleus and tibialis anterior muscles: local interactions. *Journal of Neuroscience*, **8**(6), 2142–2152.

a) State the motor unit with the largest recorded fibre diameter.
b) Distinguish between territory size in the three SOL motor units.
c) Discuss whether there is a correlation between territory size and the number of fibres in the seven motor units studied.

Solution

a) TA2, as the range of values recorded is between 28 and 120 μm—this means the largest diameter measured is 120 μm.
b) SOL2 has the largest territory percentage, followed by SOL3 and SOL1, which have almost the same value.
c) Correlation can only be determined using a graph. Nevertheless, there does not seem to be a relationship, as in SOL1 and SOL2 there is the same number of fibres but SOL1 has a smaller territory size. Similarly, TA1 and TA4 have a similar number of fibres and a different territory size. SOL1 and SOL3 have similar territory size, but the number of fibres is very different.

Assessment tip

In a discussion, you must include at least one positive argument and one negative argument in your answer.

- **Ligaments** attach bone to bone.
- **Synovial fluid** provides lubrication preventing friction.
- **Tendons** attach muscle to bone.

Functions of the skeleton

Skeletons can be inside the body as in vertebrates (endoskeleton) or outside the body as in arthropods (exoskeleton). The roles of the musculoskeletal system are movement, support and protection. Movement is achieved mainly by muscles pulling on bones, while bones provide most support and protection. Bones and exoskeletons provide anchorage for muscles and act as levers.

Synovial joints

Synovial joints are found between bones that move against each other, such as the joints of the hip. The hip joint is found between the pelvis and the femur with **ligaments** joining the bones. The heads of the bones are covered with protective cartilage. The **synovial fluid**, found in sacs next to the cartilage, helps to reduce friction during movement. **Tendons** join muscles to bones.

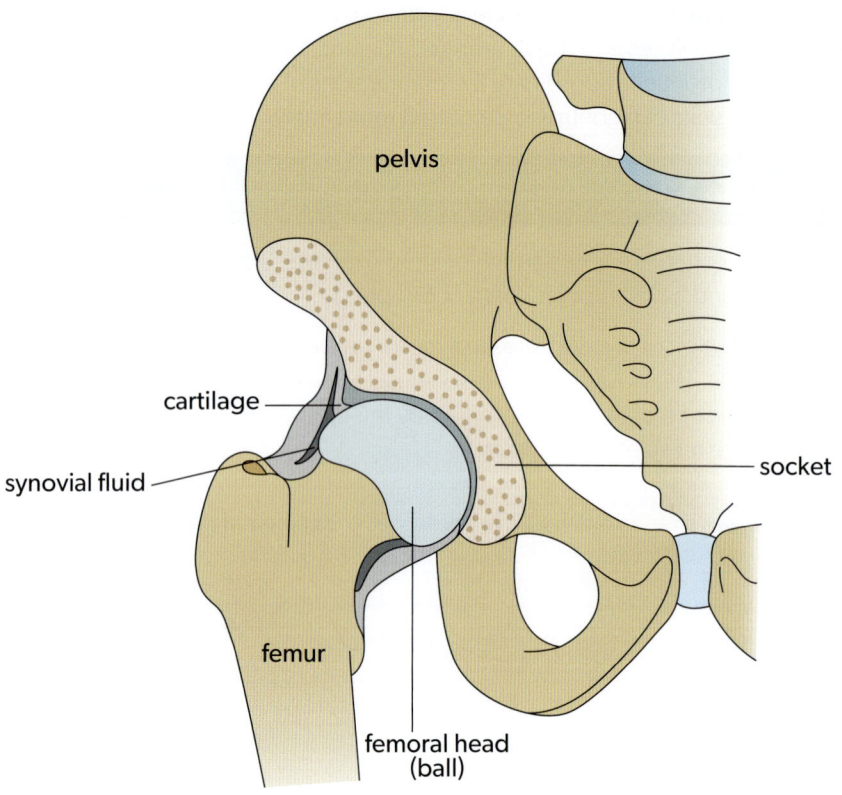

▲ Figure 9 The hip joint

Synovial joints allow certain movements but not others. The hip joint, where the femur joins the pelvis, is a ball and socket joint. This type of joint allows movement in all directions, including rotation, flexion (bending) and extension, **abduction** and **adduction**. The elbow allows less movement because it is a hinge joint, allowing flexion and extension only.

- **Abduction** is the movement of a limb away from the midline of the body.
- **Adduction** is the movement of a limb towards the midline of the body.

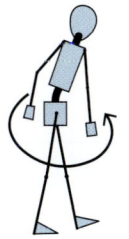 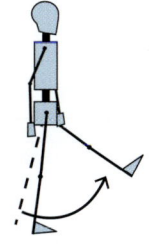 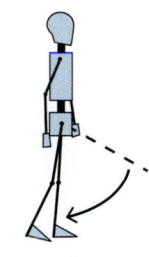 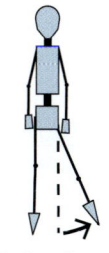 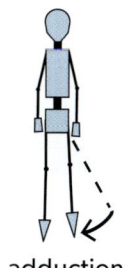

rotation   flexion   extension   abduction   adduction

▲ Figure 10 Movement at a synovial hip joint

Joint angles can be measured using computer analysis of images or a goniometer. A goniometer has a central body and a lever with a protractor to measure the angle formed between the bones.

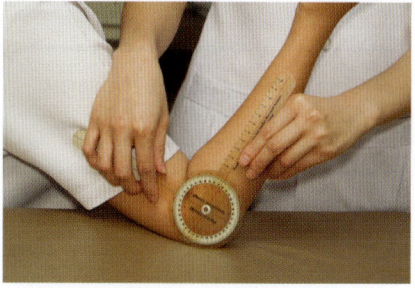

▲ Figure 11 A goniometer

Antagonistic action of intercostal muscles

Internal and external intercostal muscles take part in lung ventilation through **antagonistic muscle** action. When one contracts, the other relaxes. In inhalation, the external intercostal muscles contract, while in exhalation, they relax. In contrast, during inhalation, the internal intercostal muscles are relaxed and contract only during exhalation. This results in the moving of the ribcage in opposite directions. When one of these layers contracts, it stretches the other, storing potential energy in the sarcomere protein titin.

- **Antagonistic muscles** produce opposing movement at a joint—when one contracts, the other relaxes.

Intercostal muscles are found between the ribs, in the intercostal space. As there are 12 ribs on each side of the thorax, there are 11 intercostal spaces. The intercostal muscles join the rib bones at the back (dorsal) and front (ventral) part of the body. Each external intercostal muscle joins the lower border of the rib above and the upper border of the rib below with fibres running obliquely downward and forward. The internal intercostal muscle fibres are also directed obliquely, but they run inferiorly and posteriorly, in a direction opposite to that of the external intercostals.

You learned about the function of intercostal muscles in ventilation in subtopic B3.1.

Sample student answer

In a study, the orientations of the external and internal intercostal muscles were measured in the dorsal and ventral parts. The graphs show the angles between the ribs and the external and internal intercostal fibres in intercostal spaces 2 (between ribs 2 and 3), 4, 6 and 8. Data means ± standard error were obtained from five subjects.

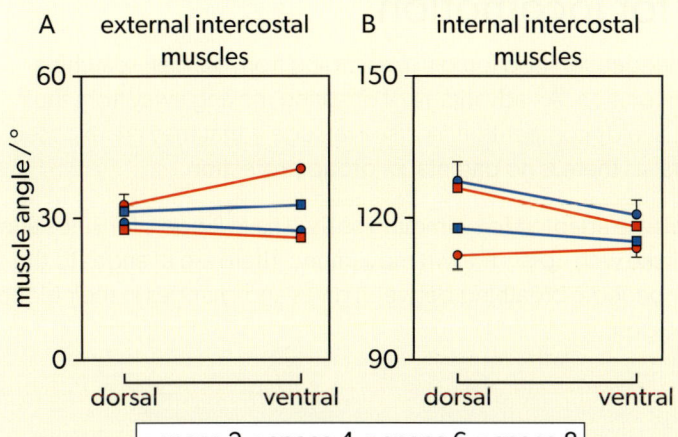

Source: Wilson, T. A., et al. (2001). Respiratory effects of the external and internal intercostal muscles in humans. *Journal of Physiology*, **530**(2), 319–330. (Pt 2):319-30. doi: 10.1111/j.1469-7793.2001.0319l.x. PMID: 11208979; PMCID: PMC2278403.

Form and function

a) State the angle of the external intercostal muscle in space 8. [1]

This answer could have achieved 0/1 mark:

> 33

▼ Although the answer is correct, there are no units (degrees), so it does not score a mark.

b) Compare and contrast the angles in external and internal intercostal muscles in space 2. [3]

This answer could have achieved 3/3 marks:

> In both muscles the ventral is larger than the dorsal, although in external the percentage difference between ventral and dorsal is greater. The internal intercostal muscles have larger angles whereas the external have smaller angles.

▲ This answer has one similarity and two differences, therefore obtaining the three marks.

c) Explain why the external have acute angles while the internal intercostal muscles have obtuse angles. [1]

This answer could have achieved 1/1 mark:

> Acute angles are smaller than 90° whereas obtuse are greater than 90°. Because the fibres in the external intercostal muscles are in the opposite orientation to those of the internal intercostal muscles, the angles are different (see diagram).

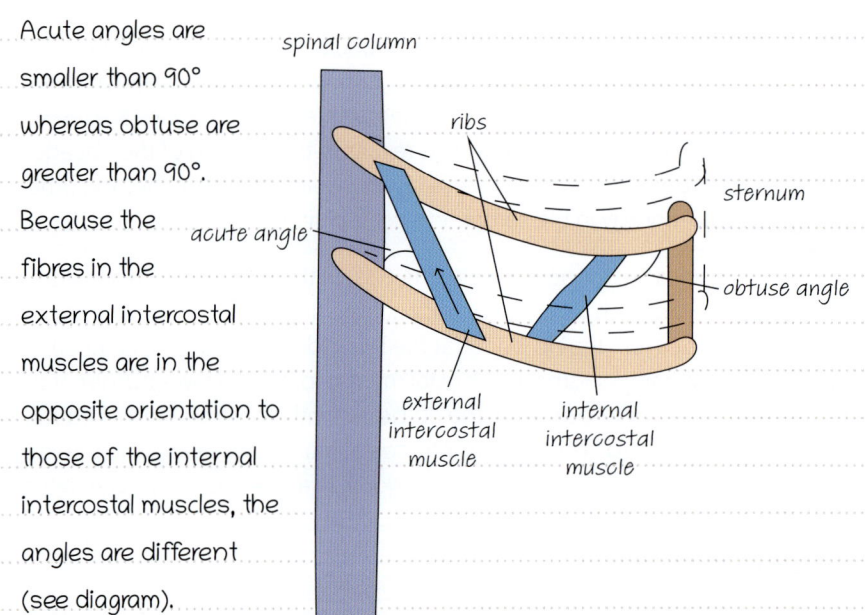

Examiner tip

Your answer can be in an annotated diagram.

Reasons for locomotion

Locomotion is needed for food foraging, escaping from danger, searching for a mate and migration. An advantage of offspring moving away from their parents is that it avoids competition. A disadvantage is that they are exposed to more dangers, as there is no parental or group protection.

 You studied adaptations to living in water in subtopic A1.1.

Marine mammals are adapted to swimming as they have flippers for limbs and the tail forms a fluke with up-and-down movement. There are changes to the airways to allow periodic breathing between dives and changes in their blood to hold more oxygen.

Example 14

Pinnipeds are marine mammals with fins or flippers and include fur seals and sea lions. Some pinnipeds forage for prey near the surface (epipelagic) while others forage on the bottom of the sea (benthic). The graph shows the foraging behaviour and the relative time spent diving while at sea for five pinniped species.

a) State the relative time the Australian fur seal spent diving while at sea.

b) Using the data in the bar chart, deduce which factor has the most significant effect on the relative time spent diving while at sea.

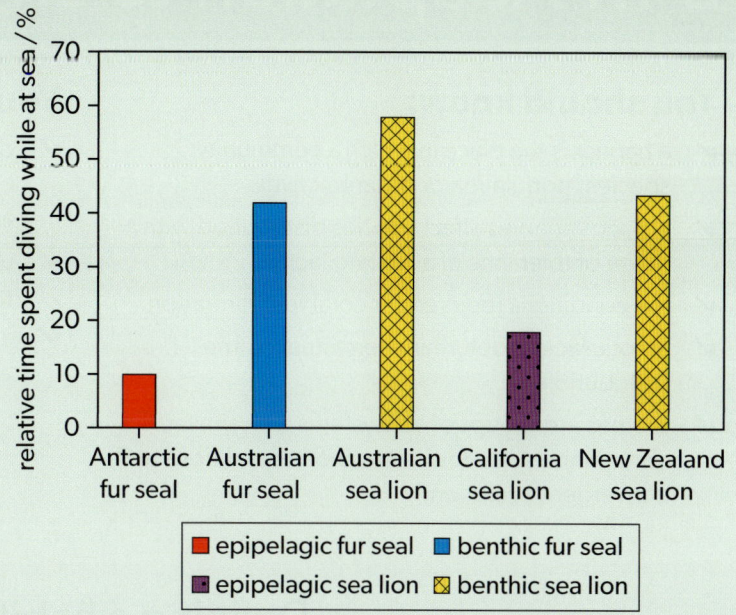

Solution

a) 42%

b) The benthic organisms in both fur seals and sea lions spend more relative time diving while at sea, so the foraging behaviour seems to be the most important.

Source: Daniel P. Costa, Carey E. Kuhn, Michael J. Weise, Scott A. Shaffer, John P.Y. Arnould, When does physiology limit the foraging behaviour of freely diving mammals?, International Congress Series, Volume 1275, 2004, Pages 359-366, ISSN 0531-5131, https://doi.org/10.1016/j.ics.2004.08.058.

Practice problems

1. Design an experiment to compare the number of stomata per unit area in the lower and upper surfaces of a leaf.

2. The transpiration rate of a leafy shoot was measured using a potometer. The distance moved by the bubble in the capillary tube was measured every 5 minutes. The experiment was performed at 25 °C at normal atmospheric conditions and was repeated later using a fan. The results are shown in the table.

Time / minutes	Distance moved by bubble / mm	
	Without fan	With fan
0	0	0
5	15	20
10	33	45
15	45	65
20	65	84
25	85	100
30	93	118

a) Sketch a graph to show the results obtained.

b) Predict the result for 35 minutes with a fan.

c) Explain the results obtained with and without a fan.

3. Discuss the composition of plasma and tissue fluid in relation to the exchange of substances between capillaries, tissue fluid and cells in tissues.

4. Explain the transport of blood under high pressure from the heart to the rest of the body.

5. Explain how the increased level of calcium ions triggers the contraction of muscle cells.

6. Explain how locomotion in animals in searching for a mate and in migration can contribute to evolution within living organisms.

- What are the advantages and disadvantages of dispersal of offspring from their parents?
- In what ways does locomotion contribute to evolution within living organisms?

Form and function

B4.1 Adaptation to environment

You should know:
- ✔ a habitat is the place in which a community, species, population or organism lives.
- ✔ abiotic variables affect species distribution with a range of tolerance of a limiting factor.
- ✔ the conditions required for coral reef formation.
- ✔ abiotic factors determine terrestrial biome distribution.
- ✔ biomes are groups of ecosystems with similar communities due to similar abiotic conditions and convergent evolution.

You should be able to:
- ✔ describe adaptations of organisms to the abiotic environment of their habitat, especially hot deserts and tropical rainforest.
- ✔ use transect data to correlate the distribution of plant or animal species with an abiotic variable.
- ✔ analyse graphs showing the distribution of biomes with temperature and rainfall.

Defining a habitat

- **Habitat** is the place in which a community, species, population or organism lives.

A **habitat** is the place in which a community, species, population or organism lives. The habitat of a species can include both geographical and physical locations, and the type of ecosystem.

Example 1

Obtaining information on movement and spatial patterns of animals and understanding the factors that influence their movements are critical in designing conservation strategies. Ecologists carried out a study of the threatened copper belly water snake (*Nerodia erythrogaster neglecta*) in northwest Ohio and southern Michigan to assess differences in movement patterns, spatial ecology and resource use. The snakes feed on fish, frogs and toads found in wetland pools. The graphs show the relative abundance of frogs, toads and fish in 2001 and 2002 for six wetlands at the study site. Graphs A, B and C correspond to temporary wetlands, and graphs D, E and F correspond to permanent wetlands.

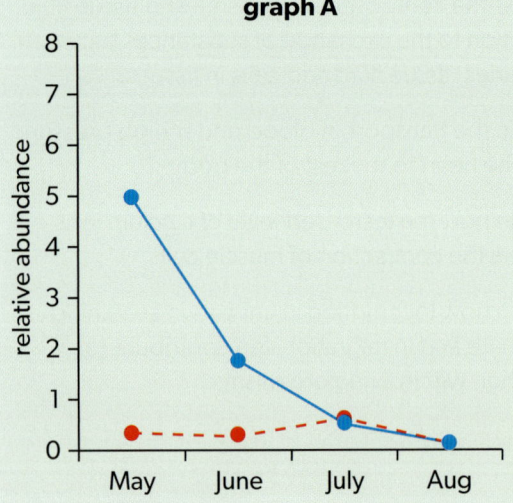
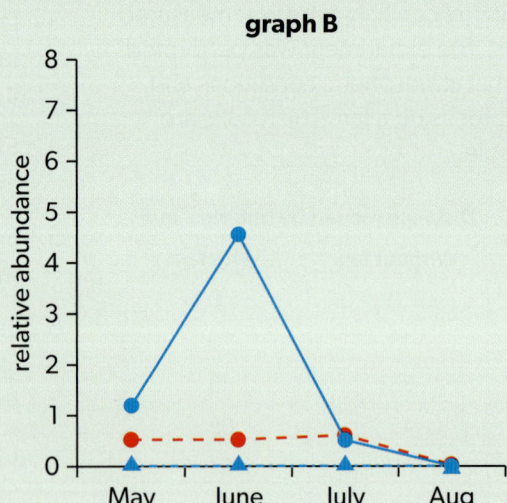

B4.1 Adaptation to environment

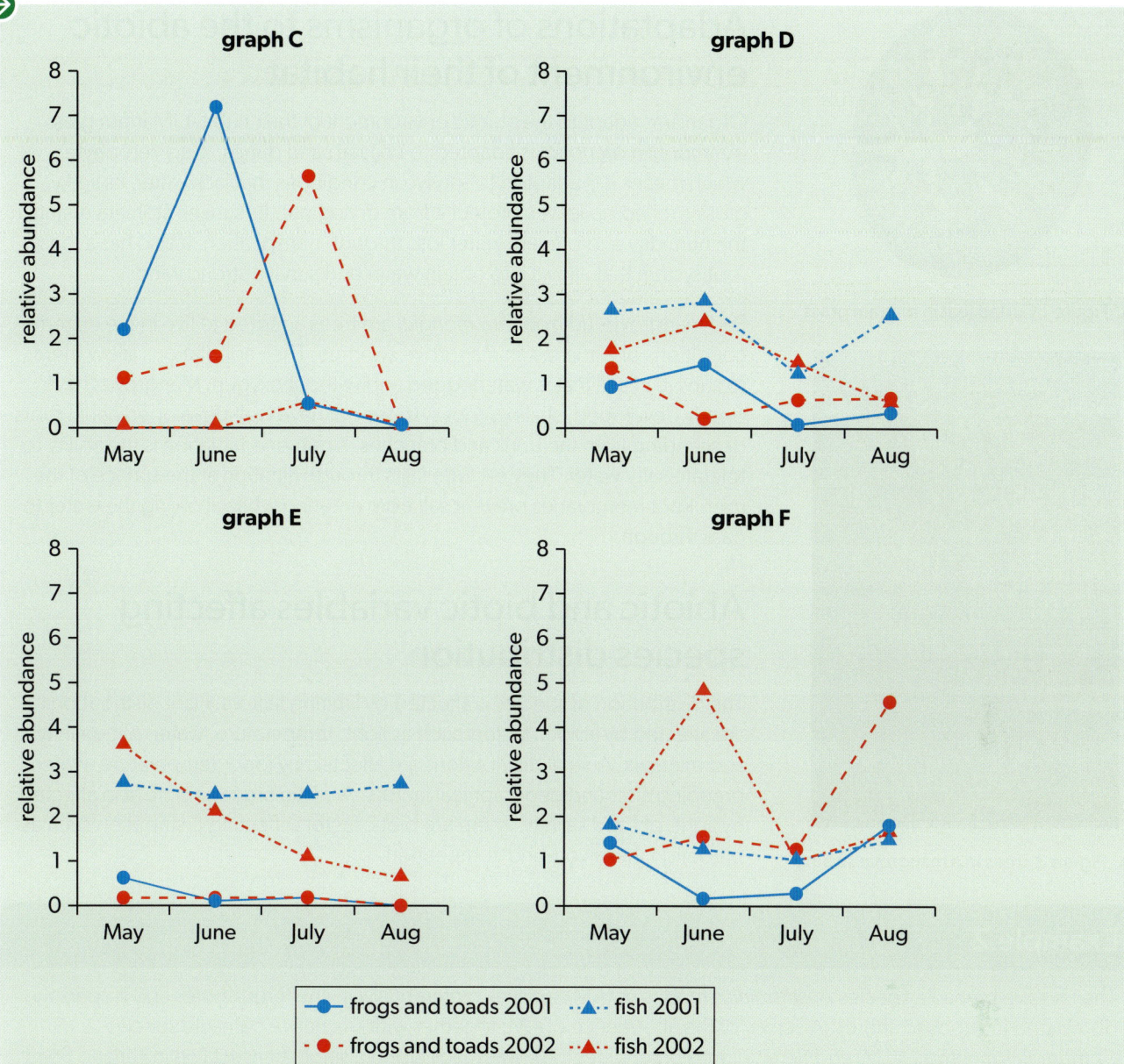

Source: J H Roe, et al. (2004). *Biological Conservation*, **118** pages 79–89.

a) Identify the type of wetland that
   (i) had the greatest frog and toad abundance during August.
   (ii) had the most stable fish population.

b) Deduce, giving a reason, which type of wetland is likely to provide the best food supply for the snakes.

c) Changing land use is decreasing the numbers of wetlands. Suggest what impacts such changes in habitat may have on *N. erythrogaster neglecta*.

Solution

a) (i) As seen in graph F, it is the permanent wetlands.
   (ii) As can be seen in all graphs, there are more fish in permanent wetlands than in temporary wetlands that hardly have any fish.

b) The permanent wetlands will provide the best food supply for the snakes as the total abundance of food over the two years is always greater in each month.

c) Temporary wetlands would be affected first and be the first to disappear. *N. erythrogaster neglecta* numbers would decrease and could be at risk of extinction. At the same time *N. erythrogaster neglecta* would have to travel farther to find food or a new habitat where there would be increased competition with other predator species.

Form and function

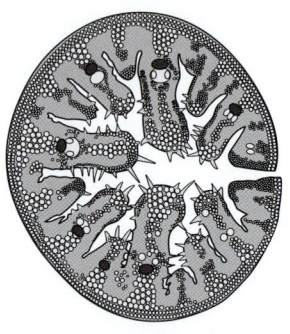

▲ Figure 1 Curled leaf in marram grass

▲ Figure 2 Trees in a mangrove swamp

Adaptations of organisms to the abiotic environment of their habitat

Organisms adapt to the abiotic environment of their habitat. Marram grass (*Ammophila arenaria*) is adapted to coastal sand dunes. It is a xerophyte, which means it is adapted to survive in conditions that lack water. Its spiky, glossy, rolled-up leaves protect it from drying out. Its sunken stomata maintain the humidity and prevent water loss through transpiration. It also has a long root system that allows it to obtain water and survive strong winds.

Red mangroves (*Rhizophora mangle*) are trees adapted to live in the extreme conditions of estuaries, in mangrove swamps. Their adaptations allow them to survive in soil that is waterlogged and without oxygen. Mangroves have poorly developed, shallow root systems but have well-developed aerial roots descending from the trunk and branches. Another adaptation is the ability to tolerate salty water. They exclude salts through filtration at the surface of the root. Root membranes prevent salt from entering while allowing the water to pass through.

Abiotic and biotic variables affecting species distribution

The distribution of species is affected by limiting factors. Plant distributions are affected by abiotic factors such as light, temperature, water, pH, salinity and minerals. Animal distributions are affected by food, temperature, water, breeding sites and geographical factors. Aquatic organisms are also affected by light, pH and salinity. There are biotic factors affecting distribution such as predation, herbivory and competition.

Example 2

The invasive barnacle species *Austrominius modestus* (AM) survives preferentially in sheltered shores and is sparsely distributed in exposed shores. A survey of the abundance of AM, together with the native barnacle species *Chthamalus montagui* (CM) and *Semibalanus balanoides* (SB), was made using a single transect line in Farland Point and in Fintry Bay in Scotland. The vertical distribution above the lowest tidal height (chart datum) of the three species is shown in the kite diagram.

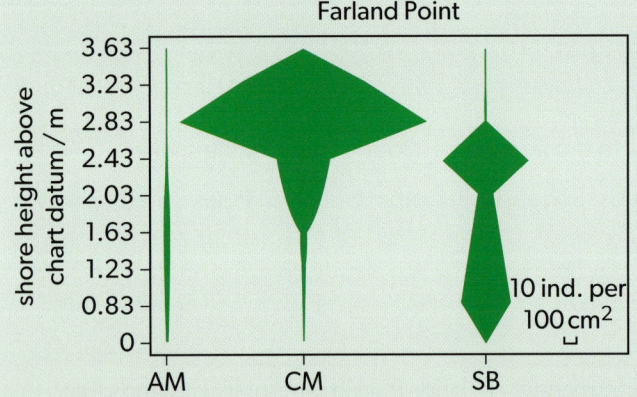
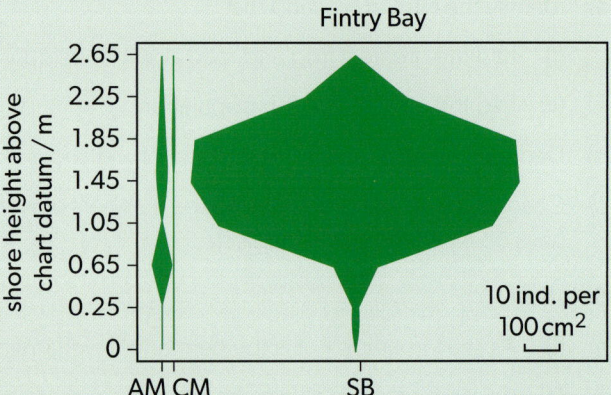

Source: Gallagher, M. C., et al. (2015). The invasive barnacle species, *Austrominius modestus*: Its status and competition with indigenous barnacles on the Isle of Cumbrae, Scotland. *Estuarine, Coastal and Shelf Science*, **152**, 134–141. www.doi.org/10.1016/j.ecss.2014.11.014

a) (i) State the number of SB barnacles per 100 cm^2 found in Farland Point at a height of 2.43 m above chart datum.

   (ii) Suggest one reason the units used for number of barnacles is individuals per 100 cm^2.

b) State the barnacle that is least abundant in

   (i) Farland Point.

   (ii) Fintry Bay.

c) Justify, with a reason, which site has more sheltered areas.

d) Suggest a factor other than shelter that could affect the distribution of barnacles.

Solution

a) (i) number of individuals per 100 cm² = $\dfrac{\text{width of the kite} \times 10 \text{ individuals}}{\text{width of scale}}$

   (ii) The sampling is done using a transect where quadrats of 100 cm² are placed periodically. The number of each barnacle in this quadrat is counted.

b) (i) AM

   (ii) CM

c) Fintry Bay, as AM thrives better in sheltered areas and is more abundant in this site.

d) There are several factors. One could be temperature—SB can live in deeper, colder waters. The hours exposed to air without water coverage—the higher the shore height value, the longer the organisms are uncovered by water. Light could be another factor. Food and nutrients can affect distribution. In this case, competition between species could be a factor, as when CM is present there is little AM and vice versa.

Coral reef formation

Coral reefs are large structures of calcium carbonate skeleton (limestone) deposited by coral polyps. Each polyp is a small invertebrate animal with a sac-like body and a mouth with stinging tentacles. Coral reefs form in warm, shallow, saline oceanic waters, with waves to bring nutrients, but not too strong to damage the corals. The temperature must be between 23 °C and 34 °C and the waters must be clean to allow light to pass. The water must also be free of pollutants, as coral reefs are formed by sensitive organisms.

▲ Figure 3 A coral reef

Example 3

Lettuce coral (*Undaria tenuifolia*) is the dominant coral species in shallow semi-exposed reefs in the Southern Caribbean Sea. Colonies of *U. tenuifolia* were classified according to colony size into small, medium and large. The numbers of colonies according to their size were recorded three times, starting in spring of the first year (1), then autumn of the first year (2) and ending in spring of the second year (3) in a location in the Southern Caribbean.

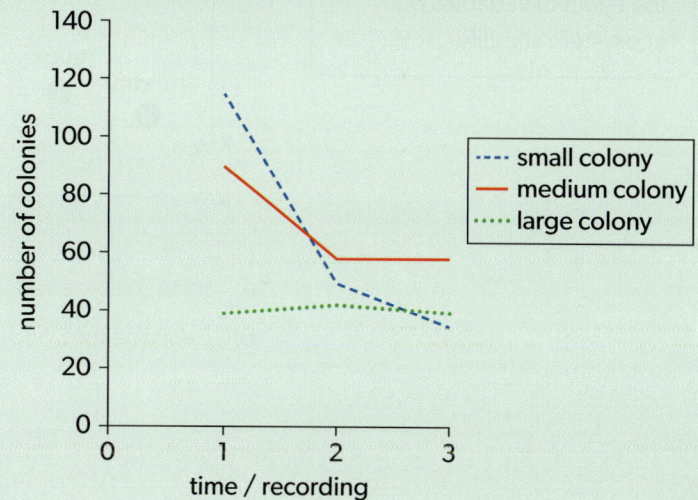

a) State the number of colonies recorded in the first spring for each colony.

b) Describe the trends in numbers of colonies.

Form and function

c) Suggest one reason for the change in population of small and medium corals.

d) Describe how bleaching is an indication of stress on corals.

Solution

a) Small: 115, medium: 90, large: 39 (a range ±2 is accepted for all values)

b) Numbers of colonies decreased in small and medium colonies. Small colonies decreased the most, while medium colonies decreased and remained constant. Large colonies had lower numbers and remained constant.

c) The number of corals could have been reduced by humans removing corals as souvenirs. Corals could have been preyed upon by natural predators (or alien species). Corals could have died due to pollution or warm temperatures. (Only one reason is needed.)

d) Bleaching is the removal of the symbiotic *Zooxanthellae* algae from corals. This could be due to lack of nutrients, warm temperatures, eutrophication or pollution. The greater the bleaching, the higher the probability the corals have suffered stress.

- **Biomes** are groups of ecosystems with similar communities due to similar abiotic conditions and convergent evolution.
- A **desert** is very dry with very hot days and cold nights.
- A **tropical rainforest** is rainy and hot.
- A **grassland** is a dry, warm biome with mainly grasses.
- A **taiga** is a temperate forest with cold winters and warm summers.
- A **tundra** is a very cold, dry biome.

Biome distribution

The type of stable ecosystem that will emerge in an area is predictable based on climate. Climographs represent the monthly average temperature and precipitation of a location. A **biome** is a group of ecosystems with similar communities due to similar abiotic conditions and convergent evolution. It can comprise many habitats. Examples of biomes are **deserts**, **tropical rainforests**, temperate forests, **grassland**, **taiga** (snow forests) and **tundra**.

You will study mesocosms in subtopic D4.2 "Investigating the effect of variables on ecosystem stability".

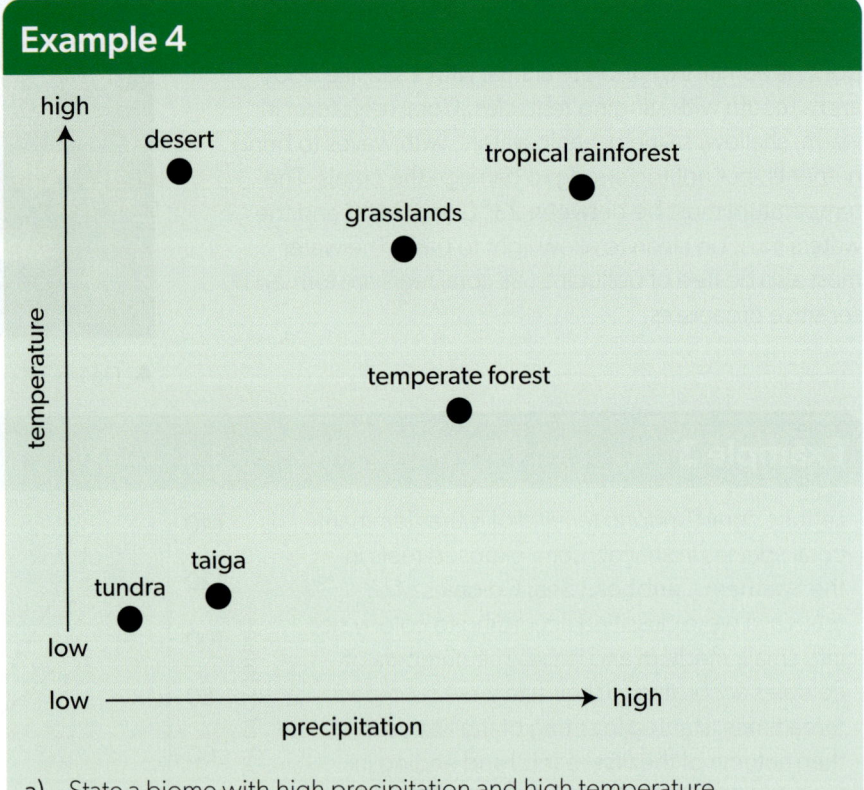

Example 4

a) State a biome with high precipitation and high temperature.

b) Describe the conditions in the tundra biome.

Solution

a) Tropical rainforest

b) The conditions are cold, frozen and dry. Temperatures are below water freezing point and there is hardly any precipitation. There is a short growing season, followed by harsh conditions, so the plants and animals in the region need special adaptations to survive. There are no trees, only ground vegetation consisting of small shrubs, grasses, mosses, sedges and lichens. The animals are adapted to extreme conditions.

Animals adapted to live in hot deserts are called xerocoles. Many desert animals have thick fur or feathers to help insulate them from the heat, while others have adapted to conserve water by sweating less and producing less urine or having a higher tolerance for salt. They sometimes have oily skins to avoid evaporation of water. They obtain most of their water from metabolic processes in breaking down their food. They tend to have a nocturnal lifestyle to avoid excessive heat and live in burrows. An example of a xerocole is the Texas banded gecko (*Coleonyx brevis*).

▲ Figure 4 The Texas banded gecko

Plants that live in deserts adapted for water conservation are called xerophytes. Deep or wide-ranging root systems increase the chances of water absorption. The thick, waxy cuticle of the leaf surface reduces transpiration. Small or no leaves or thorns instead of leaves reduce the surface area of leaves. Few stomata or the presence of stomata in pits or rolled leaves will also reduce evaporation. Hairs on the leaf surface reduce air flow near the leaf and reflect sunlight. The stomata open at night rather than during the hot daytime conditions, to reduce water loss. Stems are succulent as they can store large amounts of water and are usually photosynthetic. Plants that are adapted to live in high salt conditions are called halophytes. They have similar adaptations to xerophytes and some have structures for removing salt. An example of a plant adapted to hot deserts is the giant barrel cactus (*Echinocactus platyacanthus*).

- What are the properties of the components of biological systems?
- Is light essential for life?

▲ Figure 5 A giant barrel cactus in the hot desert

Form and function

B4.2 Ecological niches

You should know:
- the ecological niche is the role of a species in an ecosystem.
- organisms obtain nutrients by autotrophic, holozoic, mixotrophic and saprotrophic nutrition.
- teeth vary in humans according to the diet.
- herbivores and plants are adapted to herbivory.
- predators and prey are adapted to predation.
- plants are adapted to harvest light.
- the uniqueness of ecological niches depends on competitive exclusion.

You should be able to:
- compare obligate anaerobes, facultative anaerobes and obligate aerobes in terms of their tolerance to the presence of oxygen.
- distinguish between autotrophic, heterotrophic, mixotrophic and saprotrophic nutrition.
- describe nutrition in archaea.
- examine models or digital collections of skulls to infer diet from the anatomical features.
- describe named examples of adaptations to herbivory and predation.
- describe adaptations of plants to harvest light.
- distinguish between fundamental and realized niche.

Ecological niche—the role of a species in an ecosystem

- A **niche** is the role or function of an organism or species in an ecosystem.

An ecological **niche** is the role of a species in an ecosystem. Each species plays a unique role within a community because of the unique combination of its spatial habitat and interactions with other species. Interactions between species in a community can be classified according to their effect. There are biotic and abiotic interactions that influence growth, survival and reproduction, including how a species obtains food. The biotic interactions can be competition, predation, grazing, mutualism, symbiosis, parasitism and commensalism. The abiotic interactions include moisture, temperature, pH, soil composition and light availability.

Organisms can be classified according to the effect of the presence of oxygen in the environment on their metabolism and growth. Obligate anaerobes can only grow in the absence of oxygen. Many members of the human gut, such as those of the *Bacteroides* genus, are obligate anaerobes. Obligate aerobes require oxygen and do not perform fermentation, such as *Mycobacterium tuberculosis*. Facultative anaerobes such as *Saccharomyces cerevisiae* yeast can live in the presence or absence of oxygen.

Modes of nutrition in living organisms

Organisms need energy to live. Some organisms produce their own food, while others need to feed on other organisms. Each species has one specific mode of nutrition. There are only a few species that have more than one mode of nutrition.

Photosynthesis is the mode of nutrition in plants, algae and several groups of photosynthetic prokaryotes. They produce their own food. Animals, however, are **heterotrophic**—they do not produce their own food but feed on other organisms. Animals have **holozoic nutrition**, where food is ingested, digested internally, absorbed and assimilated.

Some protists, such as *Euglena*, have mixotrophic nutrition, as they are both **autotrophic** and heterotrophic. Some **mixotrophs** are obligate and others are facultative. If the trophic mode is obligate, then it is always necessary to have both types of nutrition for sustaining growth and maintenance. Otherwise, if it is facultative, it can be used as a supplementary energy source.

Some fungi and bacteria are **saprotrophs** as they obtain organic nutrients from dead organisms by external digestion. Fungi and bacteria with this mode of heterotrophic nutrition can be referred to as decomposers.

Archaea—which form one of the three domains of life—given their metabolic diversity, have a diverse mode of nutrition. Archaea species use either light, oxidation of inorganic chemicals or oxidation of carbon compounds to provide energy for ATP production. For example, methanogenic archaeans produce methane and water from carbon dioxide and hydrogen.

- **Heterotrophs** are organisms that feed on other organisms.
- In **holozoic nutrition**, food is ingested, digested internally, absorbed and assimilated.
- **Autotrophs** are organisms that produce their own food.
- **Mixotrophs** are both autotrophic and heterotrophic.
- **Saprotrophs** are heterotrophs that obtain organic nutrients from dead organisms by external digestion.

Relationship between dentition and diet

As herbivores feed on plants, they require teeth for grinding. They have smaller, sharp incisors and wider, flattened molars. In contrast, carnivores feed on meat, so they require teeth capable of tearing and biting with force. Usually, canines are pointed, incisors are large and sharp, and there are few molars. Omnivores, which eat both plants and animals, have teeth for a wide range of foods. The thickness of the enamel (protective layer in teeth) and the wear patterns observed as pits and scratches on the enamel are used to infer the type of food eaten. Trace-element studies of strontium-to-calcium ratios and fossils of butchered animal bones next to hominid fossils are indicators of a carnivore diet.

Paranthropus robustus, ancestors of humans, had teeth and jaws adapted to grinding and chewing hard and starchy plants. Their molars were large with thick enamel and their canines and incisors were small. The tooth wear patterns in *Australopithecus* show they were mainly herbivores, and that they manually pulled plant foods across their front teeth and had a jaw adapted to chewing. Their molars were large, but their incisors and canines were not as small as those of *P. robustus*. They also ate animals that were easy to catch. In *Homo floresiensis*, the teeth show signs of an omnivorous diet, with teeth used to crack open hard seeds or nuts with molars reduced in size. The molars of *Homo* became smaller with evolution, while canines and incisors became larger, adapted to their omnivorous feeding which included more meat.

Nature of science

Scientists make deductions from theories. By observing living mammals, scientists came up with theories about dentition and how it relates to herbivorous or carnivorous diets. These theories allowed scientists to work out the diet of extinct organisms from tooth wear patterns.

Form and function

Example 5

New technologies—such as dental topographic analysis—are being used to help understand how early hominids lived. This technique allows the pattern of wear of teeth over a lifetime to be analysed, revealing what types of food were eaten. Teeth from early hominids and *Australopithecus afarensis* were compared. The upper surfaces of the teeth were analysed for slope and occlusal relief (how closely the upper and lower teeth fit when brought together). The teeth examined were in groups of similar stages of wear to ensure consistency of results. The lower the slope and occlusal relief, the flatter the teeth. Flat teeth are best suited to crushing hard, brittle foods. More shaped teeth are better suited to eating elastic foods such as meat.

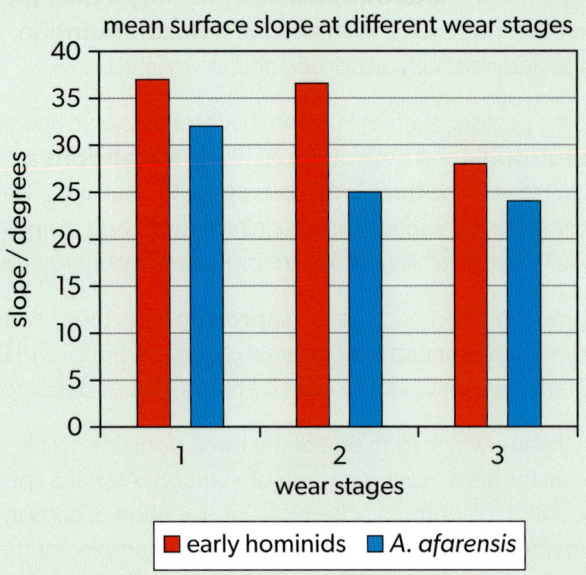

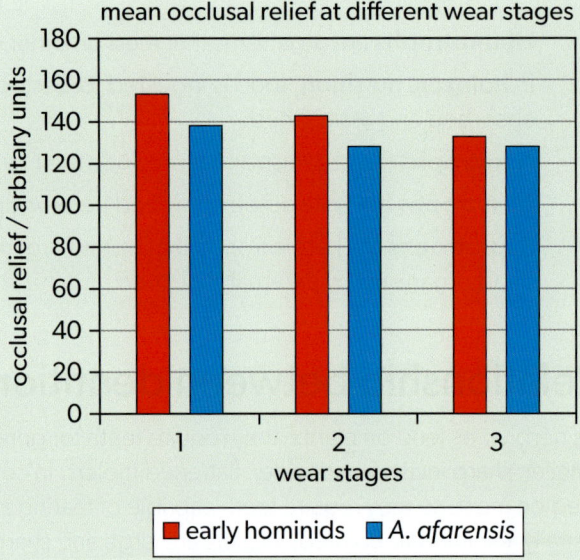

Source: Ungar, P. (2004). Dental evidence for diets of *Australopithecus afarensis* and early *Homo*. Vertebrate Paleobiology and Paleoanthropology, 121–134. www.doi.org/10.1007/978-1-4020-9980-9_11

a) (i) State what changes occurred to all teeth with wear.

   (ii) Compare the teeth of early hominids with those of *A. afarensis*.

b) Using the data, suggest how the diets of early hominids and *A. afarensis* differed.

c) Suggest what other evidence would help scientists to determine what food was eaten by early hominids.

Solution

a) (i) Slope and occlusal relief decrease with wear.

   (ii) Teeth of early hominids have a greater slope and occlusal relief at all stages than *A. afarensis*. Wear has greater effect on slope in *A. afarensis* than in early hominids, but wear has less effect on occlusal relief in both. The occlusal relief of early hominids progressively decreases whereas it remains relatively stable for *A. afarensis*.

b) Early hominids ate more elastic foods such as meat, while early *A. afarensis* ate more hard, brittle foods. The diets were similar at wear stage 3.

c) Tools found and bones of butchered animals found with early hominid fossils.

Adaptations of herbivores and of plants for resisting herbivory

Herbivores are adapted to feed on plants and at the same time plants are adapted to resist herbivory. Insects have piercing and chewing mouthparts that help them to eat leaves. Plants resist herbivory using structural barriers such as thorns or by hiding through camouflage. Another indirect method is to attract the natural predators of their pests. Plants can also avoid herbivory by producing hard, rigid leaves and stems that are difficult to chew or by producing toxins as secondary metabolites in seeds and leaves. Some animals have metabolic adaptations for detoxifying these chemicals. Monarch butterfly (*Danaus plexippus*) caterpillars store in their bodies the chemical defences cardenolides produced in milkweed (*Asclepias* spp.) to discourage predators from eating them. The cardenolides interfere with the sodium pumps of animals, but do not seem to harm these caterpillars.

Adaptations of predators and prey

There are many chemical, physical and behavioural adaptations of predators to catching their prey and of prey resisting predation. Predators have hunting adaptations that help them to catch their prey. They usually have acute senses of smell, vision and hearing. They develop individual or group hunting strategies. Predators have long, sharp teeth and claws specialized for shearing, tearing and cutting flesh to catch, kill and eat the prey. Some produce venoms to immobilize or kill their prey.

Prey can hide from predators by resembling its background (camouflage), for example having white fur in snow animals. Prey can have defence weapons, such as quills in porcupines or venom in snakes, or are adapted to run very fast to escape from predators. Also, prey must have sense of smell, sight and hearing developed to detect the presence of a predator in time to escape. Many prey use signals to warn other members of their population or to scare away predators. Some of these signals use sounds, colours, smells or poisons.

Adaptations of plants for harvesting light

Plants need to harvest light for photosynthesis. Given that a forest is densely populated by trees, plants in the lower layers—shade-tolerant shrubs or herbs—are adapted to living in low light. Lianas are vines that are rooted in the soil at ground level and use the tree as a vertical support to climb up to the canopy in search of direct sunlight. Epiphytes are plants growing on branches of trees. These plants obtain moisture and nutrients from the air, rain, water or from debris accumulating around them. They use the host only for physical support. Strangler epiphytes, for example, the figs *Ficus macrocarpa* and *F. virens*, are epiphytes that grow directly on trees. They grow from the branches down to the soil. They do not take nutrients from the sap of the host tree, but restrict the expansion of the host trunk eventually leading its death.

▲ Figure 6 (a) Liana and (b) epiphyte

Fundamental and realized niches

A niche is the role or function of an organism or species in an ecosystem. A **fundamental niche** is the potential of a species based on adaptations and tolerance limits. The **realized niche** is the actual extent of a species niche when in competition with other species. This means that each species proves to be not only physically different but also ecologically distinct in the habitat where it lives, the time of the day when it is active, what it eats and how it catches its food.

- A **fundamental niche** of a species is the potential mode of existence.
- A **realized niche** is the actual mode of existence, depending on adaptations and competition with other species.

- **Competitive exclusion** is a principle which states that two species cannot occupy the same niche.

Competitive exclusion

Competition can be for resources such as food, water, sunlight, breeding sites or territory. Intraspecific competition is between members of a species whereas interspecific competition is between different species. Two species cannot survive indefinitely in the same habitat if their niches are identical due to **competitive exclusion**.

Example 6

Two species of warbler songbirds, *Crateroscelis robusta* and *C. murina*, on Mt. Karimui in New Guinea segregate in their niches by altitude. On the chart, each circle represents one individual observation of *C. robusta* or *C. murina*.

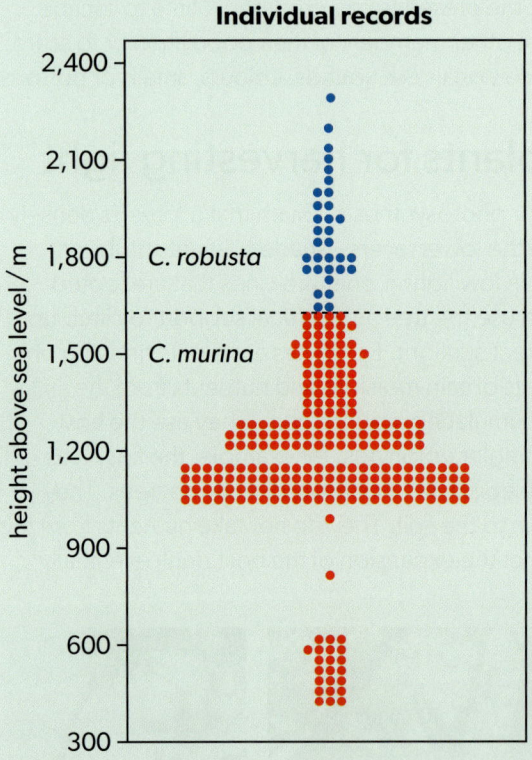

Source: Diamond, J. M. (1978). Niche shifts and the rediscovery of interspecific competition: *Why did field biologists so long overlook the widespread evidence for interspecific competition that had already impressed Darwin?* American Scientist 66(3), 322–331. www.jstor.org/stable/27848643

a) State the number of *C. murina* observations at 600 m.
b) State the lowest altitude at which *C. robusta* can be found.
c) Identify the altitude where *C. murina* is best adapted.
d) Explain the altitude niche distribution of these warblers.

Solution

a) 3
b) 1,620 m
c) From 1,050 to 1,110 m
d) These two morphologically and ecologically similar warblers inhabit non-overlapping adjoining altitudinal ranges. This is because of the competitive exclusion principle, which states that species must occupy different niches to coexist, because otherwise one species will exclude the other competitively. The species that occupy distinct niches survive by minimizing the struggle for existence.

Practice problems

1. The diagram shows a climograph.

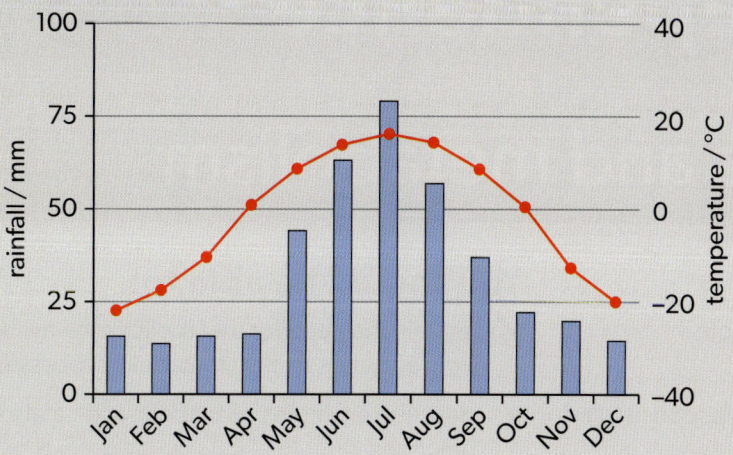

   a) State the maximum precipitation in this location.

   b) State the warmest month in this location.

   c) Identify the biome represented in the climograph.

2. What adaptation enhances resistance to herbivore predation in plants?

   A. Production of great quantities of pollen

   B. Exposed seeds

   C. Production of tasty fruit

   D. Production of secondary metabolites

- What are the relative advantages of specificity and versatility?
- For each form of nutrition, what are the unique inputs, processes and outputs?

C Interaction and interdependence

C.1.1 Enzymes and metabolism

You should know:

- enzymes are catalysts with a central role in metabolism.
- metabolic reactions are regulated in response to the cell's needs.
- metabolic pathways consist of chains and cycles of enzyme-catalysed reactions.
- enzymes are globular proteins with an active site to which specific substrates bind.
- molecular movement and collision of the enzyme's active site with the substrate are required for catalysis.
- the rate of activity of enzymes depends on temperature, pH and substrate concentration.
- enzymes can be denatured.
- enzymes lower the activation energy of the chemical reactions that they catalyse.

Additional higher level:

- examples of intracellular and extracellular enzyme-catalysed reactions.
- heat energy is generated in metabolic reactions, which mammals and birds use to maintain body temperature.
- enzyme inhibitors can be competitive or non-competitive.
- metabolic pathways can be controlled by end-product inhibition.

You should be able to:

- explain the need for metabolic pathways and how enzymes work in metabolic processes.
- describe examples of anabolic and catabolic reactions.
- describe induced-fit binding of an enzyme to a substrate.
- explain the relationships between the structure of the active site, enzyme–substrate specificity and denaturation.
- interpret graphs to show the expected effects of temperature, pH and substrate concentration on the activity of enzymes.
- calculate and plot rates of reaction from raw experimental results.
- describe that energy is required to break bonds within the substrate and that there is an energy yield when bonds are made to form the products of an enzyme-catalysed reaction.

Additional higher level:

- describe cyclical and linear pathways in metabolism.
- distinguish different types of inhibition from graphs at specified substrate concentration.
- explain the end-product inhibition of the pathway that converts threonine to isoleucine.
- explain mechanism-based inhibition using penicillin as an example.

Enzymes as catalysts

- An **enzyme** is a globular protein that acts as a biological catalyst.

Enzymes are catalysts, which means they speed up the rate of a reaction without taking part in the reaction itself. In the absence of enzymes these reactions would take too long. Cells depend on many chemical reactions and without enzymes these would be too slow to allow life processes to happen.

Enzymes in metabolism

Metabolic pathways are enzyme-mediated, interdependent and interacting biochemical reactions that lead to the synthesis or breakdown of natural **substrates** into products within a cell or tissue. The glycolysis reactions in respiration and the Calvin cycle in photosynthesis are examples of metabolic pathways.

- A **substrate** is the molecule changed by the enzyme.

Anabolic and catabolic reactions

Anabolic reactions form larger molecules from smaller ones. They include the formation of macromolecules from monomers by condensation reactions, including protein synthesis, glycogen formation and photosynthesis. Conversely, catabolic reactions break down larger molecules to smaller ones. They include hydrolysis of macromolecules into monomers in digestion and oxidation of substrates in respiration.

Induced-fit binding of substrate and active site

Enzymes are globular proteins that lower the activation energy of the chemical reactions they catalyse. An area of the protein, formed by 3 to 12 amino acids, is specific for each substrate and is responsible for the catalytic activity.

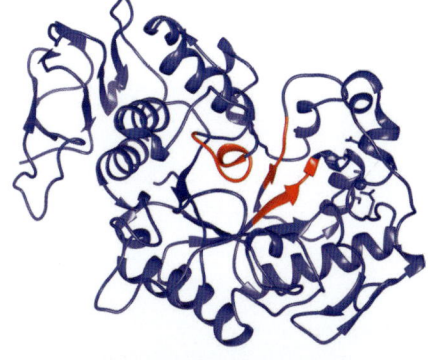

▲ Figure 1 Structure of an enzyme showing its active site in red

Movement is needed for a substrate molecule and an **active site** to come together. The enzyme and the substrate collide in a chemical reaction due to molecular motion. The greater the motion, the higher the chances a substrate will join the active site of an enzyme. The enzyme–substrate complex formation allows catalytic activity. Interactions between substrate and active site allow **induced-fit binding**, where both substrate and enzyme change shape when binding occurs. The enzyme lowers the energy needed for the reaction to occur by bringing the substrates closer to join in an anabolic reaction or by straining the molecules in a catabolic reaction.

- The **active site** is the part of the enzyme that binds to the substrate.
- In **induced-fit binding**, both substrates and enzymes change shape.

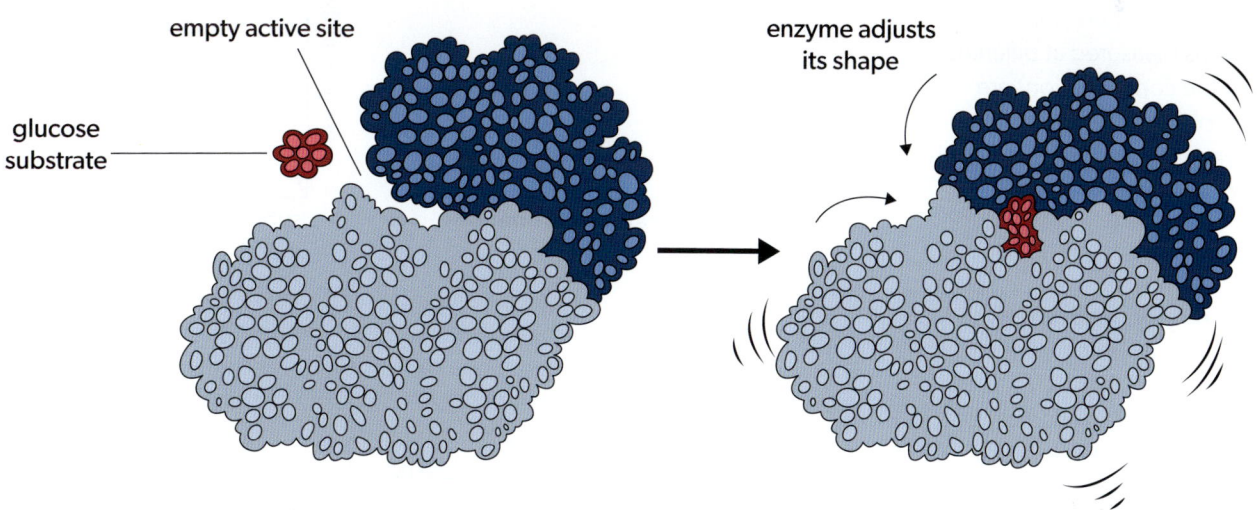

▲ Figure 2 Induced-fit binding of glucose to the enzyme glucose oxidase

Large substrate molecules or enzymes can be immobilized by being embedded in membranes. Immobilization stabilizes the enzyme by restraining the free movement of the enzyme outside the active-site region, leaving the active site flexible enough for high activity.

Interaction and interdependence

> - **Denaturation** is the loss of the tertiary structure of the enzyme.

Changes in temperature and pH can alter the spatial disposition of the protein, affecting the rate of reaction of the enzyme. If the change is extreme, the protein can denature. **Denaturation** is the reversible loss of tertiary (and even secondary) protein structure. Disulfide and other weak bonds are broken, but denaturing does not involve the breaking of peptidic bonds. Denaturing modifies the structure of the active site, which changes the enzyme–substrate specificity and leaves the enzyme unable to work.

Factors affecting the rate of enzyme activity

Enzymes and substrates collide until they form the enzyme–substrate complex, thus catalysing the formation of product. Increasing the concentration of enzymes or substrates increases the rate of reaction by increasing the frequency of successful collisions. However, an excess of one over the other will not increase the rate of reaction, as one of the molecules will be saturated.

Example 1

Which graph best represents the effect of increasing the substrate concentration on enzyme activity?

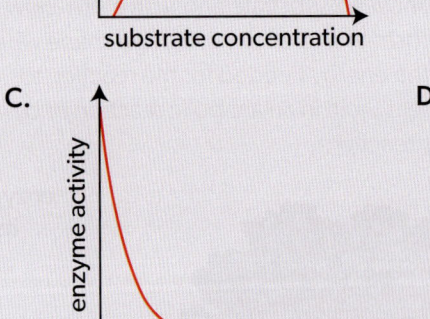
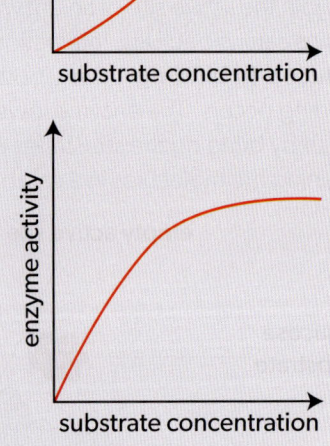

Solution

The correct answer is **D**. The enzyme activity increases until it reaches saturation point, where all enzymes have their active site full. The shape in Graph **A** is similar to the effect of pH on enzyme activity and the shape of graph **B** is a sigmoid population growth curve. Graph **C** is not showing any relationship, as at higher substrate concentration there is a higher rate of reaction.

> **Nature of science**
>
> Models can be used to describe the relationship between variables. Results from experiments where enzyme activity is measured at different substrate concentrations are used to show the relationship between these two variables in a sketched graph.

As temperature increases, the average speed and thus kinetic energy of particles also increases. The particles collide with one another more frequently. As a result, the frequency of successful collisions increases, so the **rate of reaction** increases. However, an increase in temperature above an **optimum** will start to denature the enzyme and the rate of reaction will decrease.

> - A **rate of reaction** is the speed at which a chemical reaction takes place. It can be measured according to the concentration (or amount) of product formed in a unit of time or as the concentration (or amount) of substrate that disappears in a unit of time.
> - **Optimum** describes the ideal conditions required for an enzyme to work.

C.1.1 Enzymes and metabolism

Measurements in enzyme-catalysed reactions

Example 2

Catalase is an enzyme found in living tissues that catalyses the breakdown of hydrogen peroxide into oxygen and water.

The graphs show the relative activity of catalase under different conditions of temperature (**A**), pH (**B**) and hydrogen peroxide concentrations (**C**) and the reaction time of hydrogen peroxide degradation with and without catalase (**D**). Error bars represent the mean ± standard error of the mean of three independent experiments.

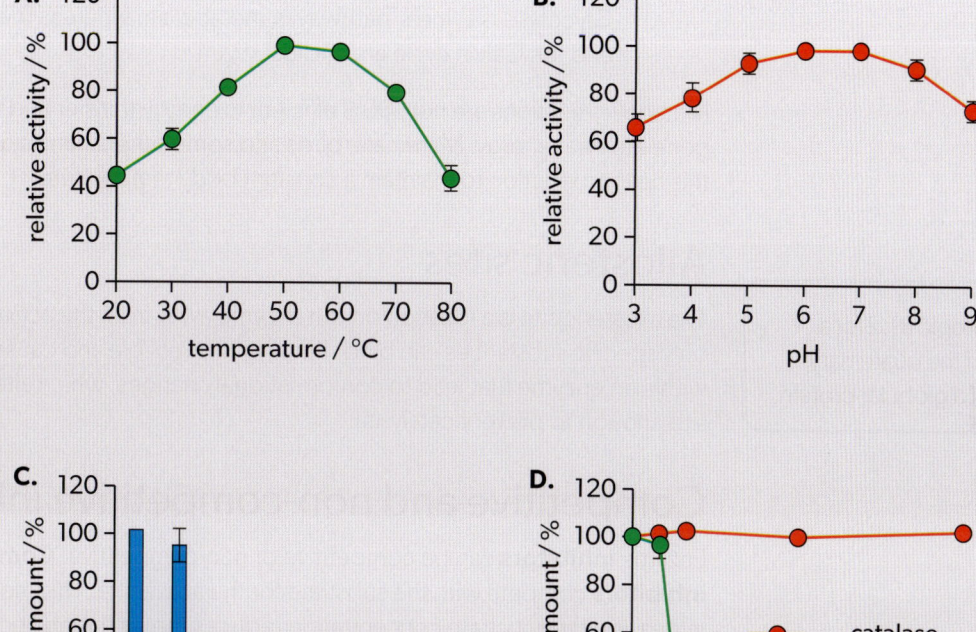

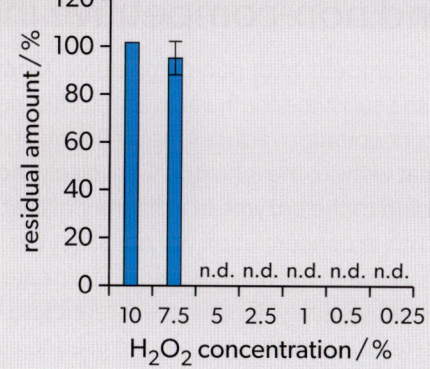

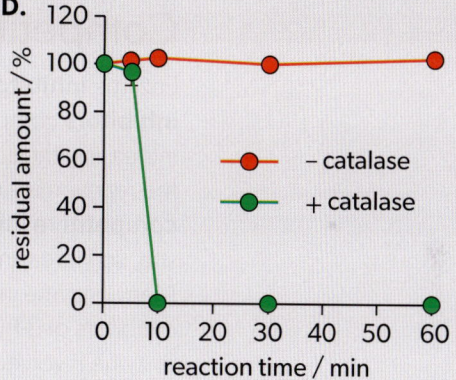

Source: Sakai, K., Okada, M. & Yamaguchi, S. (2022). Decolorization and detoxication of plant-based proteins using hydrogen peroxide and catalase. *Nature Scientific Reports*, **12**, 22432. www.doi.org/10.1038/s41598-022-26883-8 (CC BY 4.0)

a) State the optimum temperature and pH for catalase.
b) Explain the effect of hydrogen peroxide concentration on enzyme activity.
c) Explain graph **D**.
d) Suggest an experimental method for recording the catalase activity.

Solution

a) Temperature: any value between 50 and 55 °C. pH: any value between 6 and 6.5.
b) At concentrations lower than 5%, there is no hydrogen peroxide left. At concentrations of 10% and 7.5%, most hydrogen peroxide remains as it is not all broken down. The enzymes are not enough to join the excess hydrogen peroxide, therefore saturation point has been reached.
c) Without catalase, the residual amount is always 100%, which means no hydrogen peroxide is broken down. With catalase, after 10 minutes there is no hydrogen peroxide left.
d) Counting bubbles or measuring volume of oxygen (or water) or froth produced. Using probes to measure oxygen or water production or reduction of hydrogen peroxide.

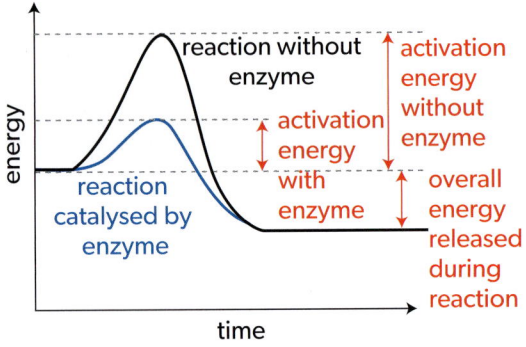

▲ Figure 3 Enzymes lower activation energy

Effect of enzymes on activation energy

Energy is required to break bonds within the substrate and there is an energy yield when bonds are made to form the products in an enzyme-catalysed reaction. Activation energy is the minimum amount of energy needed to activate molecules so that they can undergo a chemical reaction or transformation.

Metabolic pathways

Enzyme-catalysed reactions such as glycolysis and the Krebs cycle are intracellular, whereas chemical digestion in the gut is extracellular. Glycolysis is a linear pathway in metabolism whereas the Krebs cycle and the Calvin cycle are cyclical.

Metabolic reactions are not 100% efficient in energy transfer and therefore generate heat energy. Mammals, birds and some other animals depend on this heat production to maintain a constant body temperature.

Allosteric sites

> Turn to subtopic B1.2 for a reminder of the quaternary structure of a globular protein.

The allosteric site is a location on the enzyme away from the active site where only specific substances can bind. Binding is reversible and causes interactions within an enzyme that lead to conformational changes, which alter the active site enough to prevent catalysis.

Competitive and non-competitive inhibition

Enzyme **inhibitors** can be competitive or non-competitive. **Competitive inhibitors** compete with the substrate for the active site. This means that increasing the substrate concentration reduces the inhibition and eventually the reaction is similar to that without the inhibitor. Conversely, **non-competitive inhibitors** bind to the enzyme in a different place from the active site, so increasing the substrate concentration will not affect the inhibition. Many enzyme inhibitors have been used in medicine. Statins are competitive inhibitors of 3-hydroxy-3-methylglutaryl-Coenzyme A reductase (HMGR), an enzyme involved in cholesterol biosynthesis. They are used to reduce levels of cholesterol in residual amount blood.

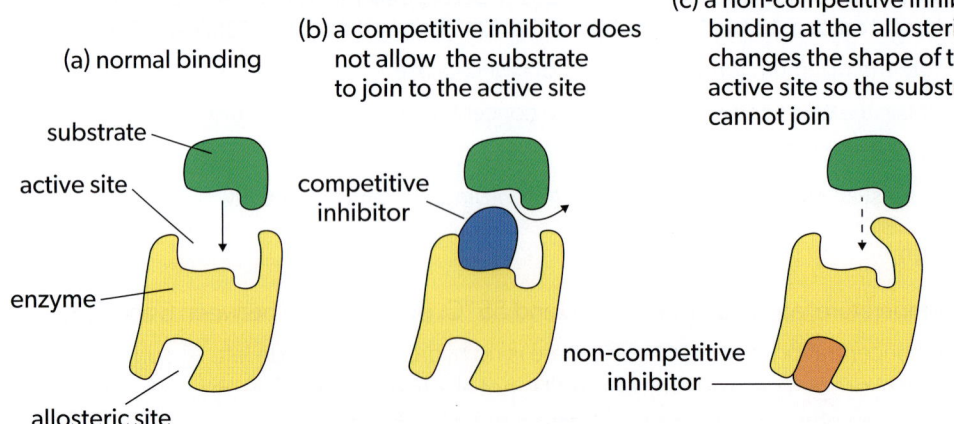

▲ Figure 4 (a) Normal binding, (b) competitive inhibition and (c) non-competitive inhibition

- An **inhibitor** is a molecule that binds to an enzyme, decreasing its catalytic activity.
- A **competitive inhibitor** joins at the active site. Its binding is affected by the substrate concentration.
- A **non-competitive inhibitor** joins away from the active site. Its binding is not affected by the substrate concentration.

Regulation of metabolic pathways by feedback inhibition

Metabolic reactions are regulated in response to the cell's needs. Metabolic pathways consist of chains and cycles of enzyme-catalysed reactions. Because enzymes lower the activation energy of the chemical reactions that they catalyse, the reactions can occur at a higher rate. Metabolic pathways can be controlled by **end-product inhibition**. This means the initial reaction is inhibited by the product of the final reaction. This avoids an excess of the product.

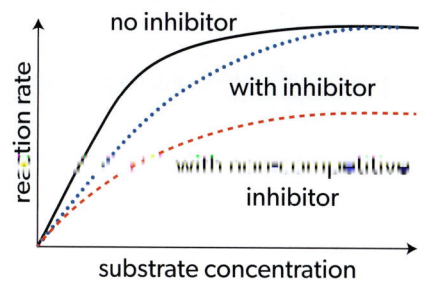

▲ Figure 5 Rate of reaction with and without inhibitors

- In **end-product inhibition** the product inhibits its own production.

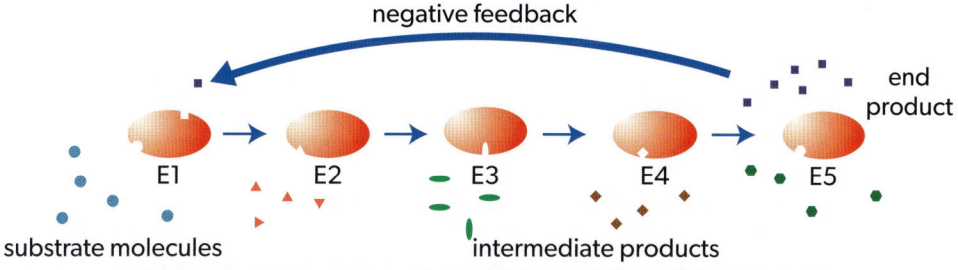

▲ Figure 6 Negative feedback by end-product inhibition

The production of isoleucine is controlled by end-product inhibition. It inhibits the pathway that converts threonine to isoleucine. Through a series of five enzyme-catalysed reactions, the amino acid threonine is converted to isoleucine. As the reaction takes place, the concentration of isoleucine increases. Isoleucine then binds to the allosteric site of the first enzyme in the chain, threonine deaminase. This enzyme is therefore inhibited. Isoleucine acts as a non-competitive inhibitor. When the level of isoleucine is low again, the enzyme will start breaking down threonine again.

Sample student answer

The enzyme α-1,4-glucosidase catalyses the breakdown of maltose to two glucose molecules. Its activation by maltose and inhibition by glucose was studied. The activity of α-glucosidase was measured in yeast which was grown in a medium containing 2% maltose only, 2% glucose only, or a mixture of 1.5% maltose and 0.5% glucose.

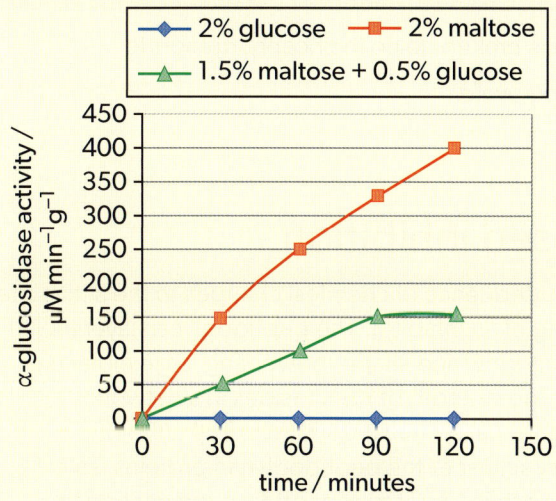

Interaction and interdependence

a) Calculate the percentage change in α-glucosidase activity at 60 minutes in yeast grown with 2% maltose compared to yeast grown with a mixture of both maltose and glucose. [1]

This answer could have achieved 1/1 marks:

> 2% maltose = 250μ min⁻¹g⁻¹
> both = 100μ min⁻¹g⁻¹
> difference = 150μ min⁻¹g⁻¹
> 250μ min⁻¹g⁻¹ is 100%
> 150μ min⁻¹g⁻¹ = $\frac{150\mu\ min^{-1}g^{-1}}{250\mu\ min^{-1}g^{-1}} \times 100\%$
> answer = 60% less for mixture

▲ The reduction is correct. The calculation could have also been done for the increase in 2% maltose with respect to the mixture, giving a 150% increase.

b) Distinguish between α-glucosidase activity in yeast incubated with only glucose and in yeast incubated with only maltose. [1]

This answer could have achieved 1/1 marks:

> Yeast incubated with glucose shows no activity, it remains at zero throughout the two hours, whereas yeast incubated with maltose shows rapid activity.

c) Suggest the effect of incubating the yeast cells with both maltose and glucose in terms of metabolic pathways. [3]

This answer could have achieved 0/3 marks:

> If the yeast cells are incubated with maltose and glucose yeast activity will be very low. The breakdown of maltose to glucose will be very slow and the regular function of metabolism of carbohydrates will be affected.

▼ The answer is too vague as it does not explain the low activity. Glucose inhibition is greater than maltose activation as glucose is a product of the breakdown of maltose. The enzyme is inhibited by negative feedback by the product of the digestion of maltose to glucose. When the yeast is grown in a medium with glucose there is no activity. Glucose represses expression of α-glucosidase.

d) Outline the metabolism of glucose during glycolysis. [2]

This answer could have achieved 2/2 marks:

> During glycolysis, glucose is broken down in phosphorylation to make two molecules of pyruvate.

▲ One mark is given for the phosphorylation of glucose and one for the formation of two pyruvates. After two phosphate groups are added to glucose (phosphorylation), glucose is transformed to hexose biphosphate, which is converted to two 3C compounds. Its oxidation forms pyruvate with a net gain of 2 ATP and 2 NADH + H.

- What are examples of structure–function relationships in biological macromolecules?
- What biological processes depend on differences or changes in concentration?

Mechanism-based inhibition

Inhibition can also be a consequence of chemical changes to the active site by the **irreversible** binding of an inhibitor. The peptidoglycan layer (PGN) in bacteria preserves cell integrity and shape, preventing large molecules from penetrating the cell. The strength of PGN resides in a network of proteins linked by enzymes called transpeptidases. The antibiotic penicillin is structurally very similar to the peptidoglycan proteins, so it can bind irreversibly to the active site of transpeptidase, preventing the formation of PGN. Bacteria with weakened walls are susceptible to death by osmotic stress. Some bacteria have developed resistance to different antibiotics through three different mechanisms: enzymatic destruction of the antibiotic, reduced binding affinity of the penicillin-binding proteins or reduced permeability.

- An **irreversible** inhibitor joins the enzyme and cannot be unbound.

C1.2 Cell respiration

You should know:

- ATP is an immediate source of energy in the cell.
- ATP supplies energy for life processes such as active transport.
- energy is released by hydrolysis of ATP to ADP and phosphate, but energy is required to synthesize ATP from ADP and phosphate.
- cell respiration is the controlled release of energy from organic compounds (such as glucose) to produce ATP.
- glucose and fatty acids are the main substrates for cell respiration.

Additional higher level:

- in glycolysis, glucose is converted to pyruvate in the cytoplasm.
- in aerobic cell respiration, pyruvate is decarboxylated and oxidized, converted into an acetyl compound and attached to coenzyme A to form acetyl coenzyme A in the link reaction.
- in the Krebs cycle, the oxidation of acetyl groups is coupled to the reduction of hydrogen carriers, liberating carbon dioxide.
- energy released by oxidation reactions is carried to the cristae of the mitochondria by reduced NAD.
- transfer of electrons between carriers in the electron transport chain in the membrane of the cristae is coupled to proton pumping.
- oxygen is needed to bind with the free protons to maintain the hydrogen ion gradient and to accept electrons from the end of the electron transport pathway, resulting in the formation of water.

You should be able to:

- distinguish between the processes of cell respiration and gas exchange.
- compare anaerobic and aerobic cell respiration in humans.
- write word equations for aerobic and anaerobic respiration in humans.
- make measurements to determine the rate of cell respiration.
- calculate the rate of cell respiration from primary or secondary data.

Additional higher level:

- state that cell respiration involves the oxidation and reduction of electron carriers.
- explain how energy is made available through respiration.
- outline how NAD is regenerated by conversion of pyruvate to lactate, to give a small net gain of ATP without the use of oxygen.
- describe the pathway of anaerobic cell respiration in yeast and its use in brewing and baking.
- outline the reactions in glycolysis, the link reaction and the Krebs cycle.
- explain the electron transport chain and the involvement of NAD in respiration.
- explain chemiosmosis and the function of ATP synthase in the production of ATP.
- analyse diagrams of the pathways of aerobic respiration to deduce where decarboxylation and oxidation reactions occur.
- distinguish between lipids and carbohydrates as respiratory substrates.

Interaction and interdependence

Properties of ATP as a distributor of energy

Adenosine triphosphate (ATP) is a nucleotide suitable for storing, using and distributing energy within cells. One phosphate is transferred to acceptor molecules, which at the same time are activated for chemical transformations. The adenosine diphosphate (ADP), which is formed after the release of phosphate, is recycled to reform ATP through energy obtained from **oxidative phosphorylation** or by photophosphorylation. ATP is used in many processes in the cell such as active transport across membranes, synthesis of macromolecules (anabolism) and movement of the whole cell or cell components such as chromosomes.

- **Adenosine triphosphate** or **ATP** is a molecule that is used as a fast source of energy. It is formed by a ribose that has a nitrogenous base joined to carbon 1 and three phosphate groups joined to carbon 5.
- **Oxidative phosphorylation** is the phosphorylation of ADP to ATP using the oxidation of molecules such as reduced NAD (NADH).
- **Anaerobic respiration** is the catabolic process through which small amounts of energy in the form of ATP are produced from glucose without the use of oxygen.
- **Aerobic respiration** is the catabolic process involving the use of oxygen through which glucose is broken down to carbon dioxide, water and large amounts of energy in the form of ATP.

Cell respiration

Most living processes require energy, which is obtained from cell respiration mainly through the breakdown of glucose or fatty acid molecules. A wide range of carbon organic compounds can also be used as substrates for respiration. In **anaerobic respiration**, glucose is broken down to produce a small amount of ATP. The end products in anaerobic respiration can be lactic acid (in muscles or bacteria) or carbon dioxide and ethanol (in yeast or bacteria). In **aerobic respiration**, large amounts of ATP are produced from the breakdown of glucose, and carbon dioxide and water are also produced. Aerobic respiration occurs in the mitochondria. Cell respiration is the chemical process occurring in cells whereas gas exchange is the process by which gases move passively by diffusion across a surface.

Example 3

a) State a word equation for anaerobic and aerobic respiration in humans.
b) Distinguish between aerobic and anaerobic respiration in humans.

Solution

a) Anaerobic: glucose → lactic acid
   Aerobic: glucose + oxygen → carbon dioxide + water

b)

	Aerobic respiration	Anaerobic respiration
Substrate	glucose	glucose
Oxygen requirement	yes	no
Relative yield	high (36 ATP)	low (2 ATP)
Waste products	carbon dioxide and water	lactic acid
Site of reaction	mitochondria	cytoplasm

C1.2 Cell respiration

Variables affecting the rate of cell respiration

Example 4

The diagram shows a respirometer.

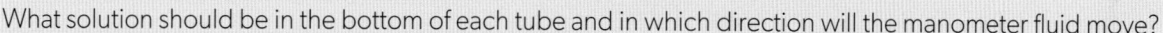

What solution should be in the bottom of each tube and in which direction will the manometer fluid move?

	Solution placed in the bottom of each tube	Direction of movement of fluid in the manometer
A.	acid	up on the left side
B.	alkali	down on the right side
C.	acid	up on the right side
D.	alkali	down on the left side

Solution

The correct answer is **B**. The solution in the bottom of the tubes is an alkali. Soda lime, which is a mixture of sodium hydroxide (NaOH) and calcium oxide (CaO), or sodium hydroxide (NaOH) or potassium hydroxide (KOH) can be used. The purpose of this solution is to absorb carbon dioxide (CO_2) from the respirometer. The manometer moves down on the right side because the small organisms (maggots in this case, but could be seeds) are respiring, therefore they are using up oxygen (O_2) from the air in the test tube and producing CO_2. As the CO_2 is absorbed by the soda lime, the consumption of O_2 causes a fall in pressure in the left-hand test tube, as the volume of air is reduced. This change in pressure pulls the liquid in the manometer up on the left so down on the right. The syringe at the top of the right tube is used to adjust the level of fluid in the manometer when a new replicate is going to be performed.

Example 5

A study was performed to see the effect of heat stress on energy metabolism in rice plants (*Oryza sativa*) of two varieties, japonica Nipponbare (NIPP) and temperature-resistant Huanghuazhan (HZ). Plants were grown at different high atmospheric temperatures (28, 34 and 38 °C) in a greenhouse and the daily respiration rate (Rd), ATP content and dry matter mass were measured.

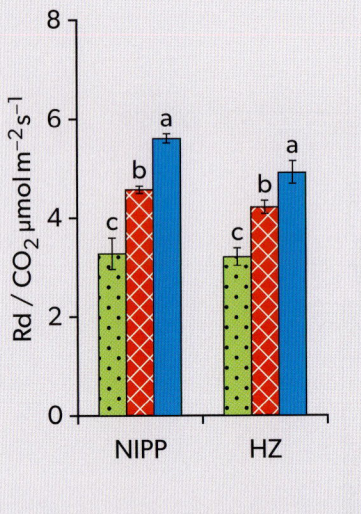 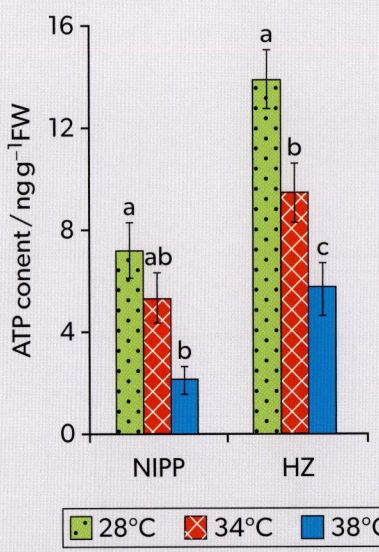 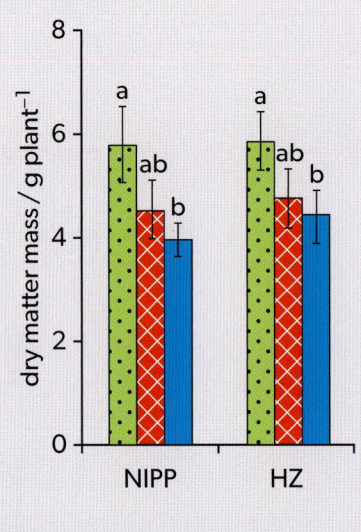

Source: Li, G., et al. (2021). Respiration, rather than photosynthesis, determines rice yield loss under moderate high-temperature conditions. *Frontiers in Plant Science*, 12, www.doi.org/10.3389/fpls.2021.678653

a) Identify the variety with the highest daily respiration rate.
b) Describe the effect of temperature on daily respiration rate.
c) Compare and contrast the effect of temperature on the ATP content in both varieties.
d) Explain the effect of heat stress on daily respiration, ATP content and biomass.

Solution

a) NIPP at 38 °C
b) As temperature increases, the daily respiration rate increases in both varieties.
c) As temperature increases, the ATP content in both varieties decreases. NIPP has lower ATP concentration than HZ at all temperatures. ATP content at 34 °C in NIPP is the same as at 38 °C for HZ.
d) Respiration is used for growth and maintenance. In growth respiration, reduced carbon compounds are metabolized to provide energy for the addition of new biomass, whereas in maintenance respiration, this energy is used for basic metabolism in cells. As temperature increases, plants require more maintenance energy than growth energy. Respiration increases using up stored carbohydrates, so biomass decreases, but ATP stores decrease as this ATP is used in maintenance reactions.

Redox reactions

Energy can be released through a series of redox reactions from the breakdown of glucose, amino acids or lipids and stored as ATP. Redox reactions involve both oxidation and reduction. Oxidation is a process of electron loss, so when hydrogen with an electron is removed from a substrate (dehydrogenation) the substrate has been oxidized. During cell respiration, the oxidation of carbon compounds produces hydrogen that is accepted by the carrier NAD. NAD is reduced to NADH when it accepts hydrogen.

Glycolysis and regeneration of NAD

In glycolysis, glucose is converted to pyruvate in the cytoplasm. Glycolysis gives a small net gain of ATP and reduced NAD without the use of oxygen. In anaerobic respiration, the pyruvate can be used in alcoholic or lactic fermentation. Alcoholic fermentation produces carbon dioxide and alcohol, while lactic fermentation in humans produces lactate. Both anaerobic fermentations are a linear sequence of reactions, each step catalysed by a different enzyme. Pathways of anaerobic respiration are the same in humans and yeasts and they include phosphorylation, lysis, reduction of NAD, with a net yield of two ATP molecules per molecule of glucose. Alcoholic fermentation in yeasts is used in brewing and baking. Production of pyruvate and regeneration of NAD in humans allows glycolysis to continue into aerobic respiration.

Link reaction and Krebs cycle

After glycolysis, pyruvate enters the mitochondrion. In aerobic cell respiration, pyruvate is decarboxylated, oxidized and converted into an acetyl compound. It is then attached to coenzyme A to form acetyl coenzyme A in the link reaction. In the **Krebs cycle** (sometimes called the citric acid cycle), the oxidation of acetyl groups is coupled to the reduction of hydrogen carriers, liberating carbon dioxide. Energy released by oxidation reactions is carried to the cristae of the mitochondria by reduced NAD. Lipids are also metabolized to form acetyl groups (2C), which are transferred by coenzyme A to the Krebs cycle.

- The **Krebs cycle** is a series of chemical reactions to reduce NAD, by oxidation of two-carbon acetyl groups. It occurs in the matrix of the mitochondrion.

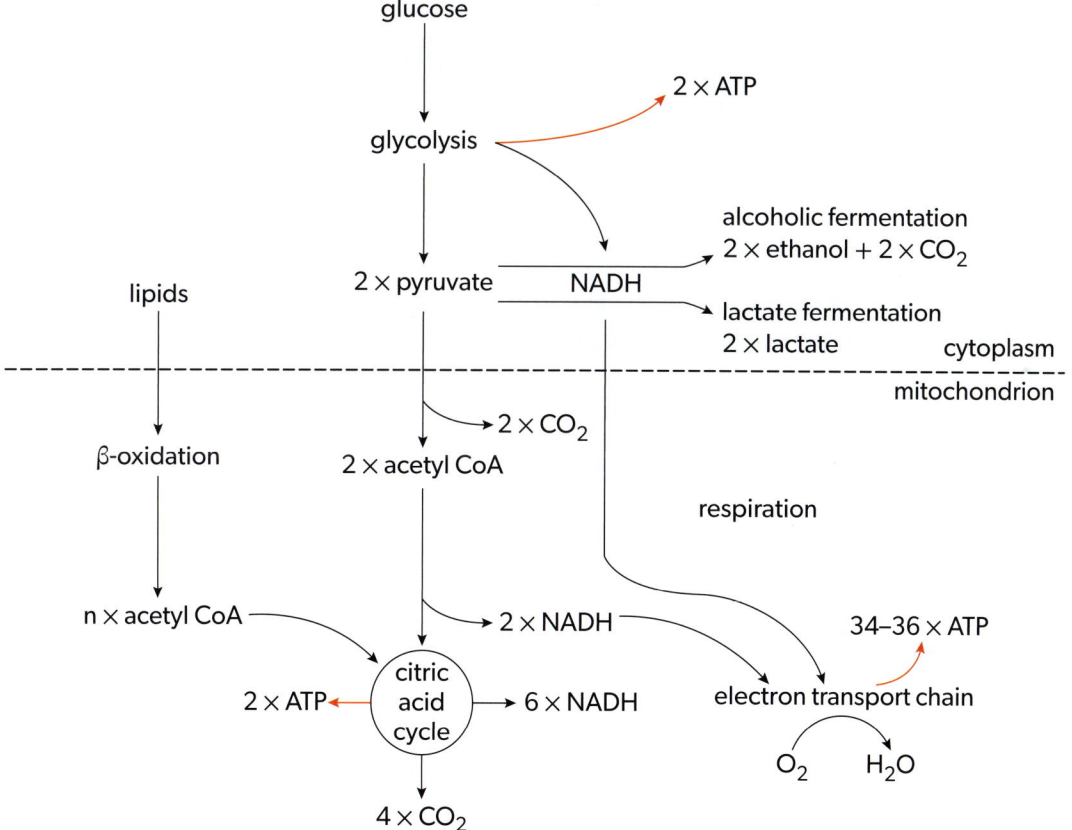

▲ Figure 7 Summary of the processes producing energy in a cell

Example 6

The diagram shows some steps of the link reaction and the Krebs cycle.

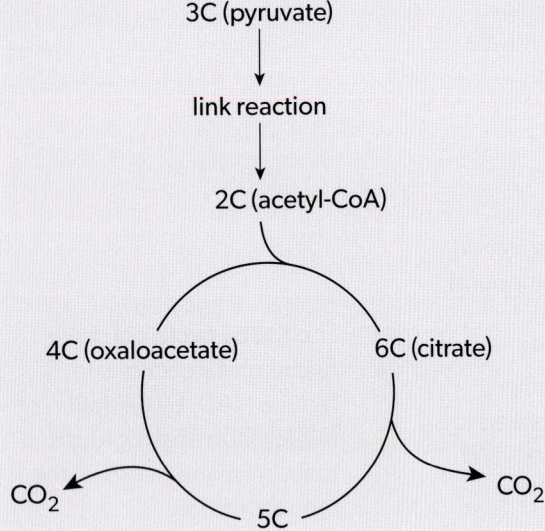

What occurs during the Krebs cycle?

A. Four oxidations and two decarboxylations to regenerate oxaloacetate

B. Two decarboxylations to form a 4C molecule and one oxidation to form a citrate

C. The product of the link reaction is decarboxylated to form citrate

D. One dehydration to form oxaloacetate and two decarboxylations to form acetyl-CoA

Solution

The correct answer is **A**. Citrate is produced by transfer of an acetyl group to oxaloacetate and that oxaloacetate is regenerated by the reactions of the Krebs cycle, including four oxidations and two decarboxylations. The oxidations are dehydrogenation reactions.

Transfer of energy to the electron transport chain

The reduced NAD produced in glycolysis, the link reaction and the Krebs cycle are carried to the cristae of the mitochondria. Energy is transferred to the electron carrier chain when a pair of electrons is passed to the first carrier in the chain, converting reduced NAD (NADH) back to NAD. Transfer of electrons between carriers in the **electron transport chain** in the membrane of the cristae is linked to proton pumping. The molecules of the electron transport chain include protein complexes such as cytochromes. Each donor will pass the electron to a more electronegative acceptor through redox reactions. This liberates energy that will be used to build up a proton (H^+) gradient across the inner mitochondrial membrane.

- The **electron transport chain** is the process by which electrons are passed from one carrier to another, liberating energy that is used to transfer protons across the inner membrane from the matrix to the intermembrane space.

- **Chemiosmosis** is the process by which protons move across the inner mitochondrial membrane, down the concentration gradient, releasing the energy needed for the enzyme ATP synthase to make ATP.

Chemiosmosis and the synthesis of ATP in the mitochondrion

The synthesis of ATP in the mitochondrion occurs through **chemiosmosis**. ATP synthase couples the release of energy from the proton gradient with the phosphorylation of ADP.

Chemiosmosis is the movement of ions across a partially permeable membrane, down their electrochemical gradient. In the mitochondria, chemiosmosis occurs when protons diffuse through ATP synthase to generate ATP. The energy to maintain the hydrogen ion gradient comes from the redox reactions of the electron transport chain. Protons are pumped out of the membrane into the intermembrane space where they build up. When they come back through the ATP synthase enzyme they enable the phosphorylation of ATP.

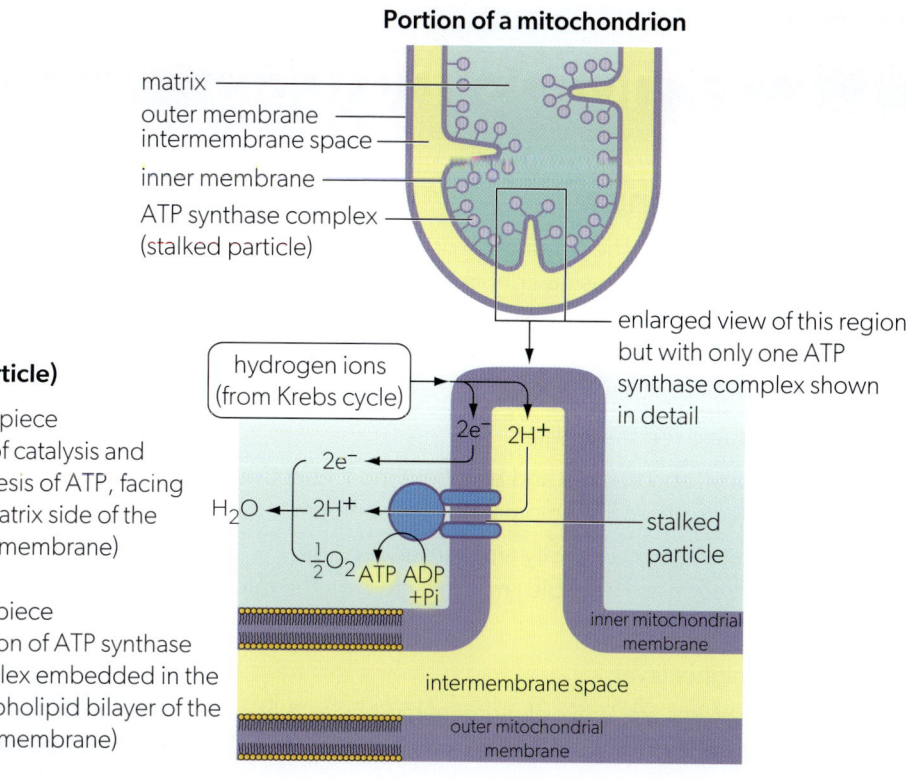

▲ Figure 8 Chemiosmosis in the mitochondrion

Oxygen is the final electron acceptor

Oxygen is the final acceptor of electrons in the transport chain. Oxygen is needed to bind with the free protons from the matrix of the mitochondrion to maintain the hydrogen ion gradient. This results in the formation of metabolic water and a continued flow of electrons along the chain.

Comparing lipids and carbohydrates as respiratory substrates

When electrons move from an atom with a low affinity for electrons, such as carbon, to one with a high affinity for electrons, such as oxygen, energy is released. Carbon atoms in fatty acids have more electrons around them so when the fatty acids are oxidized, a higher yield of energy per gram of lipids is obtained than when the same process happens to carbohydrates.

Example 7

Distinguish between lipids and carbohydrates as respiratory substrates.

Solution

Although lipids and carbohydrates are both respiratory substrates, lipids have a higher yield of energy per gram, due to less oxygen present in the molecule and more oxidizable hydrogen and carbon. Glycolysis and anaerobic respiration occur only if carbohydrate is the substrate, with 2C acetyl groups from the breakdown of fatty acids entering the pathway via acetyl-CoA (acetyl coenzyme A).

- In what forms is energy stored in living organisms?
- What are the consequences of respiration for ecosystems?

Interaction and interdependence

C1.3 Photosynthesis

You should know:

- photosynthesis is the production of carbon compounds in cells using light energy.
- oxygen is produced in photosynthesis from the photolysis of water.
- visible light has a range of wavelengths: violet is the shortest and red the longest.
- chlorophyll absorbs red and blue light most effectively and reflects green light more than other colours.
- temperature, light intensity and carbon dioxide concentration are possible limiting factors on the rate of photosynthesis.

Additional higher level:

- photosystems are arrays of pigment molecules that can generate and emit excited electrons.
- oxygen is a waste product generated by photolysis of water in photosystem II.
- excited electrons from photosystem II are used to contribute to generating a proton gradient.
- ATP synthase in thylakoids generates ATP using the proton gradient.
- excited electrons from photosystem I are used to reduce NADP.
- reduced NADP and ATP are produced in the light-dependent reactions.
- glycerate 3-phosphate is reduced to triose phosphate using reduced NADP and ATP.
- triose phosphate is used to regenerate ribulose bisphosphate and produce carbohydrates.
- ribulose bisphosphate is reformed using ATP.
- the products of the Calvin cycle and mineral nutrients are used to synthesize carbohydrates, amino acids and other carbon compounds.
- the light-dependent and light-independent reactions are interdependent.

You should be able to:

- write a simple word equation for photosynthesis.
- explain the separation of photosynthetic pigments by chromatography.
- draw and explain an absorption spectrum for photosynthetic pigments and an action spectrum for photosynthesis.
- design experiments to investigate the effect of temperature, light intensity and carbon dioxide on photosynthesis.
- describe how carbon dioxide enrichment experiments can be used to predict future rates of photosynthesis and plant growth.

Additional higher level:

- describe the steps involved in the light-dependent reactions of photosynthesis.
- explain how photosystems in the chloroplasts absorb light, generating excited electrons.
- describe the process of photolysis.
- explain the transfer of excited electrons between carriers in thylakoid membranes.
- explain how NADP and ATP are produced in the light-dependent reactions.
- state the locations of the light-dependent reactions in the intermembrane space of the thylakoids, and the light-independent reactions in the stroma.
- explain the carboxylation of ribulose bisphosphate.
- analyse diagrams of the pathways of photosynthesis to deduce where decarboxylation and oxidation reactions occur.

- **Photosynthesis** is the process by which photoautotrophs use sunlight to synthesize nutrients from carbon dioxide and water.
- **Chlorophyll** is a green pigment that absorbs light in most plants.

Overview of energy transfers in photosynthesis

Photosynthesis is the transformation of light energy into chemical energy in the form of carbon compounds occurring in plants, algae and cyanobacteria that contain **chlorophyll**. This energy transformation supplies most of the chemical energy needed for life processes in ecosystems. Photosynthesis consists of two interdependent processes—the light-dependent and light-independent reactions. A lack of light stops light-dependent reactions, and a lack of carbon dioxide prevents both the light and light-independent reactions.

C1.3 Photosynthesis

In the light-dependent reactions, photosynthesis uses light energy to split water molecules (**photolysis**). The purpose of photolysis is to release electrons that can eventually be used to reduce carbon dioxide and hydrogen ions to release energy. **Oxygen** is a by-product of the splitting of water. Energy from ATP and the hydrogen from photolysis are used to produce carbohydrates and other carbon compounds from carbon dioxide.

- **Photolysis** is the breaking down of water molecules by photons present in light.
- **Oxygen** is a by-product of photosynthesis.

Chromatography to separate photosynthetic pigments

It is possible to separate photosynthetic pigments using chromatography and identify them according to their colour and R_f.

Sample student answer

The picture shows a thin-layer chromatogram of an extract of leaves of annual meadow grass (*Poa annua*). Plastic sheets were coated with silica gel. A drop of extract was spotted 0.5 cm from the bottom of the sheet. The sheet was then placed in a beaker of solvent (75% acetone and 25% petroleum ether).

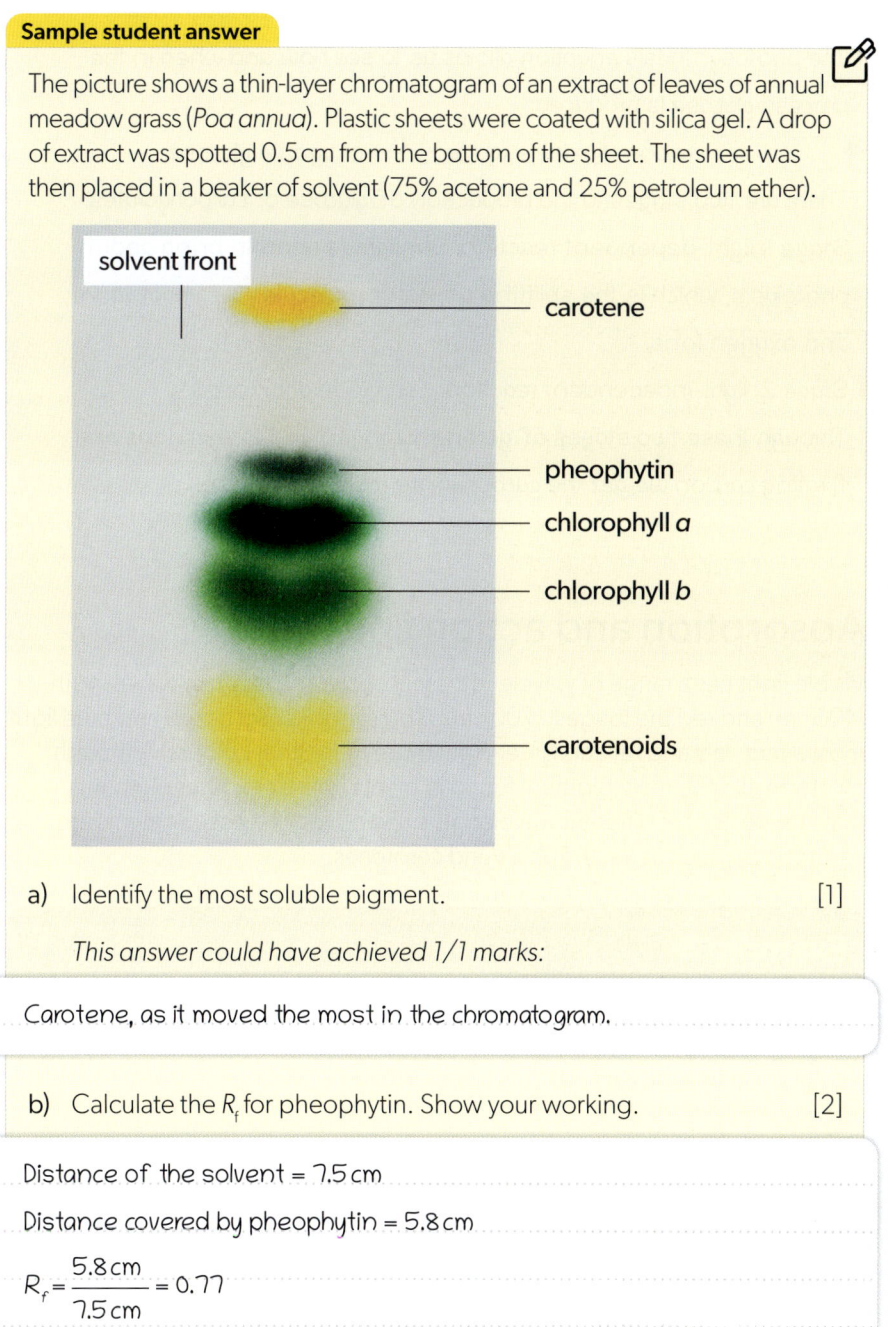

> **Examiner tip**
>
> Always check whether the answer makes sense. R_f has no units and the value is always between 0 and 1.

a) Identify the most soluble pigment. [1]

This answer could have achieved 1/1 marks:

Carotene, as it moved the most in the chromatogram.

b) Calculate the R_f for pheophytin. Show your working. [2]

Distance of the solvent = 7.5 cm

Distance covered by pheophytin = 5.8 cm

$R_f = \dfrac{5.8 \text{ cm}}{7.5 \text{ cm}} = 0.77$

123

Interaction and interdependence

c) Suggest what would have happened to the pigments if the solvent had been different. [1]

This answer could have achieved 1/1 marks:

> If the solvent is changed, the solubility will be different, and the chromatogram will be different too.

d) Outline the production of carbohydrates in photosynthesis. [4]

This answer could have achieved 3/4 marks:

> Glucose is a type of carbohydrate formed through photosynthesis. The photosynthesis equation allows us to see how and when in the process glucose is produced:
>
> carbon dioxide + water $\xrightarrow{Light}$ oxygen + glucose
>
> There are two stages to the production of glucose or carbohydrates. Stage 1: light-dependent reaction. Sunlight is used to bring about photolysis, which is the splitting of water molecules into hydrogen and oxygen ions.
> Stage 2: light-independent reaction. Carbon fixation occurs. Through these two stages of getting hydrogen and oxygen ions and fixating carbon we get the carbohydrate molecule glucose.

▲ The word equation scores one mark. Another mark is for photolysis and for mentioning oxygen and hydrogen as products of photolysis. The third mark is for carbon fixation.

▼ The answer does not mention that light is absorbed by chlorophyll and is converted to chemical energy. Some of the energy is used to produce ATP, which is needed to produce carbohydrates or starch.

Absorption and action spectra

Visible light has a range of wavelengths with violet the shortest wavelength (400 nm) and red the longest (700 nm). Chlorophyll absorbs red and blue light most effectively and reflects green light more than other colours. Although chlorophyll is the main pigment found in plants, there are other pigments that absorb light at different wavelengths. Examples of these pigments are carotenoids, such as xanthophylls and carotenes.

- An **absorption spectrum** is a graph of the amount of light absorbed by each pigment at different light wavelengths.

Example 8

The graph shows the **absorption spectrum** of three different pigments.

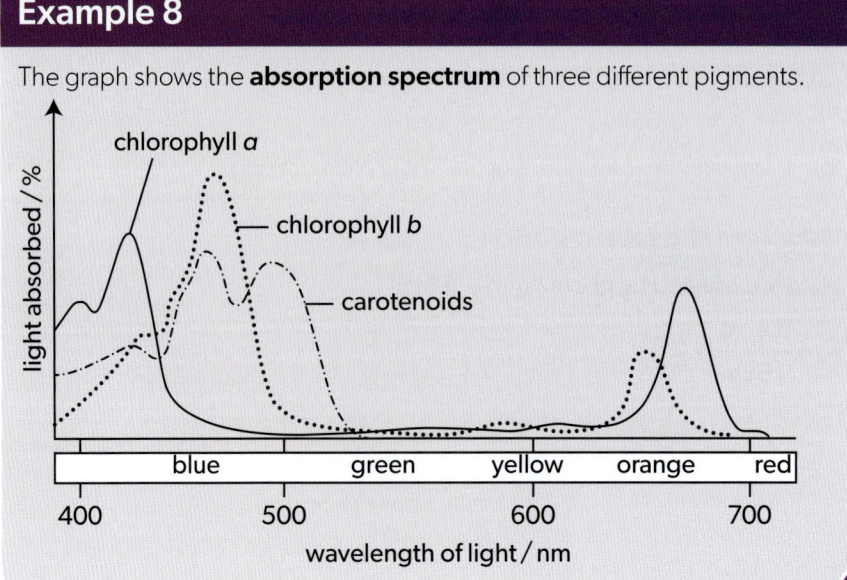

What is shown in the graph?
A. Chlorophyll b absorbs best in blue light.
B. Carotenoids absorb best in orange light.
C. Chlorophyll a and chlorophyll b both absorb green light.
D. Chlorophyll b always absorbs at the same wavelengths as carotenoids.

Solution

The correct answer is **A**, as chlorophyll absorbs most in blue light and reflects green. Carotenoids are orange and absorb more in blue and green. Chlorophyll b is green and absorbs mostly in blue light.

Example 9

The graph shows the absorption spectrum of different pigments in a green plant.

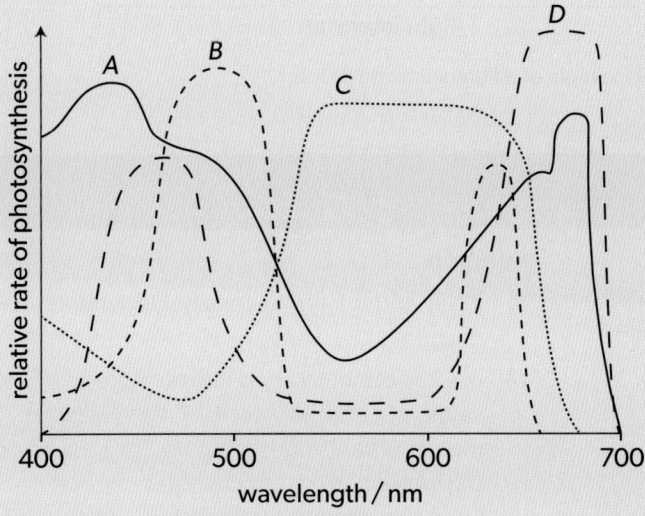

a) Which line shows the **action spectrum** in photosynthesis?
b) How was the relative rate of photosynthesis measured for varying wavelengths?
A. Volume of oxygen produced, or carbon dioxide consumed using a probe
B. Volume of water produced counting bubbles
C. Rate of discoloration of leaves by chromatography
D. Measurement of chlorophyll molecules using R_f values

- An **action spectrum** is a graph of the rate of photosynthesis performed by an organism at different light wavelengths.

Solution
a) Line **A**, as the action spectrum is the rate of photosynthesis performed by an organism at different light wavelengths. It is the sum of the action of the different pigments.
b) The correct answer is **A**, as water is consumed in photosynthesis, it is not produced. The leaves do not change colour in photosynthesis. The rate of photosynthesis could depend on the amount of chlorophyll, but it is not usually a limiting factor.

Investigating limiting factors of photosynthesis

Temperature, light intensity and carbon dioxide concentration are possible limiting factors on the rate of photosynthesis.

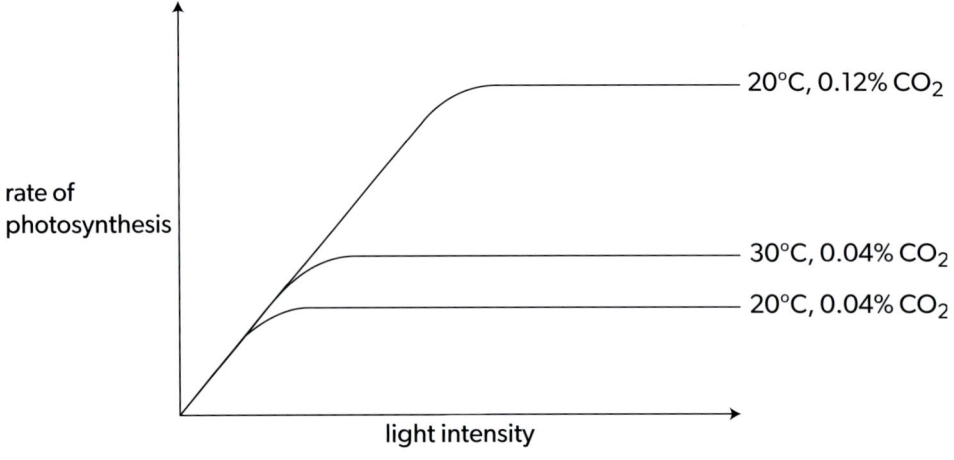

▲ Figure 9 Example of different limiting factors

Example 10

To test the effect of light intensity on the rate of photosynthesis, the following apparatus was set up in a laboratory. Light was placed at different distances and the volume of oxygen was recorded.

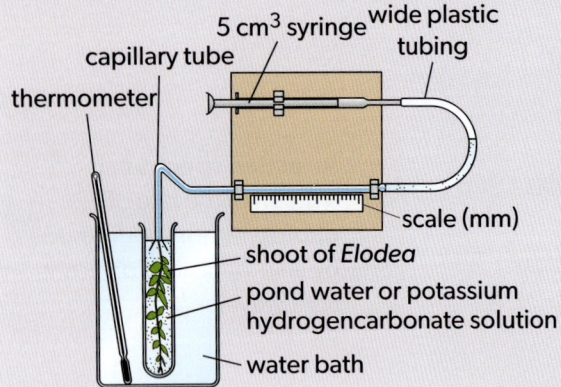

a) State the independent and dependent variables in this experiment.
b) Identify the reason a thermometer is placed in the water bath.
c) Design an experiment to measure the effect of concentration of carbon dioxide on the rate of photosynthesis using this apparatus.
d) Explain how the rate of photosynthesis was recorded in this experiment.

Solution

a) Independent: light intensity. Dependent: volume of oxygen.

b) It is important to keep temperature as a fixed value, not another variable. This will make measurements consistent. As lamps give off heat, the water bath might change temperature. To keep the temperature constant, cold water or ice can be added.

c) Add different concentrations of potassium hydrogencarbonate solution ($KHCO_3$) and record movement on the scale. When the hydrogencarbonate dissolves in water, potassium hydroxide (KOH) and carbonic acid (H_2CO_3) are formed. The latter breaks down into water (H_2O) and carbon dioxide gas (CO_2). Higher carbon dioxide concentration will increase the photosynthesis rate until saturation point, where it plateaus. Sodium hydrogencarbonate ($NaHCO_3$) can also be used in this experiment.

d) The capillary tube and plastic tubing must be filled with water at the start of the experiment. As the pondweed carries out photosynthesis, oxygen bubbles are produced. These bubbles ascend through the capillary tube, displacing the water. The volume of gas produced is recorded on the scale (in mm), as the diameter of the tube is known. It is important to take the time, as rate is measured in volume per unit of time.

Carbon dioxide enrichment experiments

Example 11

A long-term, free-air carbon dioxide enrichment (FACE) technology experiment was performed to test whether rising atmospheric carbon dioxide may lead to increased forest biomass carbon. Two experiments were conducted in deciduous forests: Rhinelander (Rhin.) and Oak Ridge National Laboratory (ORNL), and two in evergreen forests: Duke and Kennedy Space Center (KSC). The FACE experiments all increased carbon dioxide by 50% above ambient, while at KSC the increase was about 100%.

Model predictions were compared against experimental observations of the vegetation biomass change (ΔC_{veg}) in response to the 10-year carbon dioxide enrichment. In the box-and-whisker plots, grey boxes represent the observations and purple show the predicted values.

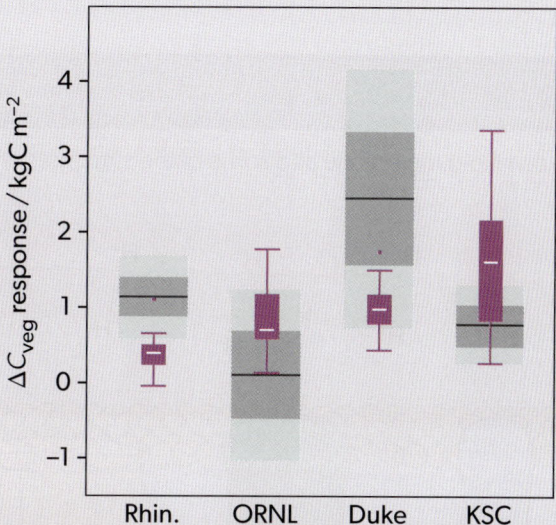

Source: Walker, A. P., et al. (2019). Decadal biomass increment in early secondary succession woody ecosystems is increased by CO_2 enrichment. *Nature Communications*, **10**, (1). www.doi.org/10.1038/s41467-019-08348-1 (CC BY 4.0)

a) State the observed mean ΔC_{veg} for Duke.
b) Outline the effect of carbon dioxide enrichment on forest biomass.
c) Compare and contrast observed results from deciduous and from evergreen forests.
d) Suggest a reason for the change in biomass with carbon dioxide enrichment.
e) Comment on the reliability of the predicting models.

Solution

a) 2.5 kg C m^{-2}

b) Carbon dioxide enrichment increased biomass. In most cases, the change is positive, as there is more biomass at the end of the experiment than at the beginning. In ORNL the mean is greater than 0, but there are negative values, which means that in some parts of the forest the biomass decreased.

c) In all forests, the mean ΔC_{veg} is positive. Negative values are only found in deciduous forests (ORNL). In deciduous forests the observed ΔC_{veg} is in general smaller than in evergreen forests, although in KSC the mean is smaller than in Rhin.

d) More carbon dioxide implies more photosynthesis. This process produces glucose that is used for growth and development, increasing vegetation biomass.

e) In general, the predictions were not so good. At Rhin. and Duke, the ΔC_{veg} response to carbon dioxide enrichment was strongly underpredicted by the model. ORNL and KSC models overpredicted biomass increment. The ranges were also not predicted as observed: in KSC the range of values was much smaller than predicted while in Duke the range was much larger than predicted.

Photosystems

Photosystems are two large membrane protein complexes that can generate and emit excited electrons. They are always embedded in membranes and can occur in cyanobacteria and in the chloroplasts of photosynthetic eukaryotes. The membrane allows these proteins to be one next to the other. Both photosystems also contain an array of hundreds of chlorophyll molecules that capture light energy to excite electrons and work synergistically. A single molecule of chlorophyll or any other pigment would not be able to perform any part of photosynthesis.

> - **Photosystems** are molecular arrays of chlorophyll and accessory pigments with a special chlorophyll as the reaction centre from which an excited electron is emitted.

Example 12

The diagram shows a section of a thylakoid membrane and some of the structures involved in the light-dependent reactions of photosynthesis.

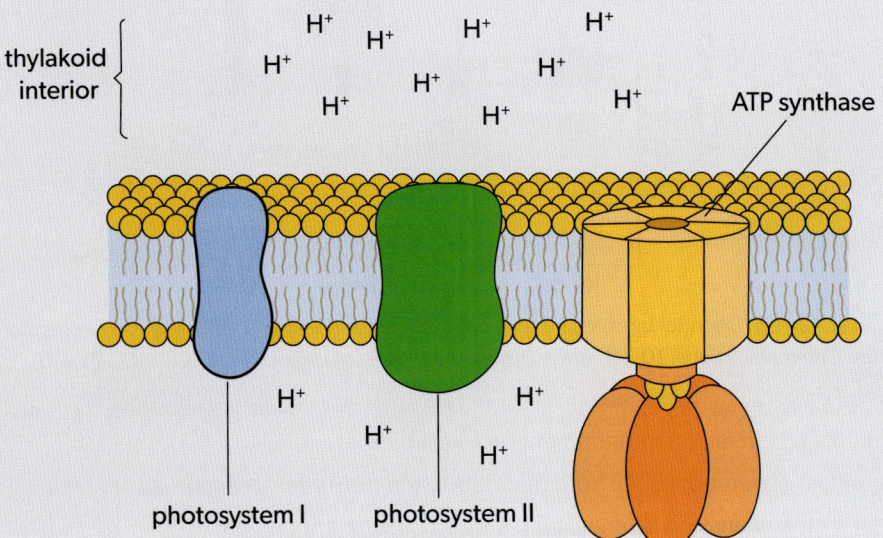

a) Explain the high proton (H^+) concentration in the thylakoid interior.
b) Describe how ATP is produced in the thylakoid membrane.

Solution

a) The electron transfer in the thylakoid membrane causes these protons to be actively pumped across the thylakoid membrane into the thylakoid space. This builds a high concentration of protons inside, creating a difference in charge across the membrane.

b) As the membrane is semi-permeable, the charged H^+ cannot leave the interior of the thylakoid directly, but they can do so through carrier proteins. These are ATP synthase, which connect the interior with the stroma. The H^+ pass through the ATP synthase by a concentration gradient. This passage is coupled with the synthesis of ATP. It liberates enough energy to phosphorylate ADP in the process of chemiosmosis.

Light-dependent reactions of photosynthesis

Light-dependent reactions take place in the intermembrane space of the **thylakoids.** The absorption of light in photosystem II excites electrons (1) that are accepted by carrier proteins (2) transferred by carriers such as

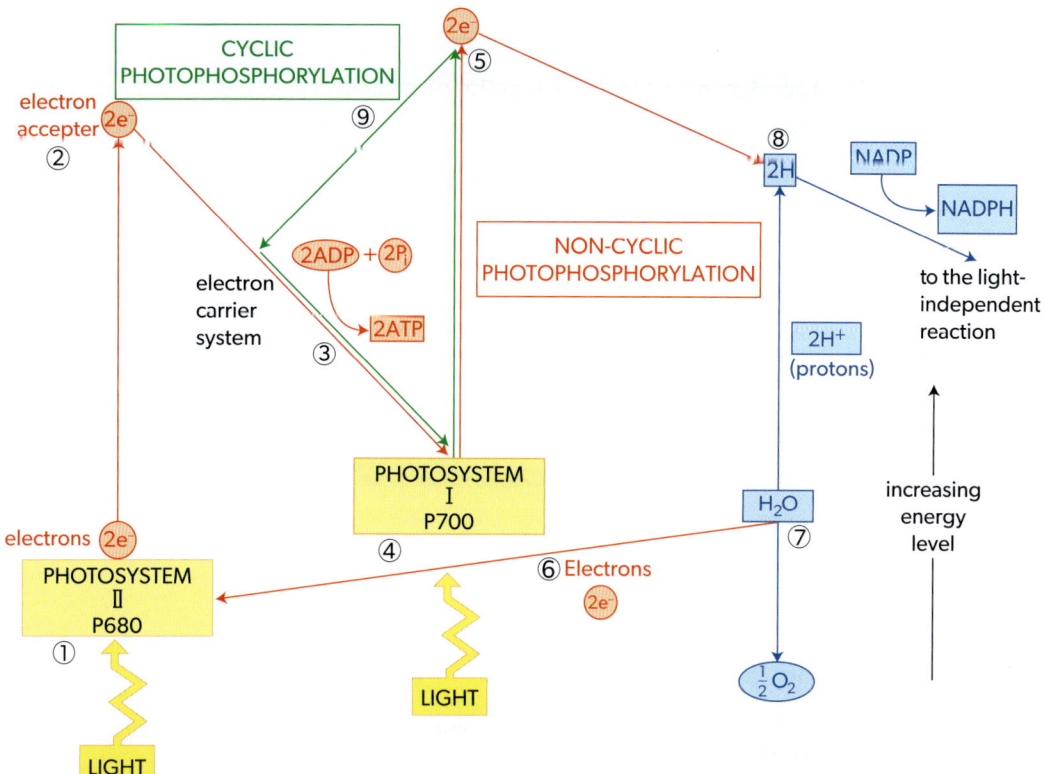

▲ Figure 10 Light-dependent reactions in photosynthesis

cytochrome in the **thylakoid membranes**. Protons are actively pumped into the **thylakoid space**. ATP is produced by chemiosmosis in ATP synthase (3). Excited electrons from photosystem I (4) are used to reduce NADP forming **reduced NADP** (two electrons per molecule) to be used in the light-independent reactions in the stroma of the chloroplast (5 and 8). The electrons lost from photosystem II are replaced (6) by the oxidation of water, which is lysed into protons and oxygen (7). Oxygen diffuses out of the chloroplasts. Cyclic photophosphorylation (9) only uses electrons from photosystem I. It occurs in prokaryotes to accomplish the rapid conversion of ADP to ATP.

The generation of oxygen as a waste product by photolysis had immense consequences for living organisms and geological processes on Earth.

Rubisco and the Calvin cycle

Ribulose-1,5-bisphosphate carboxylase oxygenase (Rubisco) is the most abundant enzyme on Earth and is involved in **carbon fixation**. High concentrations of it are needed in the **stroma** of chloroplasts because it works relatively slowly and is not effective in low carbon dioxide concentrations.

The light-independent reactions occur in the stroma of the chloroplast. The enzyme Rubisco catalyses the carboxylation of a 5C sugar, ribulose bisphosphate (RuBP). In a carboxylation, carbon dioxide is added to the molecule. The 6C sugar formed is broken down into two 3C molecules. Glycerate 3-phosphate (3C) is reduced to triose phosphate (3C) using reduced NADP and ATP. Triose phosphate is used to regenerate RuBP and produce carbohydrates using ATP. In total, five molecules of triose phosphate are converted to three molecules of RuBP, allowing the **Calvin cycle** to continue. If glucose is the product of photosynthesis, five-sixths of all the triose phosphate produced must be converted back to RuBP.

- **Thylakoids** are membrane-bound compartments inside chloroplasts, formed by a membrane surrounding a thylakoid space. They form stacks of disks called grana.

- A **thylakoid membrane** is the site of light absorption and the electron transport chain.

- A **thylakoid space** is the site of photolysis and proton build-up.

- **Reduced NADP** (NADPH) is the molecule produced in the last step of the electron transport chain. It is used as reducing power for the biosynthetic reactions in the Calvin cycle.

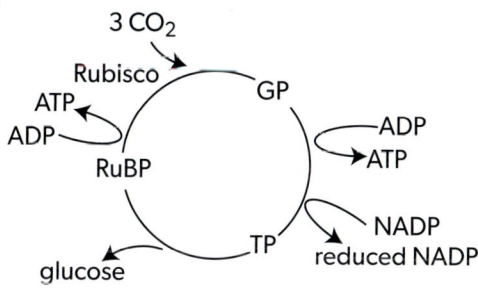

▲ Figure 11 The light-independent cycle

Interaction and interdependence

> - **Ribulose-1,5-bisphosphate carboxylase oxidase (Rubisco)** is the enzyme that catalyses the incorporation of carbon dioxide.
> - **Carbon fixation** is the addition of carbon dioxide to organic molecules.
> - The **stroma** is a fluid part of the chloroplast and is the site of the Calvin cycle.
> - The **Calvin cycle** is the light-independent stage of photosynthesis consisting of carbon fixation, reduction reactions and ribulose 1,5-bisphosphate (RuBP) regeneration.

Example 13

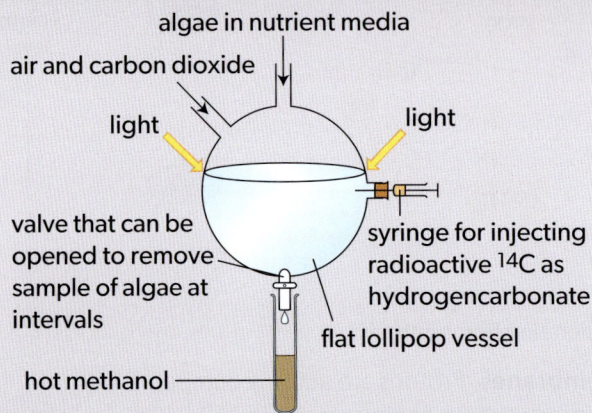

Melvin Calvin, James Bassham and Andrew Benson discovered the light-independent reactions of photosynthesis. In an experiment similar to that performed by Calvin, green algae (*Chlorella*) were placed within a thin transparent container (the lollipop container) and radioactive carbon-14 was added to the apparatus. At 10-second intervals the samples of algae were dropped into hot methanol. Samples were analysed using two-dimensional paper chromatography, which separates out the different carbon compounds. The results of these experiments showed that carbon dioxide was converted to carbohydrates during the light-independent reactions of photosynthesis.

a) Justify the use of transparent and not dark glass in the lollipop vessel.

b) The substances found at different times were recorded in a table. Identify the substances detected at 20 and 40 seconds.

Time / seconds	Radioactive molecules
0	carbon dioxide
10	glycerate 3-phosphate (GP), carbon dioxide
20	
30	glycerate 3-phosphate (GP), triose phosphate (TP), glucose
40	

Solution

a) Transparent glass allows light to pass through. Light is needed for the production of hydrogen ions to produce reduced NAD (NADH) and ATP required in the light-independent reactions.

b) At 20 seconds: glycerate 3-phosphate (GP), triose phosphate (TP). At 40 seconds: glycerate 3-phosphate (GP), triose phosphate (TP), glucose, ribulose bisphosphate (RuBP).

All carbon atoms are fixed in the Calvin cycle

All the carbon in photosynthesizing organisms is fixed in the Calvin cycle. Carbon compounds other than glucose are made by metabolic pathways that can be traced back to an intermediate in the cycle.

Practice problems

1. Explain the denaturation of enzymes.

2. What occurs during chemiosmosis?

    A. ATP synthase transports electrons to the cell cytoplasm.
    B. Protons are passed from one carrier to another, liberating energy.
    C. Electrons move across a membrane, down a concentration gradient, releasing energy.
    D. Energy stored in organic compounds is transformed into ATP.

3. Scientists wanted to evaluate the effect of temperature on the rate of photosynthesis of maidenhair tree (*Ginkgo biloba*) leaves, exposed to normal and very cold temperatures at different atmospheric carbon dioxide concentrations.

    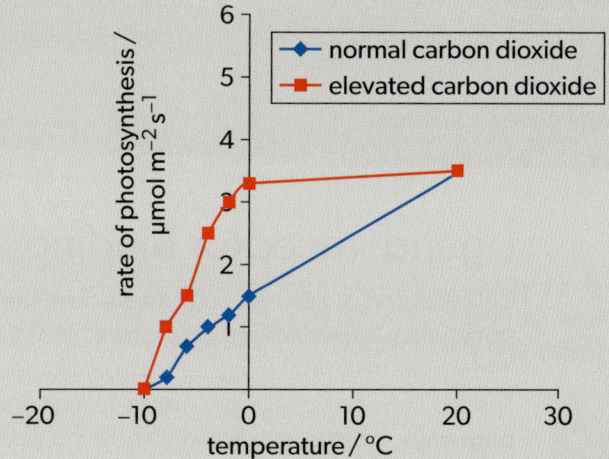

    a) Calculate the difference in rate of photosynthesis between normal and elevated carbon dioxide levels at 0 °C.

    b) (i) Outline the effect of temperature on rate of photosynthesis at different carbon dioxide levels.

    (ii) Suggest how the change in photosynthetic rate will affect plant growth.

    c) Suggest which reactions of photosynthesis could be affected by cold temperatures.

4. Describe how the rate of photosynthesis is affected by the intensity and wavelength of light.

5. Explain the limiting factors of photosynthesis.

- What are the consequences of photosynthesis for ecosystems?
- What are the functions of pigments in living organisms?

C2.1 Chemical signalling

You should know:
Additional higher level:
- receptors are proteins with specific sites where ligands bind.
- bacteria signal in quorum sensing.
- hormones, neurotransmitters, cytokines and calcium ions are signalling chemicals in animals.
- a range of substances can serve as hormones and neurotransmitters.
- neurotransmitter transmembrane receptors change the cell membrane potential by opening ion channels.
- the mechanism of action of epinephrine (adrenaline) receptors.
- binding to transmembrane receptors with tyrosine kinase activity causes autophosphorylation and sets off transduction pathways.
- intracellular receptors affect gene expression.

You should be able to:
- distinguish between types of signalling.
- explain how each type of signalling works and the effect it has on cells or organisms.
- compare signalling with hormones and neurotransmitters.
- compare transmembrane with intracellular receptors.
- describe how binding of a signalling chemical sets off a sequence of responses.
- explain signalling via G protein-coupled receptors
- describe the effects of oestradiol and progesterone on target cells.
- compare positive and negative feedback in regulating signalling pathways.

Ligand–receptor binding

Cell receptors are proteins on either side of the membrane that transmit a signal when a specific ligand molecule binds to them. The ligand is a chemical messenger released by one cell to signal either to a different cell or to itself. The binding of a signalling chemical to a receptor sets off **transduction pathways**, which are a sequence of responses within the cell. These pathways include proteins joining to gene promoters, opening of channel proteins, activation of other proteins or sending of a second messenger. All these pathways can induce changes in gene transcription, send nervous impulses or change the shape and/or function of a cell or group of cells.

- In **signal transduction pathways**, messages from ligands cause changes in biological activity of target cells.

Quorum sensing by bacteria

Bacteria can synchronize their gene expression through a mechanism called **quorum sensing**. This means one organism senses there are more organisms and only then does it express certain proteins.

Vibrio fischeri, a marine bacterium, regulates its bioluminescence through a quorum sensing mechanism by releasing diffusible small molecules (autoinducers) that accumulate in the environment as the population density increases. This accumulation of autoinducer activates transcriptional regulators for bioluminescence. These at the same time induce the production of luciferase enzymes that produce light when they oxidize their substrate. This is a positive feedback mechanism.

- In **quorum sensing**, organisms in groups synchronize their gene expression.

> ### Example 1
>
> Bioluminescent bacteria *Vibrio fischeri* reside in the light-emitting organ of the squid, *Eupryma scalopes*. The gene *luxA* encodes one of the subunits of luciferase, the enzyme responsible for bioluminescence. *LuxI* synthesizes the autoinducer AI which binds to *luxR*, the transcriptional activator for the bioluminescence genes.
>
> Scientists constructed *lux* mutants, defective either for *luxA*, *luxI* or *luxR*. The luminescence was measured in *lux* mutants and their parents with or without the addition of the AI.
>
> | Strain | Luminescence without AI addition / quanta s^{-1} $cell^{-1}$ | Luminescence after addition of AI / quanta s^{-1} $cell^{-1}$ |
> |---|---|---|
> | parental | 24 | 4,400 |
> | luxA | 0.01 | 0.01 |
> | luxI | 8.7 | 3,600 |
> | luxR | 17 | 1 |
>
> Source: Visick, K. L., et al. (2000). *Vibrio fischeri lux* genes play an important role in colonization and development of the host light organ. *Journal of Bacteriology*, 183(16), 4578–4586.
>
> a) Calculate the difference in luminescence in parental strains with and without the addition of AI.
>
> b) Deduce, with a reason, whether the strains were in large or small densities in the experiment without AI addition.
>
> c) Evaluate the effect of the different mutations on luminescence without addition of AI.
>
> d) Explain the results for *luxI*.
>
> ### Solution
>
> a) 4,376 quanta s^{-1} $cell^{-1}$
>
> b) They were in small densities, as the luminescence is low and the quorum sensing effect is not shown.
>
> c) The mutant *luxA* is the most affected, as hardly any luminescence is seen, while the others lower their luminescence to half or less than half that of the parental strain.
>
> d) The mutant *luxI* regains luminescence activity with the addition of AI. *LuxI* determines AI production, so if AI is added to the experiment, there will be no difference in luminescence compared to the parental strain.

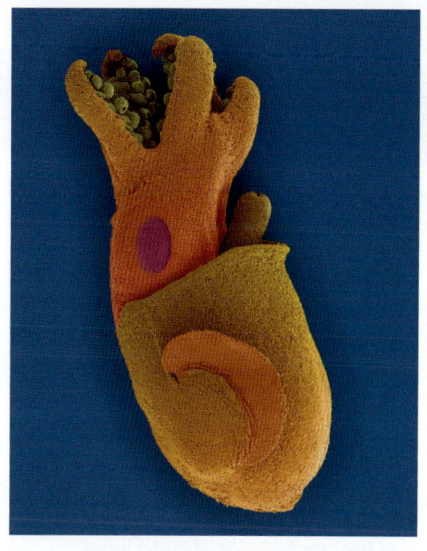

▲ **Figure 1** Bioluminescent *Vibrio fischeri* in squid, *Eupryma scalopes*

Signalling molecules and their effects

There are different types of signalling molecules—lipids, amino acids and amines, proteins, glycoproteins and ions. Hormones are amines, proteins and steroids. Substances that can serve as neurotransmitters range from amino acids, peptides and amines to nitrous oxide. Cytokines are signal proteins that control growth and immune response. Calcium ions can serve as signalling chemicals too.

Signalling can be between cells in direct contact or cells that are distant. Neurotransmitters diffuse across a synaptic gap to neighbouring cells, whereas hormones are transported by the blood system to distant target organs. Signalling can be between cells, different tissues or organs of the same organisms or even between organisms of the same species (for example, pheromones) or of different species (odours).

Interaction and interdependence

▲ One mark is awarded for comparing distance of transport, although rather incomplete. The category "chemical signal" also could have obtained one mark, although in some cases the chemical name is given and others the type. Another mark could be achieved for the effect.

▼ Answering comparative questions in a table is allowed, but here it did not help. The answer is incomplete and vague. There is no similarity mentioned. The transport column is incomplete—it should have included start and end points. There is no mention of the type of receptor involved in each case. The effects could have been more detailed—for example, cytokines are involved in activation of T-cells, neurotransmitters can initiate or inhibit nerve impulses. "Gene expression altered" is also vague—binding to promoters or operators and an example of each functional category could have been given.

Sample student answer

Compare and contrast signalling through hormones, cytokines, neurotransmitters and calcium ions. [8]

This answer could have achieved 3/8 marks:

	Transport	Chemical signal	Effect
Hormones	distant	steroids, proteins, amines	gene expression altered
Cytokines	distant	peptides	gene expression altered
Neurotransmitters	close	amino acids, nitric oxide, gabba, acetylcholine	nerve impulse
Calcium ions	close	calcium ion	muscle contraction

Comparing transmembrane and intracellular receptors

Receptors can be transmembrane or intracellular. There are different types of cell-surface receptors: ion channels, G proteins and enzyme-linked receptors. Transmembrane receptors are found across the plasma membrane allowing the signalling chemical to penetrate the cell. As in all transmembrane proteins, the hydrophilic amino acids of the globular protein are found in the interior of the channel and in the parts in contact with the exterior and interior of the cell, while the hydrophobic amino acids are found in contact with the phospholipid tails of the membrane. Intracellular receptors are found in the cytoplasm or nucleus of the cell. Signalling causes changes in other proteins of the membrane and the signalling chemical does not enter the cell—it remains outside the cell.

Acetylcholine receptor signalling

Acetylcholines are neurotransmitters—chemicals that cross the synaptic gap in mammals. Once acetylcholine reaches the postsynaptic neuron, it attaches to a receptor, inducing an action potential by the opening of sodium ion channels. This allows positively charged ions to diffuse into the cell, changing the voltage across the plasma membrane. The acetylcholine remains outside the cell. It is then broken down by the enzyme acetylcholinesterase and the resting potential is recovered.

You learned about the structure of proteins in subtopic B1.2 and how transport occurs through membranes in subtopic B2.1.

In subtopic C2.2 you will study the effects of acetylcholine signalling in detail.

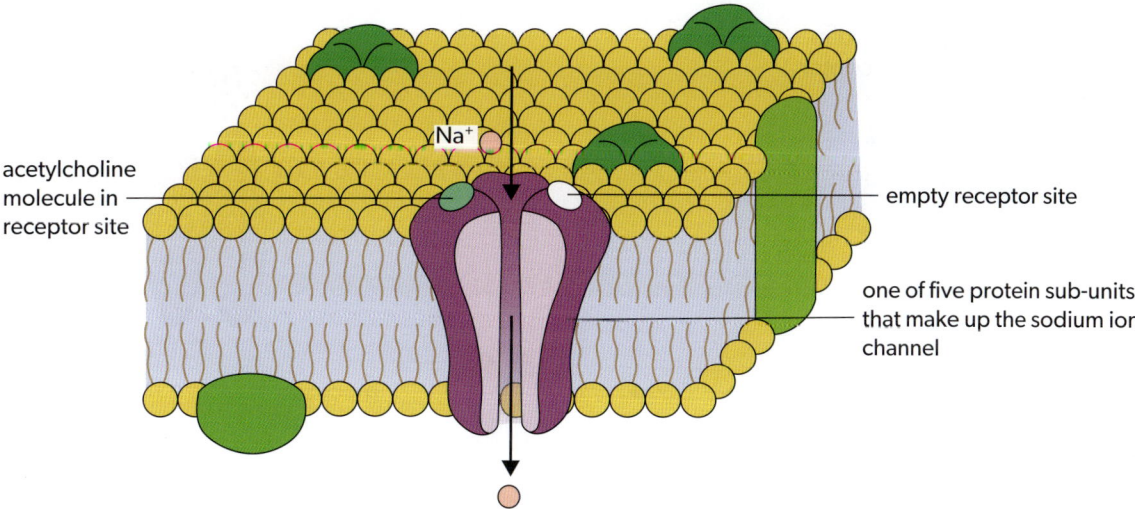

▲ Figure 2 An acetylcholine receptor

G protein receptor signalling

There are many transmembrane receptors in humans, for example, epinephrine (adrenaline) receptors, coupled to a G protein that convey a signal into cells. There are many ligands that can join these **G protein-coupled receptors**, including hormones, odours, rhodopsin, neurotransmitters, peptides and proteins. When a ligand joins the receptor, it causes a change in the coupled G protein. This, at the same time, activates adenylate cyclase enzyme to produce cyclic AMP (cAMP) from ATP. The cAMP acts as the **second messenger**. Increasing cAMP activity results in enzyme action. For example, with epinephrine, it induces heart muscle contraction and smooth muscle relaxation. There are several substances that can act as second messengers apart from cAMP, for example cGMP, nitric oxide, carbon monoxide and calcium ions.

- **G protein-coupled receptor** signals involve second messengers.
- A **second messenger** is an effector signal that is synthesized within a cell in response to an external signal (first messenger).

Nature of science

Naming conventions are an example of international cooperation in science for mutual benefit. Both "adrenaline" and "epinephrine" were coined by researchers and are based on production of the hormone by the adrenal gland. "Adrenaline" comes from Latin *ad* = at and *ren* = kidney, and "epinephrine" comes from old Greek *epi* = above and *nephros* = kidney. Unusually, both these terms are still widely used in different parts of the world.

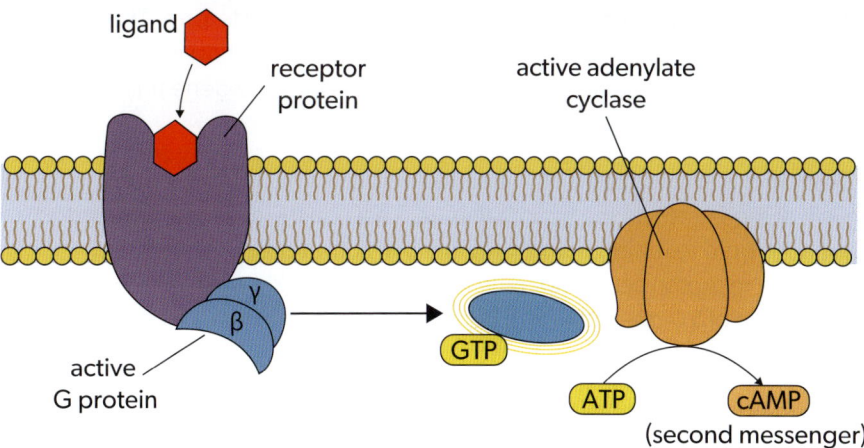

▲ Figure 3 G protein-coupled receptor signalling

Transmembrane receptors with tyrosine kinase activity

The **receptor tyrosine kinases (RTKs)** are transmembrane proteins that play an active role in signalling, growth, differentiation and motility. When a ligand joins the receptor on the outside part of the cell membrane, it activates the tyrosine kinase enzyme inside the cell. This enzyme subsequently phosphorylates tyrosine amino acids, activating the intracellular signal proteins.

- The **receptor tyrosine kinases (RTKs)** are transmembrane proteins that undergo autophosphorylation, activating tyrosine kinases.

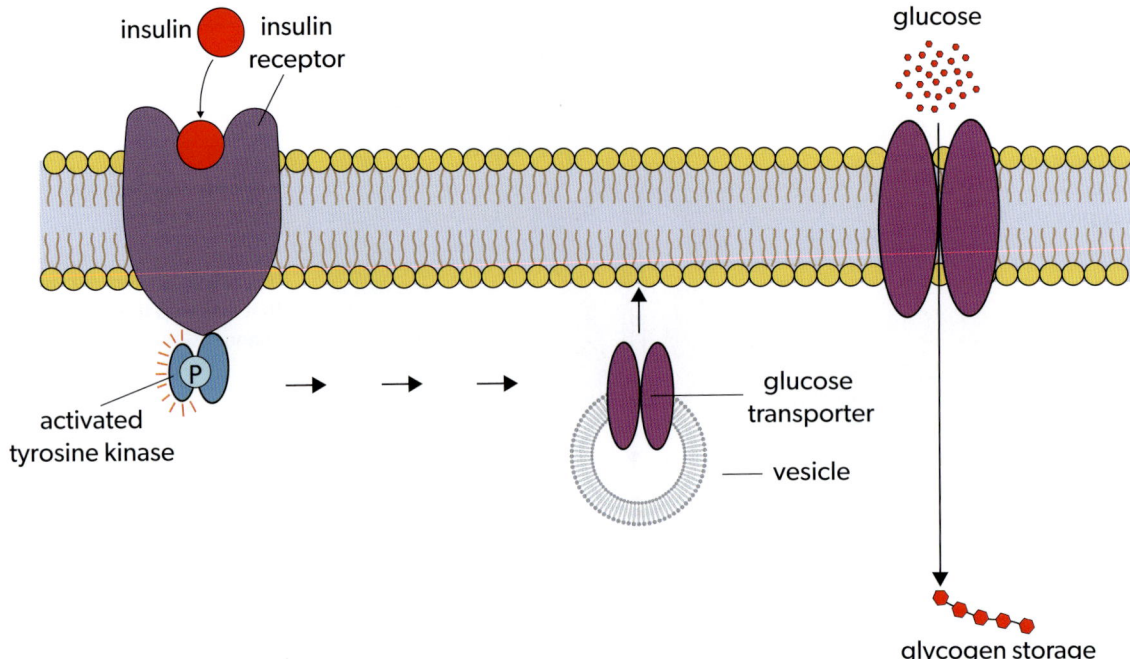

▲ **Figure 4** Mechanism of action of receptor tyrosine kinases

The insulin receptor is a tyrosine kinase receptor that is activated by the insulin hormone. The insulin RTK pathway regulates cellular responses to insulin, such as the movement of vesicles containing glucose transporters to the plasma membrane for glucose uptake.

Intracellular receptors that affect gene expression

There are **intracellular receptors** that affect gene expression, for example steroid hormones oestradiol, progesterone and testosterone. The signalling chemical binds to a site on a receptor, activating it. The activated receptor binds to specific DNA sequences to promote gene transcription in target cells.

Oestradiol is an oestrogen hormone that regulates the reproductive cycle of the female. In the hypothalamus, it acts on the cells that produce gonadotropin-releasing hormone (GnRH). It activates a nuclear steroid hormone receptor, activating gene transcription. GnRH stimulates the production of follicle-stimulating hormone and luteinizing hormone in the pituitary gland.

Progesterone is a lipophilic hormone that readily crosses cell membranes by diffusion and interacts at the nuclear level. The hormone binds to progesterone receptors and regulates expression of genes. It regulates the development and function of the endometrium and induces changes in cells essential for implantation and the establishment and maintenance of pregnancy. The expression of endometrial progesterone receptors changes during the various phases of the menstrual cycle. Withdrawal of progesterone receptor signalling triggers menstruation and parturition.

- **Intracellular receptors** bind chemical signals such as oestrogen and progesterone and affect gene expression.

Example 2

The diagram demonstrates the action of steroid and peptide hormones in a section of cell and adjacent capillary.

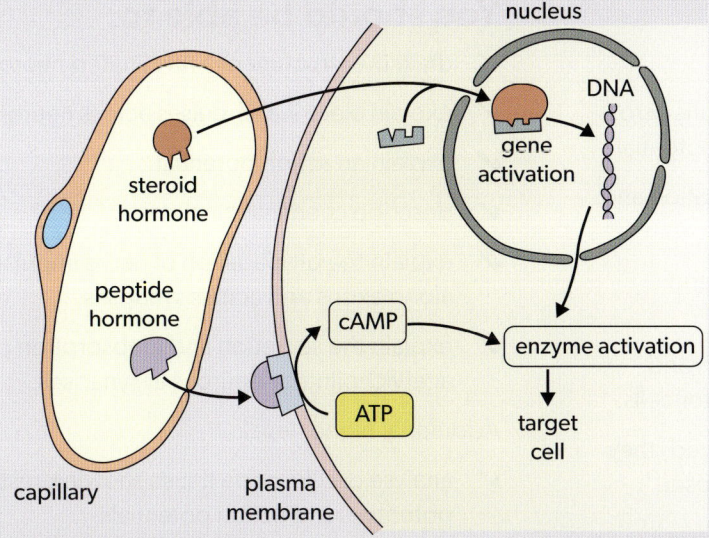

a) On the diagram label a:
   (i) second messenger.
   (ii) gene regulatory protein.
b) Outline one characteristic of steroid hormones that allows them to readily diffuse through cell membranes.

Solution

a) (i) Cyclic AMP (cAMP)
   (ii) Protein in the cell cytoplasm (shown in grey) that joins the steroid hormone to produce gene activation.
b) Steroid hormones are lipids that are lipid-soluble (lipophilic) molecules and are small enough to diffuse through membranes.

Regulation of cell signalling pathways by positive and negative feedback

Within cells, signals propagate through an integrated network with positive and negative feedback loops. In positive feedback, a signal induces its own expression, or of another molecule that amplifies the initial signal, and this serves to amplify or prolong the signalling. An example is the blood clotting cascade, where upon injury, there is release of signal chemicals. These signals activate signals in the platelets, causing a rapid cascade and the formation of a blood clot. In negative feedback, a signal induces the expression of its own negative regulator or inhibitor, so that when a threshold has been reached, the signal ceases. Negative regulators in receptor tyrosine kinase (RTK) and G protein-coupled receptor (GPCR) signalling are those that reversibly modify proteins and other signalling molecules. This can be done either by phosphorylation or dephosphorylation or by binding of guanosine triphosphate (GTP) versus guanosine diphosphate (GDP) or by promoting receptor internalization.

- What patterns exist in communication in biological systems?
- In what ways is negative feedback evident at all levels of biological organization?

Interaction and interdependence

C2.2 Neural signalling

You should know:
- neurons transmit electrical impulses.
- neurons pump sodium and potassium ions across their membranes to generate a resting potential.
- an action potential consists of depolarization and repolarization of the neuron.
- speed of transmission depends on size and myelination of neurons.
- synapses are junctions between neurons and between neurons and receptor or effector cells.
- when presynaptic neurons are depolarized, they release a neurotransmitter into the synapse.

Additional higher level:
- a nerve impulse is initiated only if the threshold potential is reached.
- propagation of nerve impulses is the result of local currents that cause each successive part of the axon to reach the threshold potential.
- perception of pain is in nerve endings.
- consciousness is an emergent property.

You should be able to:
- draw the structure of a myelinated neuron.
- explain electrical impulses across neurons.
- explain an action potential.
- describe chemical synapses.
- explain the propagation of nerve impulses along axons and across synapses.
- explain the secretion and reabsorption of acetylcholine by neurons at synapses.

Additional higher level:
- analyse oscilloscope traces showing resting potentials and action potentials.
- explain the effect of myelination on conduction.
- explain the blocking of synaptic transmission at cholinergic synapses in insects by binding of neonicotinoid pesticides to acetylcholine receptors.
- explain the effect of cocaine on presynaptic reabsorption of neurotransmitter.
- explain hyperpolarization in neurons.
- explain summation of excitatory and inhibitory potentials in postsynaptic neurons.

Properties of neurons

Neurons are nerve cells that have differentiated to perform a function. They transmit electrical impulses. There is a gap between neurons called a synapse. The message must cross this gap using chemicals called neurotransmitters. Neurons transmit the message and synapses modulate the message. Some neurons have a myelin sheath covering the axon.

Resting membrane potential and nerve impulses

Nerve impulses are **action potentials** propagated along the axons of neurons. Propagation of nerve impulses is the result of local currents that cause each successive part of the axon to reach the threshold potential. A nerve impulse is electrical because it involves movement of positively charged ions. The **resting potential** is the potential across the membrane caused by the exchange of three sodium ions moving out of the cell and two potassium ions into the cell by the sodium–potassium pump, with the use of one ATP molecule. An action potential consists of depolarization and repolarization of the neuron.

Membrane potentials in neurons can be measured by placing electrodes on each side of the membrane.

- An **action potential** is the depolarization and repolarization of the neuron allowing impulse transmission.
- **Resting potential** is the membrane potential as long as there is no perturbance. It is around $-70\,mV$.
- **Membrane potential** is the difference in voltage between the outside and inside of the cell membrane.

Example 3

The diagram shows a motor neuron.

Which identifies the labels on the diagram?

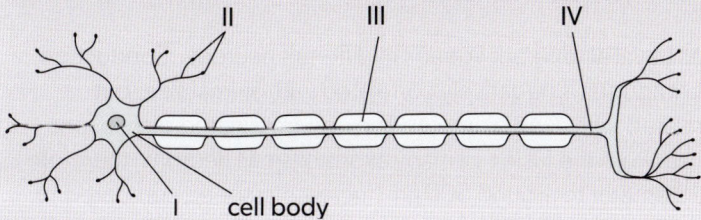

	I	II	III	IV
A.	nucleus	axon	dendrites	myelin sheath
B.	nucleus	dendrites	myelin sheath	axon
C.	dendrites	nucleus	axon	myelin sheath
D.	myelin sheath	dendrites	nucleus	axon

Solution

The correct answer is **B**. I is the cell nucleus, II is dendrites, III is myelin sheath and IV is the axon.

Example 4

Ions move across the plasma membrane of a neuron during an action potential. The trace shows voltage changes generated in a neuron during three action potentials.

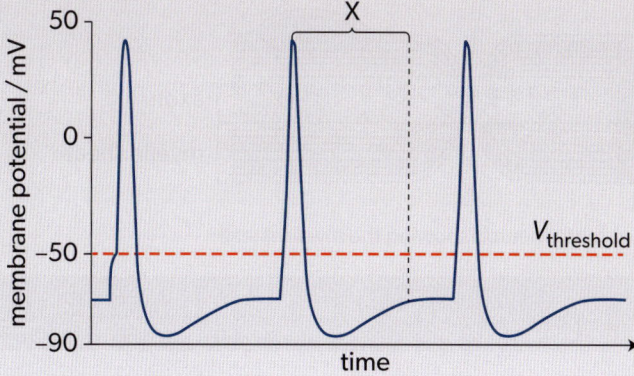

a) Describe depolarization.

b) Explain the movement of ions which causes the voltage changes observed during the interval labelled **X** on the graph.

Solution

a) Depolarization is caused by the opening of voltage-gated sodium channels, allowing sodium ions to diffuse into the neuron down a concentration gradient.

b) At the peak, sodium (Na^+) channels close and potassium (K^+) channels open. The membrane potential is approximately +45 mV. Potassium ions flow out, causing repolarization to occur. The membrane potential reaches approximately −70 mV.

Assessment tip

The question refers to a section labelled **X**, so do not answer referring to what occurs before or after the section labelled **X**, as no marks will be given, or marks can be lost due to contradictions.

Interaction and interdependence

Myelination affects speed of nerve impulses

The myelin sheath is produced by Schwann cells around axons. It causes the concentration of ion channels, including voltage-gated sodium channels, in the nodes of Ranvier. Saltatory conduction in myelinated neurons is much faster than in non-myelinated neurons.

Nerve impulses travel at different speeds in different neurons. Conduction speed of nerve impulses is negatively correlated with animal size, but positively correlated with axon diameter. The non-myelinated giant axons of squid originate from the fusion of many axons of small neurons, therefore increasing the diameter to around 500 μm, increasing speed compared to smaller non-myelinated nerve fibres.

Assessment tip

A correlation coefficient (r) between 0.00 and 0.10 is negligible, between 0.11 and 0.39 is weak, between 0.40 and 0.69 is moderate, between 0.70 and 0.89 is strong, and between 0.90 and 1.00 is very strong.

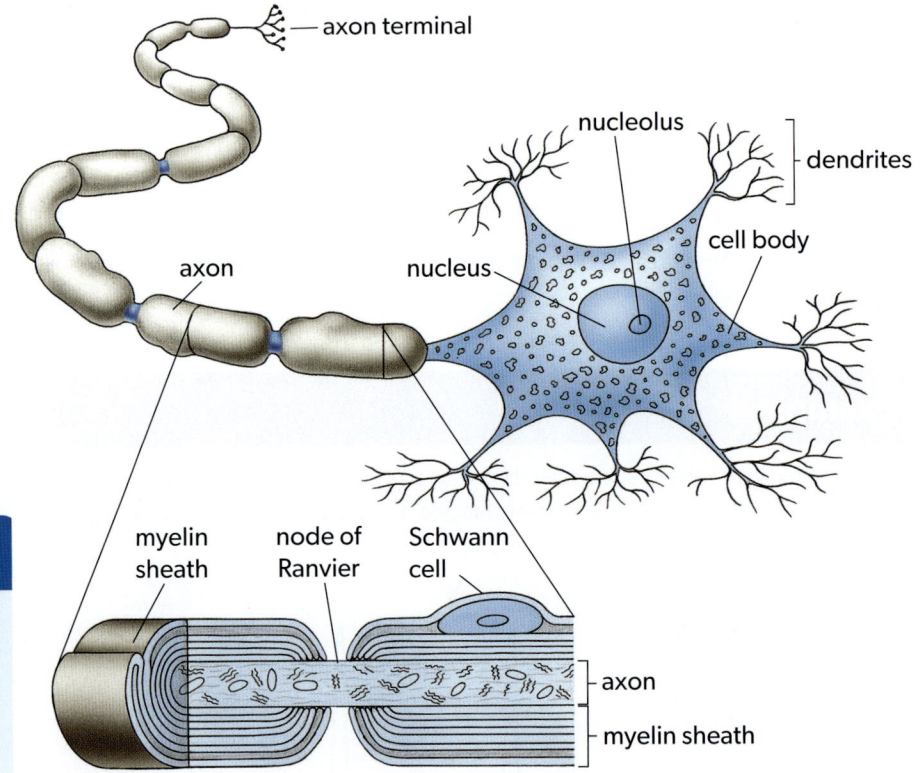

▲ Figure 5 A motor neuron showing the myelin sheath

Example 5

Brain neurons of pre-adolescent and adolescent male rats were isolated to measure the difference in the mean velocity of action potentials propagating along the axons. The response was measured at different distances from the stimulation source. Students' *t*-tests were used to assess the difference in mean velocity.

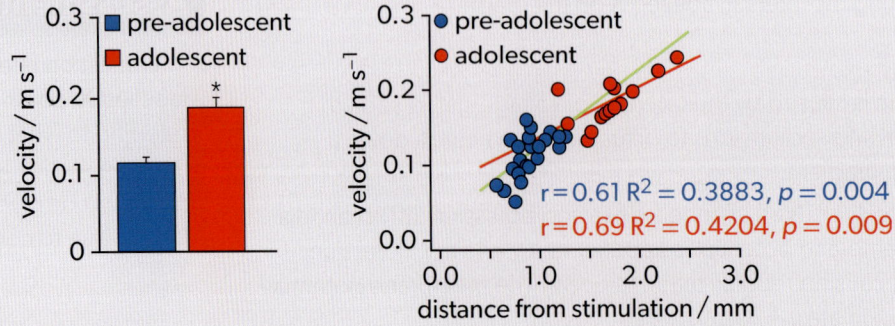

Source: McDougall, S., et al. (2018). Myelination of axons corresponds with faster transmission speed in the prefrontal cortex of developing male rats. *eNeuro*, 5(4), e0203–18.2018. www.doi.org/10.1523/eneuro.0203-18.2018 (CC BY 4.0)

a) Calculate the difference in mean velocity between the different age groups.
b) Results of the *t*-test for mean velocity at different age groups were $t = 6.47$, $p < 0.001$. Explain these results.
c) The diameter of the axons did not vary in the different age groups. Suggest a reason for the difference in velocity at different distances from stimulation according to age.
d) Outline the relationship between distance from the stimulation and velocity.

Solution

a) $0.18 - 0.11 = 0.07 \text{ m s}^{-1}$
b) As the *p*-value reported from the *t*-test is less than 0.05, the difference in velocity between both age groups is statistically significant.
c) Myelination of axons increases with age therefore increasing the conduction velocity. Another reason could be calcium availability, neurotransmitter release, more channels or faster opening of channels in older rats.
d) Positive correlation. As the distance increases, the conduction velocity increases for both age groups. The correlation coefficient determines the strength of this correlation. Because the correlation coefficient *r* is less than 0.7 it is a moderate correlation (strong is between 1 and 0.7). The correlation is slightly stronger for adolescents. In a regression, the coefficient of determination R^2 measures how well the regression line approximates the real data points. Ideally, the data should all be close to the linear regression line. The better the linear regression fits the data, the closer the value of R^2 is to 1. Values should be above 0.7 to have an acceptable regression. In this case, values between 0.3 and 0.4 are showing that the data is too far from the linear regression trend line to be acceptable.

Synapses

Synapses are the junctions between neurons or between neurons and effectors such as muscles or glands. When presynaptic neurons are depolarized, they open calcium membrane channels. The entering of calcium ions acts as a chemical signal that triggers the release of a neurotransmitter into the synapse. Neurotransmitters are chemicals used to cross the synaptic cleft (a gap around 20 nm wide) that travel in only one direction. Acetylcholine is a neurotransmitter that exists in many types of synapses, including neuromuscular junctions in mammals. Once acetylcholine reaches the postsynaptic neuron, it attaches to a receptor, inducing an action potential by the opening of sodium ion channels. Acetylcholine is then broken down by enzymes and the resting potential is recovered.

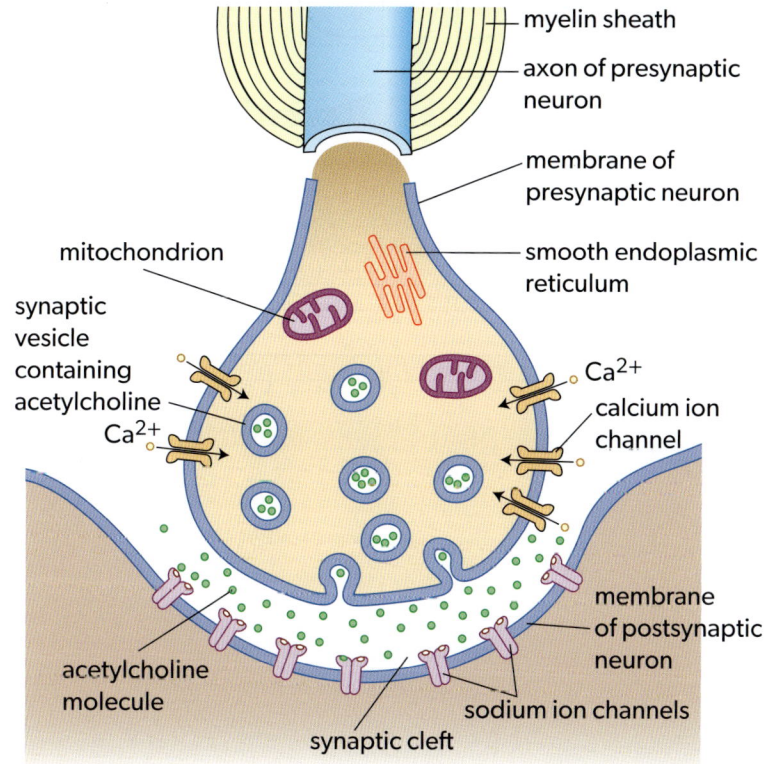

◀ Figure 6 An acetylcholine synapse

Interaction and interdependence

Propagation of an action potential

Action potentials are fast, sequential changes in the voltage across the membrane of axons along a nerve fibre because of local currents. The membrane potential (or voltage) depends on the relative amounts of ions and the permeability of the neuron to these ions. In the resting potential, the inside of the neuron has a negative charge. During an action potential there is depolarization and repolarization.

- **Depolarization** is the opening of sodium ion channels, increasing the membrane potential.
- **Repolarization** is the closing of sodium ion channels and opening of potassium ion channels restoring the low membrane potential.

Depolarization is when the inside of the neuron becomes less negatively charged and can only occur if a threshold is reached. Diffusion of sodium ions (Na^+) both inside and outside an axon can cause the threshold potential to be reached. Depolarization occurs when the voltage-gated sodium channels open, letting sodium ions in, and potassium ions (K^+) diffuse out of the neuron. Depolarization ends with the closing of the sodium ion channels.

Repolarization occurs with the opening of the voltage-gated potassium channels. The sodium–potassium ATPase pump induces the movement of sodium ions out of the neuron and potassium ions into the neuron, re-establishing the resting potential.

An oscilloscope can be used to measure these changes in potential across the cell membrane during resting and action potentials. In myelinated fibres, the impulses are faster because saltatory conduction occurs between the nodes of Ranvier where ion pumps and channels are clustered.

You learned about the structure of gated ion channels in neurons in subtopic B2.1.

Example 6

The diagram shows the results obtained with an oscilloscope attached to a neuron.

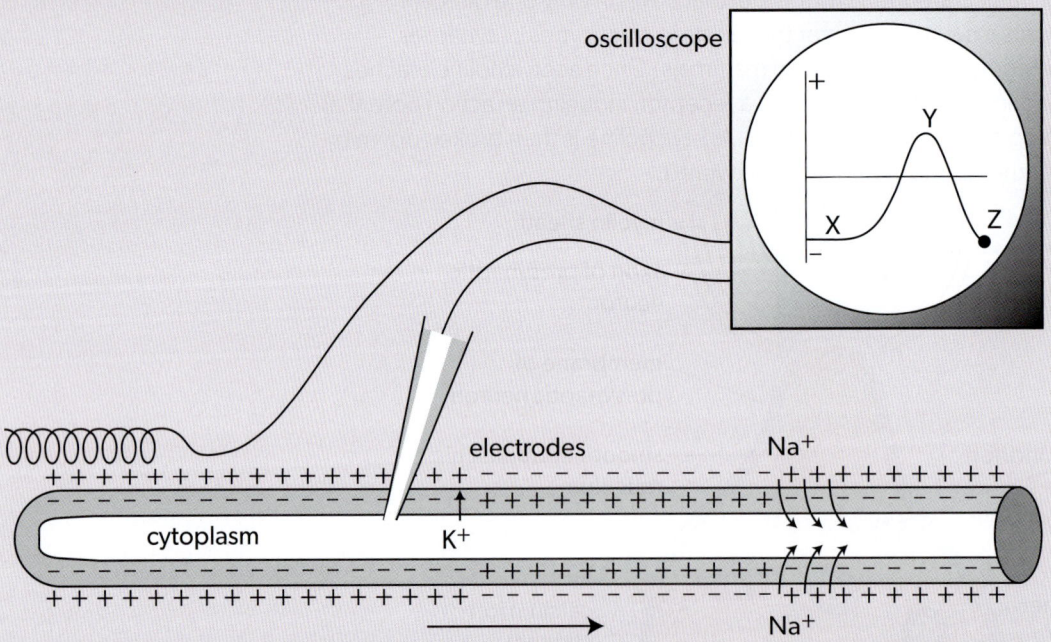

Source: adapted from www.topbiomedical.com.

a) Why is the change in the oscilloscope occurring between X and Y?

A. Hyperpolarization
B. Hypopolarization
C. Repolarization
D. Depolarization

C2.2 Neural signalling

b) Which statement describes the movement of sodium and potassium ions in the resting potential?
   A. Sodium ions are actively pumped out and some re-enter by facilitated diffusion.
   B. Potassium ions leave by simple diffusion and are actively pumped in.
   C. Three sodium ions are exchanged by simple diffusion with two potassium ions.
   D. Sodium ions diffuse out of the cell along with potassium ions.

Solution

a) The correct answer is **D**. During depolarization the cell undergoes a shift in electric charge distribution, resulting in less negative charge inside the cell compared to the outside. The resting potential is negative (approximately $-70\,\text{mV}$) because the inside of the cell is more electronegative than the outside. In an action potential, the sodium voltage-gated channels open, with an influx of Na^+ ions, giving rise to depolarization. The membrane potential is approximately $+45\,\text{mV}$. Ion channels open, altering membrane permeability to Na^+ and K^+, causing an influx of Na^+ and efflux of K^+. As the channel becomes closed/inactivated, the depolarization ends. The delay in the closing of the potassium ion channels causes a more negative charge in **hyperpolarization**.

b) The correct answer is **A**, as the generation of the resting potential is by pumping to establish and maintain concentration gradients of sodium and potassium ions. Energy from ATP drives the pumping of sodium and potassium ions in opposite directions across the plasma membrane of neurons. Sodium then returns through protein channels by facilitated diffusion.

Effects of exogenous chemicals on synaptic transmission

Exogenous chemicals can affect synaptic transmission. Neonicotinoids are synthetic chemicals used as insecticides. They have a similar composition to nicotine. They bind irreversibly to the acetylcholine receptor in cholinergic synapses in the central nervous system of insects. Because the receptors are blocked, acetylcholine is no longer capable of binding to the postsynaptic membrane, inhibiting synaptic transmission. This causes paralysis and death of insects.

- **Hyperpolarization** is when the membrane potential is at its lowest. This is caused by the delay in closing of the potassium ion channels.

Example 7

Cholinergic synapses use acetylcholine as their neurotransmitter. They are widespread in the body, passing on signals to muscle cells. These synapses are affected by neonicotinoid pesticides.

The diagrams, which are not drawn to scale, show the synapse between two neurons and a detail of the synaptic gap.

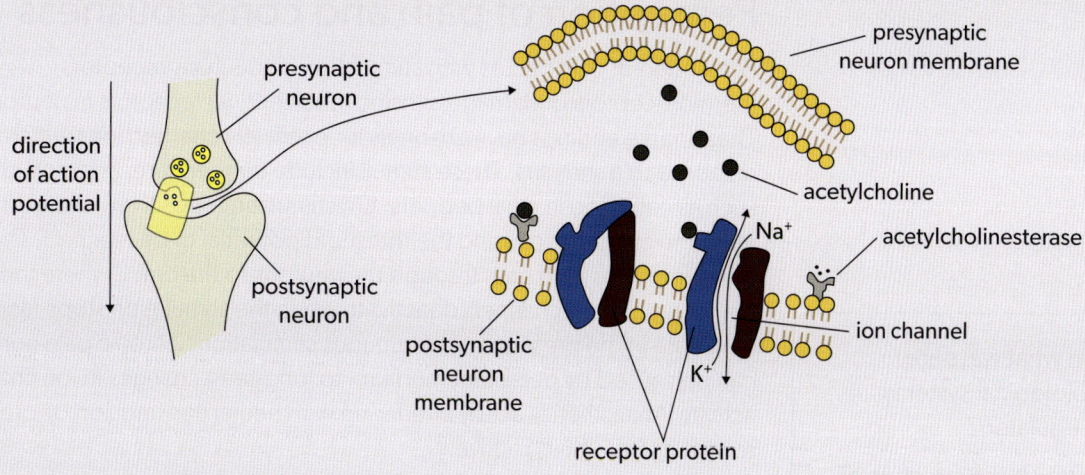

a) On the diagrams, label:
   (i) with a letter H the hydrophilic end of a phospholipid.
   (ii) with a letter E a vesicle involved in exocytosis.
   (iii) with a letter P a location where a neonicotinoid pesticide could bind.

b) Explain how acetylcholine initiates an action potential in a postsynaptic membrane.
c) State the action of the enzyme acetylcholinesterase.

Solution

a) (i) Hydrophilic end of phospholipid shown at any circle of the membrane.
   (ii) In the presynaptic neuron, the circles around black spots show the vesicles.
   (iii) Neonicotinoids bind to the receptor protein.
b) The binding of acetylcholine to the receptor channel protein causes a change in the disposition of the alpha helices of this protein, therefore a change in the tertiary structure of the protein in the channel. The core of the receptor protein is unblocked, so the transport of ions through the channel protein is allowed.
c) Acetylcholinesterase breaks down acetylcholine to acetyl and choline for recycling.

Psychoactive drugs affect the brain by either increasing or decreasing postsynaptic transmission. Stimulant drugs, such as cocaine, mimic the stimulation provided by the sympathetic nervous system by blocking the reuptake of the neurotransmitter. Dopamine is released by a neuron into the synapse, where it can bind to dopamine receptors on neighbouring neurons. Dopamine is then recycled back into the transmitting neuron by a specialized protein called the dopamine transporter. When cocaine is present, it attaches to the dopamine transporter and blocks the normal recycling process. Dopamine accumulates in the synapse to produce an amplified signal to the receiving neurons, causing euphoria and pleasure.

Excitatory and inhibitory neurotransmitters

Some neurotransmitters excite nerve impulses in postsynaptic neurons and others inhibit them. Excitatory postsynaptic potentials increase the likelihood of a postsynaptic action potential occurring while inhibitory decrease this likelihood. What determines postsynaptic excitation or inhibition is the reversal potential in relation to the threshold level. When the reversal potential is more positive than the threshold, excitation results. Nerve impulses are initiated or inhibited in postsynaptic neurons as a result of the summation of all the excitatory and inhibitory neurotransmitters received from presynaptic neurons. Multiple presynaptic neurons interact with all-or-nothing consequences in terms of postsynaptic depolarization.

Perception of pain and consciousness

Perception of pain occurs with stimulation of mechanoreceptors, thermoreceptors or damage-sensing (nociceptive) receptors in neurons. These nerve endings have channels for positively charged ions which open in response to a stimulus. These stimuli include temperature, acid and chemicals such as capsaicin in chilli peppers. On stimulation, positively charged ions enter the channels, causing the threshold potential to be reached. The nerve impulses then pass through the neurons to the brain, where the pain is perceived. These channels deactivate when the stimulation stops, and many inactivate in the presence of continuous stimulation. Chronic pain conditions can be caused by persistent medium- to long-term changes in ion channel activity. Anaesthetics act by interfering with neural transmission by preventing receptors from being active.

- In what ways are biological systems regulated?
- How is the structure of specialized cells related to function?

Emergent properties such as consciousness are a consequence of interaction. Conscious experience is the result of interactions between large-scale neuronal networks in the brain which derive from information processing.

Practice problems

1. Compare and contrast the mechanisms of action of peptide and steroid hormones.
2. Four electrical stimuli of increasing size are applied to a neuron at times t1 to t4. The graph shows the effects of each electrical stimulus on the membrane potential of the neuron at the point where the stimulus is applied.

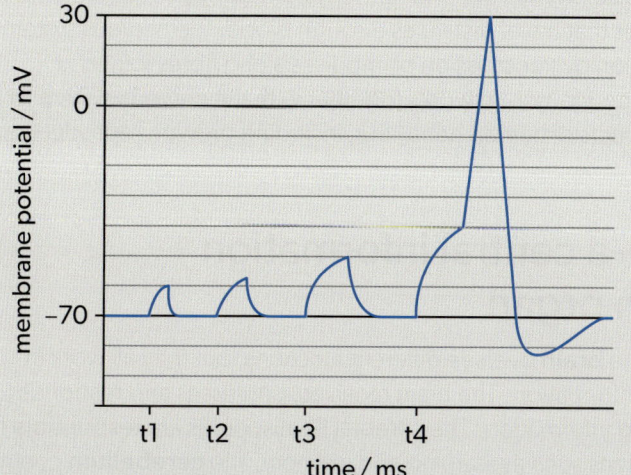

What can be concluded from the graph?

A. The stronger the stimulus, the greater the action potential.

B. Only the stimulus at t4 causes a change in membrane potential.

C. The stimulus at t4 caused a total increase of membrane potential of 30 mV.

D. The threshold potential is approximately −40 mV.

C3.1 Integration of body systems

You should know:

- ✔ coordination allows different parts of a system to collectively perform an overall function.
- ✔ emergent properties in multicellular organisms arise from integration.
- ✔ the brain is the central information integration organ.
- ✔ the spinal cord integrates unconscious processes.
- ✔ input to the spinal cord and brain is through sensory neurons and output through motor neurons.
- ✔ nerves are bundles of sensory and motor neurons.
- ✔ epinephrine prepares the body for vigorous activity.
- ✔ heart rate is controlled by feedback mechanisms through input from baroreceptors and chemoreceptors.

Additional higher level:

- ✔ seedlings have tropic responses such as positive phototropism.
- ✔ phytohormones are signalling chemicals that control growth, development and response in plants.
- ✔ auxin efflux carriers maintain concentration gradients of phytohormones.
- ✔ auxin promotes cell growth.

You should be able to:

- ✔ distinguish between the roles of the nervous system and endocrine system in sending messages.
- ✔ explain the pain reflex arc.
- ✔ state that the cerebral hemispheres are responsible for higher order functions.
- ✔ describe the role of the cerebellum in coordinating skeletal muscle contraction and balance.
- ✔ describe how melatonin regulates sleep patterns.
- ✔ explain how the hypothalamus and pituitary glands control the endocrine system.
- ✔ describe feedback control of ventilation through chemoreceptors.
- ✔ describe how the central and enteric nervous system control peristalsis.

Additional higher level:

- ✔ design experiments for tropisms.
- ✔ explain the interaction between cytokines and auxins in shoot and root growth.
- ✔ explain positive feedback of ethylene production in fruit ripening.

Integration of body systems

Cells, tissues, organs and body systems are all integrated and coordinated in a multicellular living organism to perform different functions. This integration is responsible for emergent properties. It is obtained through nervous and chemical signals that can travel through the body by blood or plasma or directly from cell to cell. While the endocrine system carries out its function mainly by hormones that travel distances through blood, the nervous system carries out its function by transmission of impulses along fibres joined at synapses. The endocrine system works together with the nervous system to influence many aspects of human behaviour, including growth, reproduction and metabolism.

The brain—a central information integration organ

Different parts of the **brain** oversee different functions, but they all work in the integration of information. The brain receives, processes and responds to different stimuli and information. The cerebral hemispheres are responsible for higher order functions such as learning and memory. The **cerebellum** controls equilibrium and posture through coordination of skeletal muscle contraction. The hypothalamus regulates metabolic processes and hormones while the pituitary gland secretes hormones. The medulla oblongata controls breathing and reflexes such as swallowing, coughing, sneezing and vomiting.

- The **brain** is a central information integration organ.
- The **cerebellum** is the part of the brain that coordinates skeletal muscle contraction and balance.

Spinal cord as an integrating centre for unconscious processes

The autonomic nervous system controls and integrates unconscious processes in the body using centres located mainly in the brainstem. Unconscious processes are not susceptible to voluntary self-regulation whereas conscious processes are in response to stimuli when a person is awake and aware. Sensory neurons convey messages from receptor cells to the **spinal cord** and cerebral hemispheres in the central nervous system. Output from the cerebral hemispheres to stimulate muscles to contract is through motor neurons. Several bundles of myelinated and unmyelinated nerve fibres enclosed in a protective sheath make up the peripheral **nerve.**

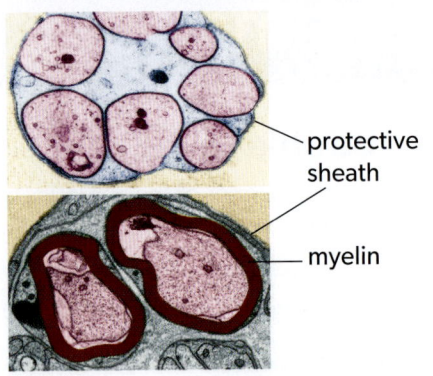

▲ Figure 1 Unmyelinated and myelinated nerve fibres

You studied the structure of myelinated neurons in subtopic C2.2.

- The **spinal cord** integrates unconscious processes.
- **Nerves** are bundles of nerve fibres of both sensory and motor neurons.

Pain reflex arcs

Autonomic and involuntary responses are referred to as reflexes. Reflex arcs comprise the neurons that mediate reflexes. The **pain reflex arc** is an example of an involuntary response with the skeletal muscle as the effector. The pain sensation is felt in the sensory neuron (pain receptor) of the hand. This sends a message to the interneuron which joins the sensory neuron and the motor neuron in the grey matter of the spinal cord. The motor neuron then sends impulses to the muscle effector to contract to remove the hand from the source of pain.

- **Pain reflex arcs** are involuntary responses with the skeletal muscle as the effector.

C3.1 Integration of body systems

Example 1

The diagram shows the reflex arc. Name the parts labelled 1 to 4.

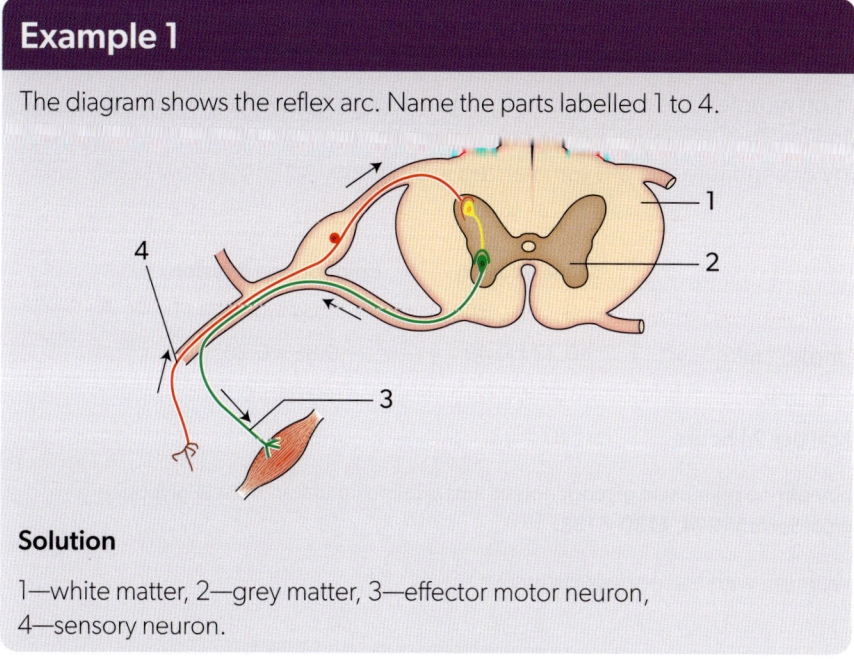

Solution

1—white matter, 2—grey matter, 3—effector motor neuron, 4—sensory neuron.

Melatonin and circadian rhythms

Melatonin is a hormone derived from the amino acid tryptophan. It is produced in the pineal gland of the brain and travels through blood to target cells in different organs. Melatonin has many functions, including modulation of daily wake and sleep patterns as part of circadian rhythms. Melatonin secretion has a diurnal (daily) pattern where it increases soon after the onset of darkness, peaks in the middle of the night and gradually falls again. Blue light suppresses melatonin biosynthesis.

- **Melatonin** is a hormone secreted by the pineal gland involved in circadian rhythms.

AHL only: You studied chemical signalling in subtopic C2.1.

Sample student answer

Glands are organs that secrete and release chemical substances. Melatonin is an important hormone secreted by the pineal gland in the brain. Describe its role in mammals. [2]

This answer could have achieved 2/2 marks:

> Melatonin controls the circadian rhythms in mammals as it sets the pattern of sleep by inducing sleep when it is night-time (and dark) and waking the mammal up during the daytime when it is bright.

▲ The student scored one mark for mentioning that melatonin regulates the circadian rhythm (or biological clock). The other mark is for saying that it affects the sleep–wake cycles. Other points that would have scored a mark include stating that production is controlled by the amount of light detected in the retina and that there is a high level of secretion during the night and a low level during the day—this means secretion is directly proportional to night-time duration.

Epinephrine action and endocrine control

Epinephrine (adrenaline) is an amine hormone that increases the heart rate to prepare for vigorous physical activity by facilitating intense muscle contraction. It is produced by the adrenal glands and travels through blood to the target organs. It produces a rapid rise in blood systolic pressure by direct stimulation of cardiac muscle, which increases the strength of ventricular contraction, thus increasing the heart rate, and also causes constriction of arterioles. It increases sugar metabolism by breakdown of glycogen, stimulating the secretion of glucagon and inhibiting the secretion of insulin.

Interaction and interdependence

Example 2

In a study to see the effect of epinephrine on muscles, the experimental group was injected with epinephrine infusion while the control group with saline infusion. Changes for the mean respiratory quotient (RQ), mean rates of glucose and lipid oxidation, and energy expenditure rate are summarized in the table. Asterisks show significant differences at $p < 0.005$.

	Saline infusion	Epinephrine infusion
ΔRQ	-0.33 ± 0.28	0.04 ± 0.01*
ΔRate of glucose oxidation / mg kg^{-1} min^{-1}	-0.33 ± 0.49	0.86 ± 0.28*
ΔRate of lipid oxidation / mg kg^{-1} min^{-1}	0.26 ± 0.12	-0.04 ± 0.10
ΔEnergy expenditure rate / kcal per 24 hours	87 ± 23	326 ± 64*

Source: Laurent, D., et al. (1998). Effect of epinephrine on muscle glycogenolysis and insulin-stimulated muscle glycogen synthesis in humans. *The American Physiological Society*, **274**, E130–E138.

Compare and contrast the effect of epinephrine with the control group.

Solution

There is a decrease of RQ in the control group but (slight) increase in the epinephrine group. There is increased glucose oxidation and decreased (very little change) lipid oxidation in the epinephrine group while the opposite in the control. The energy expenditure increased a lot in the epinephrine group and less in the control.

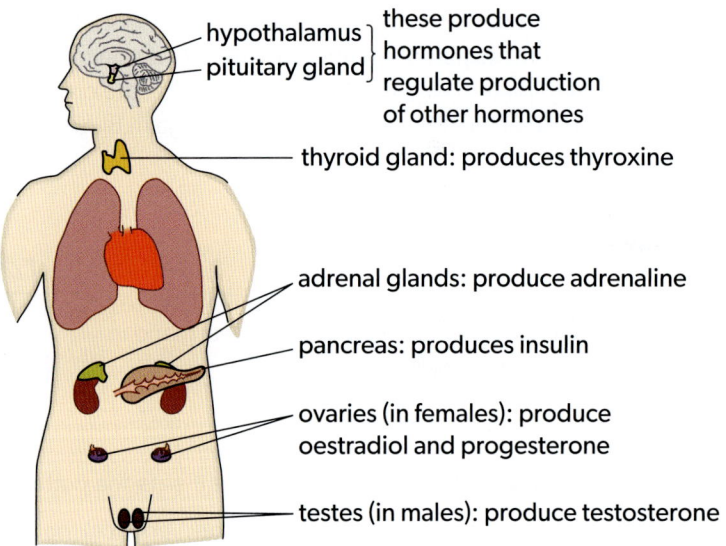

▲ Figure 2 Main endocrine glands

- The **hypothalamus** and **pituitary** are glands that control the endocrine system.

The **hypothalamus** controls hormone secretion by the anterior and posterior lobes of the **pituitary** gland. Hormones secreted by the pituitary gland control growth, developmental changes, reproduction and homeostasis. The cells in the hypothalamus secrete antidiuretic hormone (ADH) and oxytocin, which are transported down to the posterior pituitary, where they are released into the bloodstream. The hypothalamus has neurons that control the endocrine functions of the anterior pituitary. The anterior pituitary gland produces prolactin, involved in milk production. It also produces follicle-stimulating hormone (FSH) and luteinizing hormone (LH), involved in the growth of follicles and ovulation, respectively, in ovaries in females, and in sperm formation in testes in males. Thyroid-stimulating hormone (TSH) is a hormone that induces the thyroid gland to produce thyroxine and other thyroid hormones. Somatotropin or growth hormone (GH) stimulates growth, cell reproduction and regeneration in bones and soft tissues.

Feedback control of heart rate and ventilation

Baroreceptors and chemoreceptors are sensory neurons that send information to the medulla oblongata in the brainstem to induce reflex effects both on the respiratory and cardiovascular system.

Baroreceptors are mechanoreceptors in the aorta and the carotid artery. They provide information about blood volume and pressure to the medulla oblongata by detecting how much the walls of arteries have stretched. As blood volume increases, vessels are stretched causing the baroreceptors to send more impulses. On the other hand, if vessels contract in response to reduced blood volume, baroreceptors decrease impulses. Arterial baroreceptor stimulation, as occurs when pressure in the carotid and aorta arteries is raised, produces a feedback control reflex which slows down the heart (bradycardia), reduces heart **stroke volume** and causes vasodilatation, which returns arterial pressure to normal.

Chemoreceptors are sensory neurons that detect chemical changes. During exercise or when there is lack of oxygen (hypoxia), for example, at high altitude, the carbon dioxide levels in blood increase. These high levels of dissolved carbon dioxide react with water in the presence of the enzyme carbonic anhydrase to form hydrogen ions (H$^+$) and carbonic acid ions (HCO_3^-). The increase in H$^+$ decreases the pH level in blood. In the aorta and the carotid artery, there are chemoreceptors that detect the levels of oxygen and carbon dioxide in the blood. Additionally, the carotid receptors contain acid-sensing ion channels that detect H$^+$ and therefore changes in the pH of blood. Blood pH is regulated to stay within the narrow range of 7.35 to 7.45. The respiratory control centre in the medulla oblongata controls the rate of ventilation in response to changes in pH by sending nervous impulses to the external intercostal muscles and diaphragm through the intercostal nerve. During exercise or hypoxia, the rate of ventilation increases in response to the high amount of carbon dioxide in the blood. The heart rate also increases (tachycardia) to send more blood to oxygenate tissues.

- **Stroke volume** is the volume of blood pumped from the (left) ventricle per beat.
- **Chemoreceptors** are sensory nerve cells that monitor pH to control ventilation rate, whereas baroreceptors control pressure.

Control of peristalsis

The nervous control of digestion is by the **enteric nervous system** (ENS) in the walls of the specialized organs that make up the digestive system, the sympathetic and parasympathetic nervous systems in the spinal cord, and the central nervous system (CNS) in the brain. Initiation of swallowing of food and egestion of faeces is under voluntary control by the CNS but peristalsis between these points in the digestive system is under involuntary control by the ENS. The ENS has sensory neurons to detect changes in thermal, chemical or mechanical stimulus energy. These send impulses to interneurons connected by synapses that process this sensory information and control the behaviour of motor neurons connected to effectors. The intestinal musculature behaves as self-excitable with cells that function as pacemakers integrated to muscles, generating forces from one muscle to another in a coordinated manner, causing peristalsis. The vagus nerves connect the brain to the digestive tract. The CNS can modulate, but not entirely control, the motor activity by sending instructions via the two components of the extrinsic autonomic nervous system: the sympathetic and parasympathetic nervous system.

- The **enteric nervous system** is a series of nerves involved in the involuntary control of peristalsis.

Tropic responses in plants

Plant shoots respond to the environment by tropisms. Shoots have positive phototropism and negative gravitropism.

Interaction and interdependence

Example 3

In three experiments, seedlings were kept in the dark for two days and then two shoots were cut and placed on a gel for a few hours. One shoot was kept intact and the other was split in half. The concentration of auxin in the gel was measured in nmol g^{-1} fresh weight in all three experiments.

Experiment 1: Seedlings kept in the dark and the auxin concentration measured.

Experiment 2: The measurement was repeated after the shoots were exposed to unilateral light (shown with an arrow).

intact shoot
25.5

split shoot
24.1

auxin concentration / nmol g^{-1}

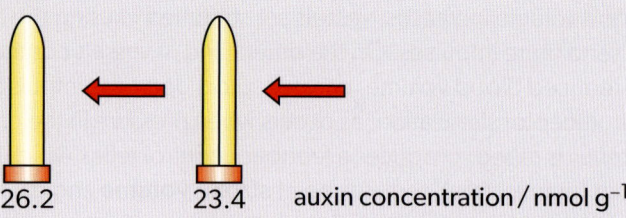
26.2   23.4   auxin concentration / nmol g^{-1}

Experiment 3: The gel itself was cut in half and the concentrations on the left and right sides of the gel were recorded after unilateral light exposure.

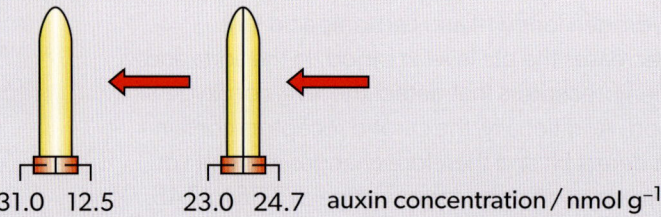
31.0 12.5   23.0 24.7   auxin concentration / nmol g^{-1}

a) Compare and contrast the results for intact and split shoots on the concentration of auxin measured in experiments **1** and **2**.

b) Calculate the concentration difference between the left and the right sides of the gel in the intact shoot and in the split shoot in experiment **3**.

c) Explain the results obtained in experiment **3**.

d) In reality, the intact shoot in experiment **3** will look different from how it is shown here. Describe the resulting shape of this shoot.

Solution

a) The intact shoots and split shoots without light have approximately the same auxin concentration (slightly higher in the intact).

b) Intact:
   31.0 → 12.5 = 18.5 nmol g^{-1} fresh weight

   Split:
   23.0 → 24.7 = 1.7 nmol g^{-1} fresh weight

c) Auxin accumulates on the dark side and is lower in the lit side when the shoot is intact, but when it is divided the auxin concentration is approximately the same. This suggest that normally auxin is transported across the shoot in unilateral light from the lit side to the dark side.

d) The intact shoot will bend to the right.

Phytohormones and their mechanisms of action

Phytohormones are signalling chemicals controlling growth, development and response to stimuli in plants. There is a great variety of phytochemicals,

> **Assessment tip**
>
> Remember the units.

including auxins, abscisic acid, salicylic acid, jasmonates, cytokinins and ethylene.

Auxin efflux carriers are transmembrane proteins in the plasma membrane that transfer auxin between the cytoplasm and extracellular space. They help to maintain concentration gradients of phytohormones. Auxin can diffuse freely into plant cells but not out of them. Auxin efflux export carriers are polarly localized, resulting in a net flux of auxin through the cells. If all cells coordinate to concentrate these carriers on the same side, auxin is actively transported from cell to cell through the plant tissue and becomes concentrated in part of the plant.

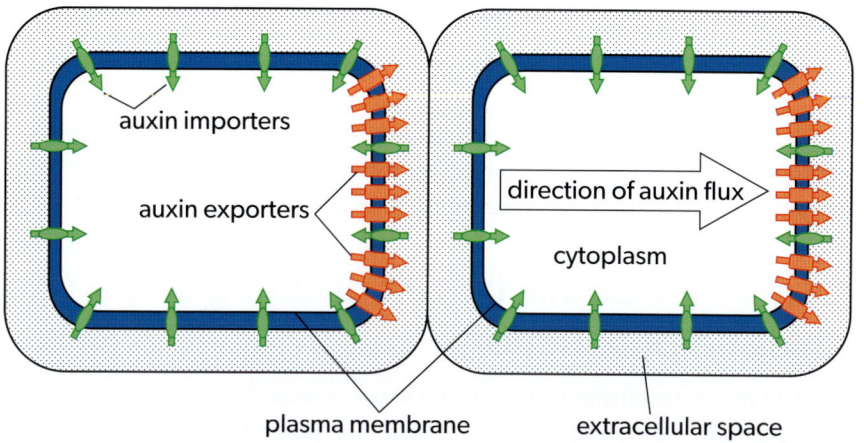

▲ Figure 3 Auxin transport

Source: Smith, R. S. (2008). The role of auxin transport in plant patterning mechanisms. *PLoS Biology*, **6**(12), e323. www.doi.org/10.1371/journal.pbio.0060323 (CC BY 4.0)

Plant hormones control growth in the shoot apex. Auxins, especially indole 3-acetic acid (IAA), are involved in plant growth behaviours. Auxin acts as a chemical signal. It influences cell growth rates by changing the pattern of gene expression, which promotes hydrogen ion secretion into the apoplast, acidifying the cell wall and thus loosening cross-links between cellulose molecules and facilitating cell elongation. Auxin efflux carriers can set up concentration gradients of auxin in plant tissue. Concentration gradients of auxin cause the differences in growth rate needed for phototropism.

Nature of science

In experiments, qualitative and quantitative observations can be made. Qualitative observations in tropism experiments could include direction of growth and state of the seedlings. The precision, accuracy and reliability of measurements in tropism experiments can be improved using quantitative measurements. Quantitative observations could include measurement of auxin concentrations (like in this case) or angle of growth under different conditions of seedlings. The precision of angle measurements and their accuracy depend on the apparatus used to measure this angle, for example a compass will be less precise than a protractor.

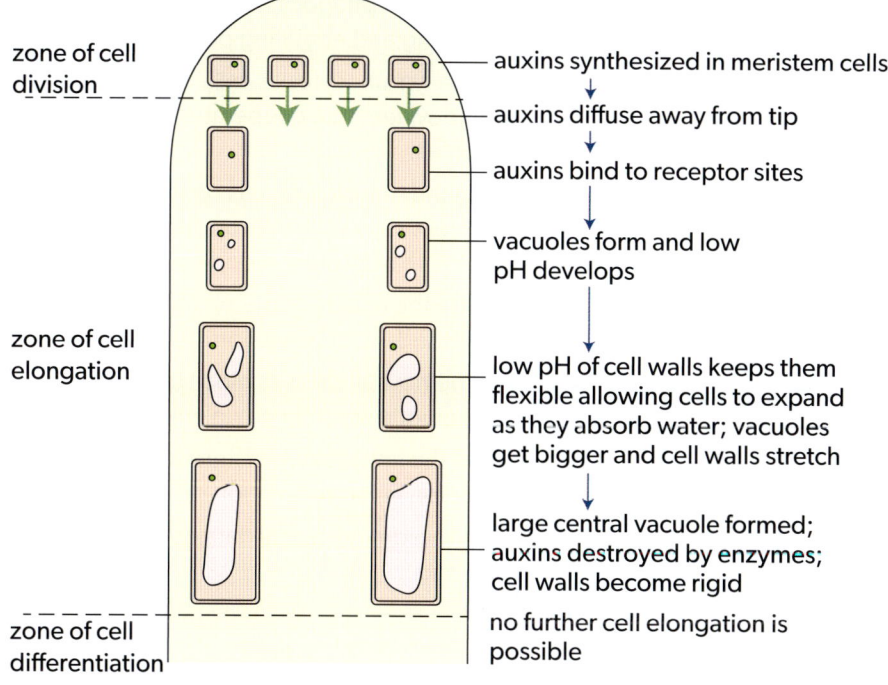

▲ Figure 4 The effect of auxin on apical shoot growth

Interaction and interdependence

Interactions between cytokinin and auxin in growth regulation

Cytokinin (CK) synthesized in the root cap and auxins are key hormones that regulate shoot growth and root development, vascular differentiation, lateral root initiation and root gravitropism. Root tips produce cytokinin, which is transported to shoots, and shoot tips produce auxin, which is transported to roots. Cytokinin and auxin regulate the synthesis of each other, demonstrating a mechanism for mutual feedback and feedforward control of auxin and cytokinin levels. When roots are turned to the horizontal position, both CK and auxins are transported laterally downwards and become concentrated in the lower root side. The high CK concentration inhibits elongation of the lower side at the distal elongation zone (closely behind the root cap). The movement of auxins from the shoot to the root tip inhibits elongation. The asymmetric distributions of both CK and auxins inhibit the lower root side and promote elongation of the upper side, resulting in downward root bending.

- What are examples of branching (dendritic) and net-like (reticulate) patterns of organization?
- What are the consequences of positive feedback in biological systems?

Positive feedback in fruit ripening and ethylene production

The plant hormone ethylene (ethene) is a key regulator of the ripening process of fruit. It is involved in the expression of genes involved in ethylene biosynthesis, so ethylene stimulates its own biosynthesis by signal transduction pathways in a positive feedback loop, which ensures that fruit ripening is rapid and synchronized.

C3.2 Defence against disease

You should know:

- ✔ skin and mucous membranes form a primary defence against pathogens that cause infectious disease.
- ✔ clotting factors are released from platelets, allowing cuts in the skin to be sealed by blood clotting.
- ✔ antigens are recognition molecules that trigger antibody production.
- ✔ B-lymphocytes are activated by helper T-lymphocytes.
- ✔ activated B-cells multiply to form clones of plasma cells, which secrete antibodies
- ✔ immunity depends upon the persistence of memory cells.
- ✔ some strains of bacteria have evolved with genes that confer resistance to antibiotics and some strains of bacteria have multiple resistance.
- ✔ zoonoses can be transferred to humans.
- ✔ vaccines contain antigens that trigger immunity but do not cause the disease.

You should be able to:

- ✔ compare the innate and adaptive immune systems.
- ✔ describe how phagocytes control infection by recognizing and engulfing pathogens.
- ✔ describe lymphocytes as cells in the adaptive immune system that cooperate to produce antibodies.
- ✔ explain the reason antigens on the surface of red blood cells stimulate antibody production in a person with a different blood group.
- ✔ describe the effects of HIV on the immune system and methods of transmission.
- ✔ describe how antibiotics block processes that occur in prokaryotic cells but not in eukaryotic cells.
- ✔ explain the importance of herd immunity in a population in the prevention of epidemics.
- ✔ analyse epidemiological data on the COVID-19 pandemic.

Defence against disease

Pathogens are organisms that cause disease. Some pathogens are species specific, but others can affect more than one species. There is a broad range of disease-causing organisms that can infect humans, including viruses, bacteria, fungi and protists. Archaea are not known to cause any diseases in humans. The human body has structures and processes that resist the continuous threat of invasion by pathogens. Skin and mucous membranes form a primary defence against pathogens that cause infectious disease, not allowing organisms to enter the body. To avoid organisms entering through a wound, clotting factors are released from platelets, allowing cuts in the skin to be sealed by blood clotting.

Comparing the innate and adaptive immune system

The innate immune system is the defence system with which an organism is born and includes physical, chemical and cellular defences against pathogens. Adaptive immunity (acquired immunity) is immunity specific to an acquired pathogen. Innate immunity is fast but does not last long, whereas acquired immunity takes longer but lasts for much longer. The innate system responds to broad categories of pathogens and does not change during an organism's life, whereas the adaptive system responds in a specific way to pathogens and builds up a memory of pathogen encounters, so the immune response becomes more effective.

> **Nature of science**
>
> Careful observations during 19th-century epidemics of childbed fever (due to an infection after childbirth) in Vienna led to breakthroughs in the control of infectious disease. In 1847, Ignaz Semmelweis proposed hand washing with chlorine and lime after observation that puerperal fever was infectious and that it could be carried from one patient to another. It was not until the work of Pasteur, who discovered the causative organisms of these diseases, that final proof of their nature could be established.

Example 4

What sequence of reactions occurs during blood clotting?

A. Fibrinogen is broken down into trypsin by fibrin.
B. Fibrin is broken down into fibrinogen by trypsin.
C. Fibrinogen is broken down into fibrin by thrombin.
D. Fibrin is broken down into fibrinogen by thrombin.

Solution

The correct answer is **C**. Fibrinogen is a globular, soluble protein found in blood plasma. In the event of a cut, clotting factors are released from platelets. In the subsequent cascade pathway, an enzyme that is found in an inactive form is transformed into active thrombin. Thrombin breaks down part of the soluble fibrinogen protein, now transforming it into a smaller but insoluble fibrous protein called fibrin. Molecules of fibrin form cross-linking bonds resulting in a large polymer which attaches to platelets, and thus clots blood.

White blood cells

Phagocytes and lymphocytes are white blood cells. Phagocytes are cells in the innate immune system. They have amoeboid movement from blood to sites of infection, where phagocytes recognize pathogens, engulf them by endocytosis and digest them using enzymes from lysosomes. Lymphocytes are cells in the adaptive immune system that cooperate to produce **antibodies**. Lymphocytes both circulate in the blood and are contained in lymph nodes. An individual has a very large number of B-lymphocytes that each make a specific type of antibody.

- **Antibodies** are proteins produced by plasma cells that attack antigens.

Interaction and interdependence

- An **antigen** is a molecule found on the surface of the cells of pathogens.
- **B-lymphocytes** are white blood cells that mature into either plasma cells or memory cells.
- **T-lymphocytes** are white blood cells that activate B-lymphocytes.
- **Plasma cells** are mature B-lymphocytes that produce antibodies.
- **Memory cells** are mature B-lymphocytes that remember how to produce the antibodies.

Antigens trigger antibody production

Immunity is based on the recognition of self and the destruction of foreign material. Every organism has unique molecules on the surface of its cells called **antigens**, which are recognized by specific antibodies. Antibodies are proteins that attach to the antigens. Most antigens are glycoproteins or other proteins (or parts of proteins) that are usually located on the outer surfaces of pathogens. Antibodies are produced when the white blood cells encounter the antigen.

In mammals, **B-lymphocytes** are activated by **T-lymphocytes**. Activated B-cells multiply to form clones of plasma cells and memory cells. **Plasma cells** secrete antibodies that aid in the destruction of pathogens. **Memory cells** are mature B-lymphocytes that remember how to produce the antibodies in the event of a second encounter with the antigen. Immunity depends upon the persistence of memory cells. Once marked by antibodies, foreign matter such as bacteria is destroyed by phagocytes. Phagocytes are non-specific, as they destroy any foreign matter.

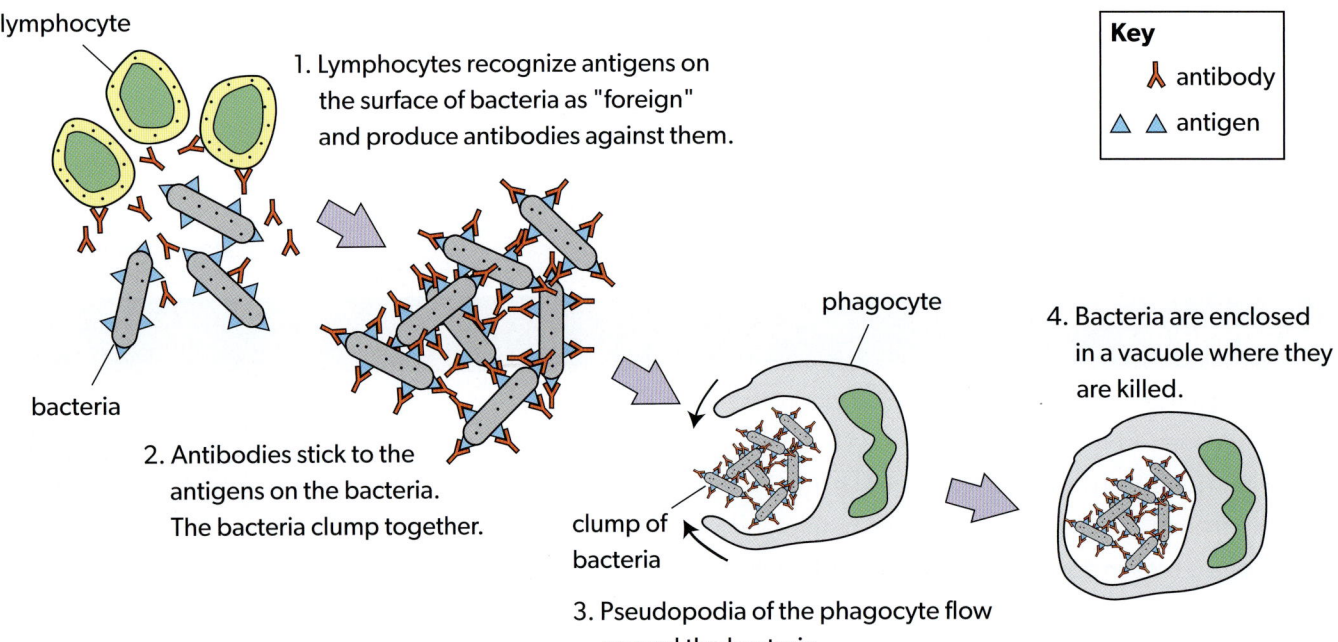

▲ Figure 5 Defence against disease

Example 5

Compare and contrast antigens and antibodies in different blood groups.

Solution

Feature	Blood type			
	A	B	AB	O
Membrane antigens	A	B	AB	O
Plasma antibodies anti-blood group produced	anti-B	anti-A	none	anti-A and anti-B
Blood type that can be received in a transfusion	A and O	B and O	A, B, AB and O	only O

Antigens on the surface of red blood cells (erythrocytes) stimulate antibody production in a person with a different blood group. This will determine the different blood donor possibilities. Blood is tested before a transfusion to avoid an incompatibility reaction. This reaction includes symptoms such as fever, chills, breathing difficulty, muscle ache, blood in urine due to kidney damage and jaundice due to liver damage. It can lead to death as blood clotting may occur throughout the body, shutting off the blood supply to the main organs.

Immunity

Immunity is the ability to eliminate an infectious disease from the body. B-cells produce antibodies and become memory cells only when they have been activated. Activation requires both direct interaction with the specific antigen and contact with a helper T-cell that has also become activated by the same type of antigen. Activated B-lymphocytes multiply to form clones of antibody-secreting plasma cells. To produce enough antibodies, activated B-cells first divide by mitosis to produce large numbers of plasma B-cells that can produce the same type of antibody. Memory cells are lymphocytes that survive for a long time and can make the specific antibodies needed to fight the infection.

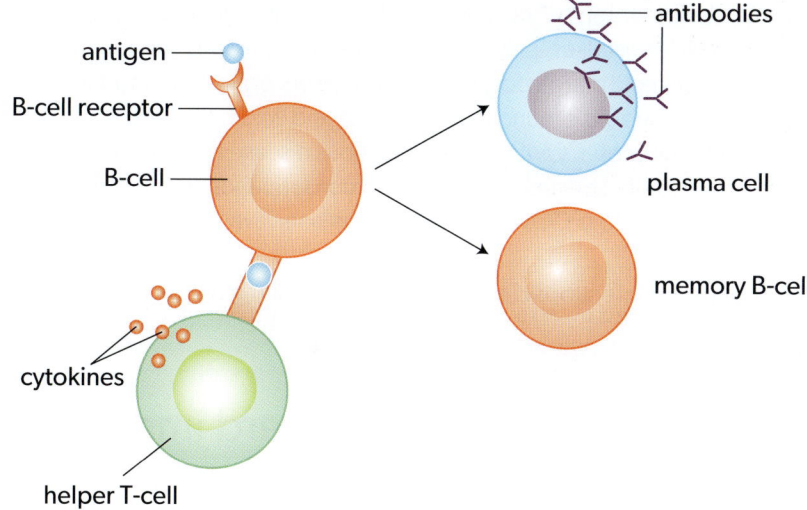

▲ Figure 6 Production of memory cells for immunity

Transmission and consequences of HIV

Human immunodeficiency virus (HIV) is transmitted through body fluids such as blood, saliva, semen and breast milk. HIV acts on the immune system, reducing the number of T-lymphocytes and causing a loss of the ability to produce antibodies, leading to the development of acquired immune deficiency syndrome (AIDS). This disease is characterized by immunosuppression that gives rise to opportunistic infections, tumours and central nervous system degeneration.

- **HIV** is a virus that causes the disease AIDS.

Antibiotics and antibiotic resistance

Antibiotics are chemicals produced by living organisms such as fungi or which can be produced synthetically. Antibiotics block processes that occur in prokaryotic cells but not in eukaryotic cells. Some block the formation of the cell wall or plasma membrane, others inhibit replication or transcription of the microorganism's genetic material. Some strains of bacteria have evolved with genes that confer resistance to antibiotics, and some strains have multiple resistance. Viruses lack a metabolism, so they cannot be treated with antibiotics.

Nature of science

The development of new techniques can lead to new avenues of research. For example, the recent technique of searching chemical libraries is yielding new antibiotics.

Interaction and interdependence

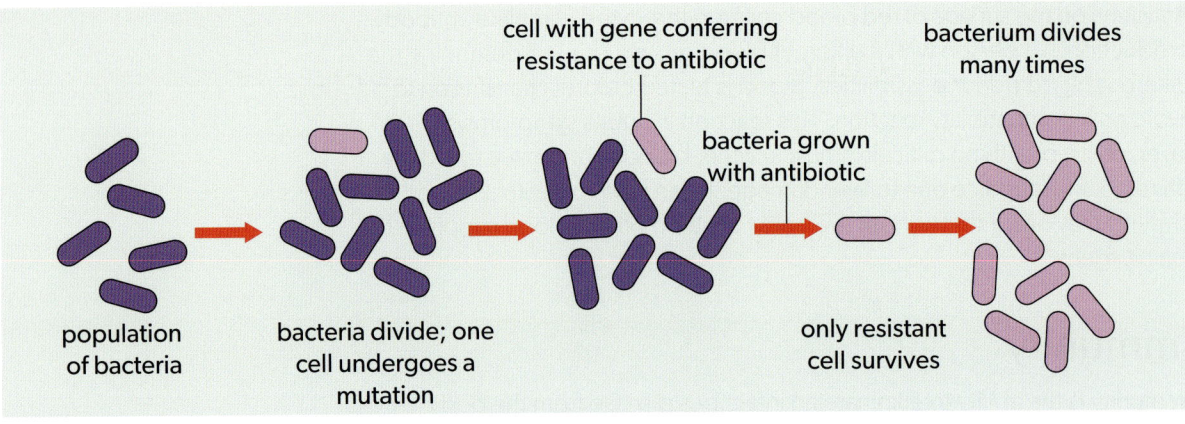

▲ Figure 7 Antibiotic resistance in bacteria

> You studied the causative agent of COVID-19 in subtopic A2.3.

Zoonosis

Zoonoses are animal infections that can reach humans through different modes, for example tuberculosis, rabies and Japanese encephalitis. Tuberculosis is caused by the bacterium *Mycobacterium tuberculosis* and can spread through water droplets liberated while coughing. Cows have a form of tuberculosis caused by *M. bovis* that can be transmitted to humans through dairy products. Rabies, caused by lyssaviruses, produces encephalitis in humans and mammals and can be transmitted through dog bites. Japanese encephalitis is an infection of the brain caused by a virus found in pigs, horses and wild birds, and is spread to humans by mosquitoes. COVID-19 is an infectious disease that has recently transferred from another species, with profound consequences for humans. Since 11 December 2020, a COVID-19 vaccine has been available.

- **Zoonoses** are infectious diseases that can transfer from other species to humans.

Nature of science

The pragmatic theory of truth is that a proposition is true when it corresponds to verified facts. Research paradigms are continually debated—there is no certainty in truths. The current paradigm holds as true that agreed by a specific group of renowned scientists. Scientists have concluded that the benefits of vaccination outweigh any minimal risk of side effects.

Vaccines, immunization and herd immunity

Immunization vaccines contain antigens, or nucleic acids (DNA or RNA) with sequences that code for antigens. Antigens stimulate the development of immunity to a specific pathogen without causing the disease. Herd immunity occurs when the proportion of the people in an area that are immune is enough to avoid transmission. Building herd immunity is important as it will reduce the probability of infection, reinfection and hospitalization, and therefore avoid epidemics. However, the time frame to achieve herd immunity depends upon completing the required and booster vaccine doses, and controlling mutations in vaccination. In global immunization, many countries are arranging vaccination campaigns, and vaccines have been administered exponentially.

Assessment tip

$$\% \text{ difference} = \frac{(\text{initial} - \text{final value})}{(\text{initial} + \text{final value})/2} \times 100$$

$$\% \text{ increase} = \frac{(\text{initial} - \text{final value})}{\text{initial value}} \times 100$$

Example 6

Estimated deaths above normal (excess deaths) are calculated as actual deaths minus normal deaths. The graphs show excess deaths per week in two regions of the USA during the COVID-19 pandemic, from 1 March to 25 July 2020.

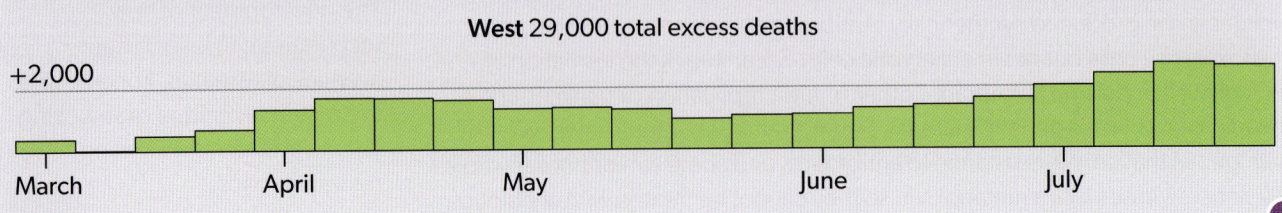

C3.2 Defence against disease

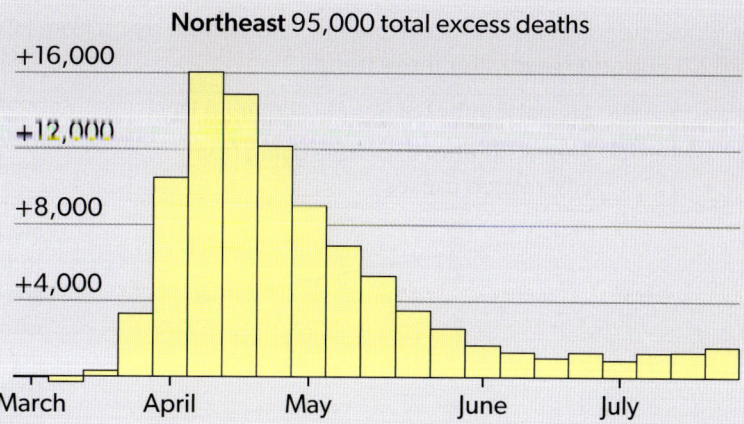

Source: US Centres for Disease Control and Prevention

a) Calculate the percentage difference in total excess deaths between the West and Northeast regions during the period shown.

b) Calculate the percentage increase in deaths between the last week in March and the first week in April 2020 in the Northeast region.

c) Suggest one reason for the trend in the Northeast region.

Solution

a) Difference = 95,000 − 29,000 = 66,000

$$\text{Mean} = \frac{(95{,}000 + 29{,}000)}{2} = 62{,}000$$

$$\% \text{ difference} = \frac{66{,}000}{62{,}000} \times 100 = 106\%$$

b) Difference = 10,500 − 3,000 = 7,500 deaths

$$\% \text{ increase} = \frac{7{,}500}{3{,}000} \times 100 = 250\%$$

c) There was a peak in April–May, as many people did not know about the virus and high population in cities favoured transmission. Herd immunity was reached by July. The use of face masks is another possible answer. Although the vaccine did not appear until December that year, it would have been a good suggestion.

Practice problems

1. The graphs show the relationship between the circadian rhythms of plasma melatonin and waking in sleep.

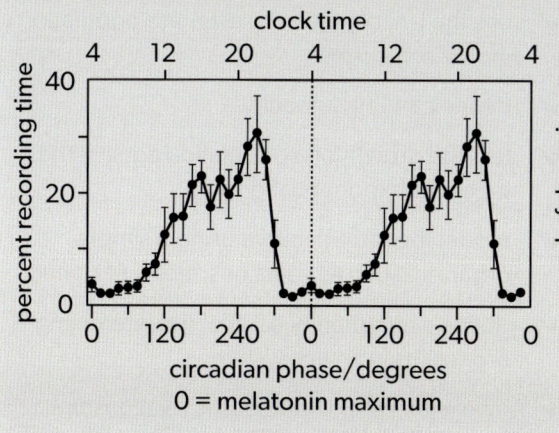 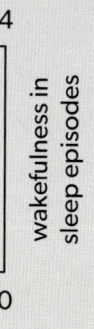 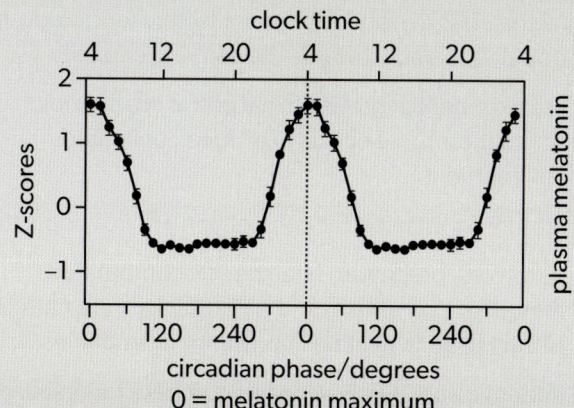

Source: Cajochen, C., et al. (2003). Role of melatonin in the regulation of human circadian rhythms and sleep. *Journal of Neuroendocrinology*, 15(4), 432–437. www.doi.org/10.1046/j.1365-2826.2003.00989.x

a) Outline the trend for plasma melatonin with time of day.

b) Describe the relationship between plasma melatonin and waking in sleep.

c) Doctors suggest taking melatonin at 18:00 clock time of the new destination in case of jet lag. Justify this statement.

Interaction and interdependence

2. The graph shows the global trends in people acquiring HIV and people dying from HIV-related causes between 2010 and 2022. Suggest reasons for these trends.

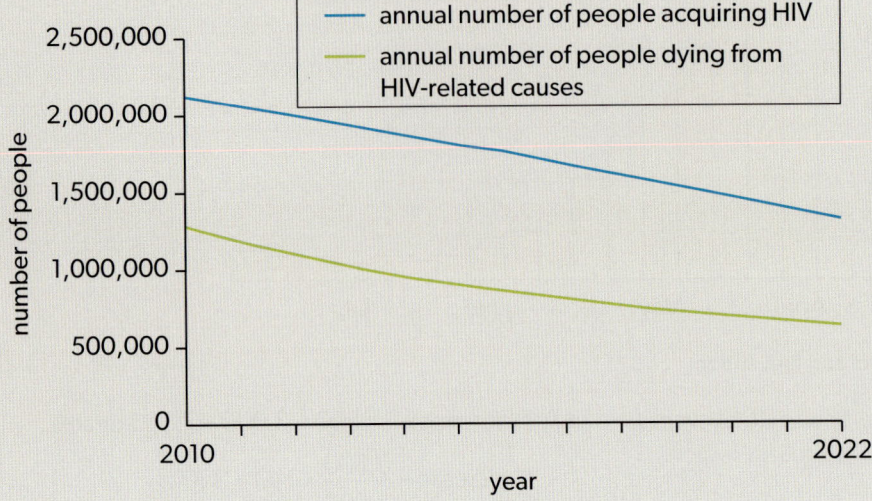

Source: HIV statistics, globally and by WHO region, 2023. (n.d.). https://cdn.who.int/media/docs/default-source/hq-hiv-hepatitis-and-stis-library/j0294-who-hiv-epi-factsheet-v7.pdf?sfvrsn=5cbb3393_7 Data from: Avenir Health using 2025 targets and UNAIDS/WHO epidemiological estimates, 2023 (CC BY-NC-SA 3.0 IGO)

- How do animals protect themselves from threats?
- How can false-positive and false-negative results be avoided in diagnostic tests?

C4.1 Populations and communities

You should know:

- ✓ populations are interacting groups of organisms of the same species living in an area.
- ✓ sampling techniques are used to estimate population size.
- ✓ population may fluctuate due to density-independent factors, but density-dependent factors tend to push the population back towards the carrying capacity.
- ✓ the exponential growth initial phase in a sigmoid population curve occurs in an ideal, unlimited environment.
- ✓ reasons for intraspecific competition within a population.
- ✓ herbivory, predation, interspecific competition, mutualism, parasitism and pathogenicity are categories of interspecific relationship within communities.
- ✓ endemic and invasive species compete for resources.
- ✓ predator–prey relationships are an example of density-dependent control of animal populations.
- ✓ limiting factors can be top-down or bottom-up.
- ✓ allelopathy and secretion of antibiotics deter competition.

You should be able to:

- ✓ explain standard deviation of a mean.
- ✓ describe the use of the capture–mark–release–recapture method to estimate the population size of an animal species.
- ✓ use the Lincoln index to estimate population size for motile organisms.
- ✓ define carrying capacity.
- ✓ test the growth of a population against the model of exponential growth.
- ✓ model the growth curve using a simple organism such as yeast or species of *Lemna*.
- ✓ define a community as all of the interacting organisms in an ecosystem.
- ✓ give examples of mutualistic relationships.
- ✓ use the chi-squared test to determine association between two species.
- ✓ describe approaches for testing for interspecific competition.

Populations

Populations are interacting groups of organisms of the same species living in an area. Members of a population normally breed. Reproductive isolation is used to distinguish one population of a species from another.

Estimating population size

Population size can be estimated by different procedures. Measuring an entire population by counting each organism takes a long time, so methods such as random sampling can be used. This method is ideal for counting sessile organisms. In random quadrat sampling, a 1 m² square is randomly thrown and the organisms of each species inside the square are counted and recorded. This is repeated many times in different locations—the more times it is done the better. The population density of each species is calculated as the mean per quadrat (individuals per m²) plus/minus the standard deviation. This gives an idea of the variation and how evenly the population is spread.

The **Lincoln index** is calculated from the capture–mark–release–recapture method.

$$\text{population size estimate} = M \times \frac{N}{R}$$

where M is the number of individuals caught and marked initially, N is the total number of individuals caught in the second sample and R is the number of marked individuals recaptured.

Assumptions need to be made when using this index: the population does not have migrations; the time between sampling is shorter than the lifetime of the organisms; and the marked individuals will distribute randomly after being released. Sampling must be random, so all organisms have the same chance of being caught. The mark should not affect the animal in any way, for example, it should not make the animal easier to be seen by predators or the person taking the samples. The marks must be permanent, so they are not lost.

Carrying capacity

Carrying capacity is the maximum number of individuals of a species that can exist in a habitat without affecting other species in that habitat. It is determined by the competition with other species for limited resources such as food, water and cover, and by predators.

Negative feedback control of population size

Numbers of individuals in a population may fluctuate due to density-independent factors such as food limitation, pollutants and climate extremes. Populations tend to increase due to reproduction, but density-dependent factors such as competition, predation and disease tend to push the population back towards the carrying capacity by negative feedback.

Population growth curves

Exponential population growth occurs in an ideal, unlimited environment. The phases shown in the **sigmoid growth curve** can be explained by relative rates of natality, mortality, immigration and emigration. In the sigmoid curve, population growth is exponential and then slows until the population reaches the carrying capacity of the environment.

- **Populations** are interacting groups of organisms of the same species living in an area that usually interbreed.

▲ Figure 1 Random sampling using a quadrat

- The **Lincoln index** uses data obtained from the capture–mark–release–recapture method to estimate the size of a population.

Nature of science

Sampling methods give estimates of the population size, not exact values. The difference between the estimate of the population size and the true size of the whole population is the sampling error. One way to reduce this error is by using large random samples.

- **Carrying capacity** is the maximum population an environment can hold due to food, territory or breeding sites.

- A **sigmoid growth curve** shows changes in population over time.

Interaction and interdependence

Nature of science

Population growth curves show population size, using a logarithmic scale, on the vertical axis and time on the horizontal axis. They represent idealized graphical models that are often simplifications of complex systems.

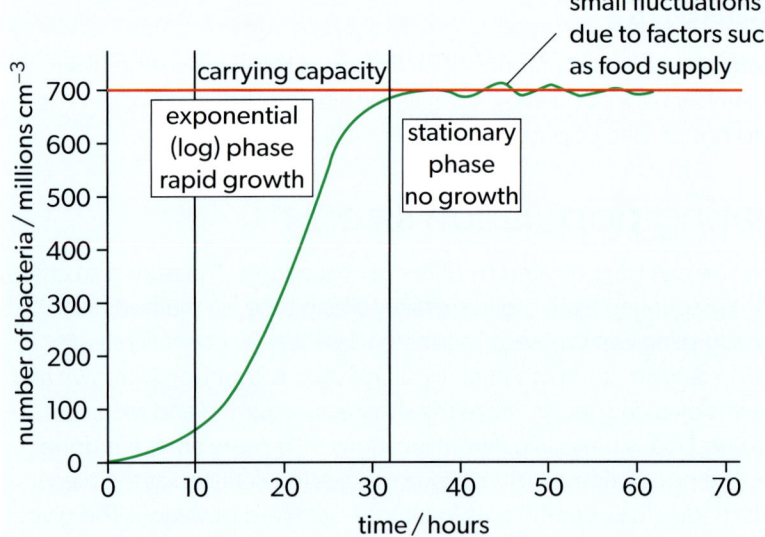

▲ Figure 2 Sigmoid growth curve

Example 1

What can cause a log phase in a population growth curve?

A. Adjusting to a new environment.
B. Reaching the carrying capacity.
C. Excess of food in the environment.
D. There is metabolic activity but not growth.

Solution

The correct answer is **C**, as the population is growing exponentially due to excess of food. Carrying capacity is reached in the stationary phase. When organisms adjust to a new environment, they take time to grow.

Example 2

Duckweed (*Lemna aequinoctialis*) is a small floating plant that sometimes forms a continuous cover over the surface of ponds. In an experiment, it was grown either in nutrient-rich medium or in sewage water for four weeks. The biomass obtained per square metre at different times is shown in the population growth curves.

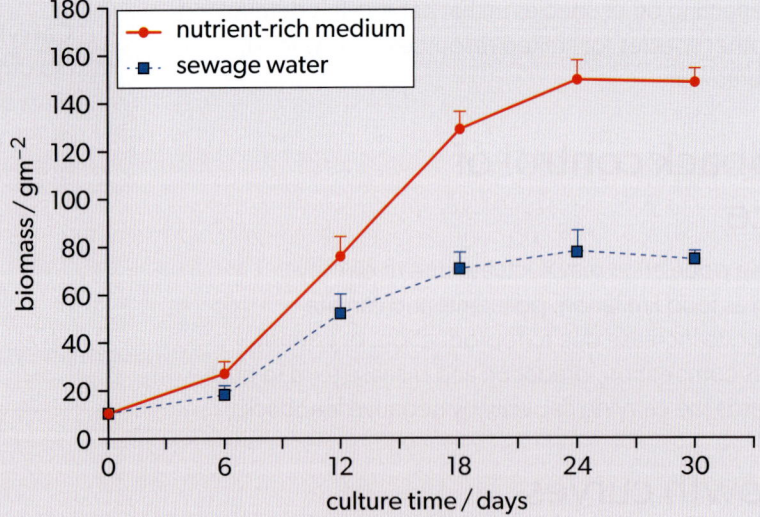

a) Label two phases on the nutrient-rich medium population growth curve.
b) Compare the population growth curve in nutrient-rich medium and sewage water.
c) State two factors that limit the duckweed population size after 24 days.

Solution

a) "Exponential/log" between 6 and 18 days. "Plateau" or "stationary growth" between 24 and 30 days.

b) Both are sigmoid curves. At all times the graph for nutrient-rich medium is higher than for sewage, showing that duckweed grows more in nutrient-rich medium than in sewage water. The final growth is higher for nutrient-rich medium as growth reaches a plateau later in nutrient-rich medium (earlier in sewage water).

c) The factors that could limit duckweed population size are light, space, food, pH, temperature, oxygen or carbon dioxide, disease and predation (by herbivores).

Intraspecific relationships

Populations of a species can **compete** or cooperate with each other. **Intraspecific competition** is when organisms of the same species compete for food, water, sunlight, breeding sites or territory. **Cooperation** is when organisms help each other, for example, the flycatcher bird (*Ficedula hypoleuca*) alerts other flycatchers by screeching when a predator is near. When other flycatchers hear the alert they get together to chase the predator away.

Interspecific relationships

Each species plays a unique role within a **community** because of the unique combination of its spatial habitat and interactions with other species. Interactions between species in a community can be classified according to their effect. These interactions can be **interspecific competition**, **predation**, **herbivory**, **mutualism**, **parasitism** and **pathogenicity**. Two species cannot survive indefinitely in the same habitat if their niches are identical, due to competitive exclusion.

- **Competition** is when one organism or species contends for the same resources as another.
- **Intraspecific competition** is between organisms of the same species.
- **Cooperation** is when organisms work together for mutual benefit.
- **Communities** comprise all the interacting organisms in an ecosystem.

- **Interspecific competition** is between organisms of different species.
- **Predation** is when an organism (predator) kills prey to feed on it.
- **Herbivory** is when herbivores feed on plants or algae.
- **Mutualism** is when both organisms benefit; e.g. birds feeding on ectoparasites of cattle.
- **Parasitism** is when an organism lives on/in a host and feeds from it, causing the host harm; e.g. worms in animal intestines.
- **Pathogenicity** is the ability of an organism to cause disease.

Example 3

Describe three mutualistic relationships.

Solution

Organisms	Relationship
Fabaceae (legumes) and *Rhizobium* bacteria	*Rhizobium* produce nitrogen and roots of legumes give protection.
orchids and mycorrhizae fungus	Mycorrhizae supply nutrients to germinate and grow, and orchids provide carbohydrates through photosynthesis.
corals and zooxanthellae	Corals provide habitat and zooxanthellae produce sugars through photosynthesis.

Example 4

Many traits vary with the presence of neighbours within plant communities. To explore the relationship between trait responses and competitive outcomes, scientists grew 15 species alone, 15 in a monoculture (mono) and 15 in a mixture (mix). They selected species to represent a range of functional types, including three grasses, three legumes and nine forbs. They measured (a) average leaf area, (b) root to shoot ratios and (c) specific root length. Error bars represent one standard error.

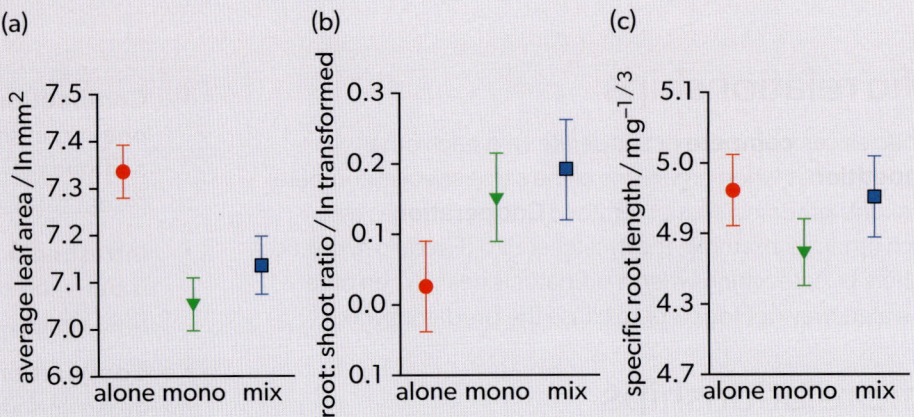

Source: Bennett, J. A., et al. (2016). The reciprocal relationship between competition and intraspecific trait variation. *Journal of Ecology*, **104**, 1410–1420. www.doi.org/10.1111/1365-2745.12614. Data from: Dryad Digital Repository http://dx.doi.org/10.5061/dryad.hg578 (CC0 1.0)

a) State the mean average leaf area for the monoculture group.
b) Compare and contrast the root : shoot ratio in the three experimental groups.
c) Identify the experimental group that tests intraspecific differences.
d) Switching to more stress-tolerant strategies, such as increasing the diameter of plant roots, reduced interspecific competition. Comment on this statement.

Solution

a) 7.05 ln mm^2
b) Alone has a smaller ratio than mono and significantly smaller than mix. Mix is slightly higher than mono, but not significantly different.
c) The mono group, as organisms of the same species are grown together.
d) The mix group gives an idea of interspecific competition. The root : shoot ratio is highest in the mix group, but the root length is similar to alone, therefore the increase in the ratio is probably due to an increase in diameter.

Resource competition and testing for interspecific competition

A typical case of interspecific competition is when invasive species compete with endemic species for resources. One of the most well-known invasive species was introduced as a biological control. The cane toad (*Rhinella marina*) was introduced to Australia to control the cane beetle (*Dermolepida albohirtum*). The cane toad and their tadpoles are highly toxic to most animals if ingested. As they do not have natural predators, they have become a competitor for food resources and have spread throughout the country. Decreases in the populations of certain species of frogs (*Limnodynastes peronii*), large lizards (*Bellatorias major*, *Intellagama lesueurii* and *Varanus varius*) and snakes (*Pseudechis porphyriacus*) occurred following the introduction of *R. marina*.

C4.1 Populations and communities

Example 5

Paramecium aurelia and *Paramecium caudatum* are single-cell organisms. They were grown separately and together. The growth curves are shown as a function of volume according to time.

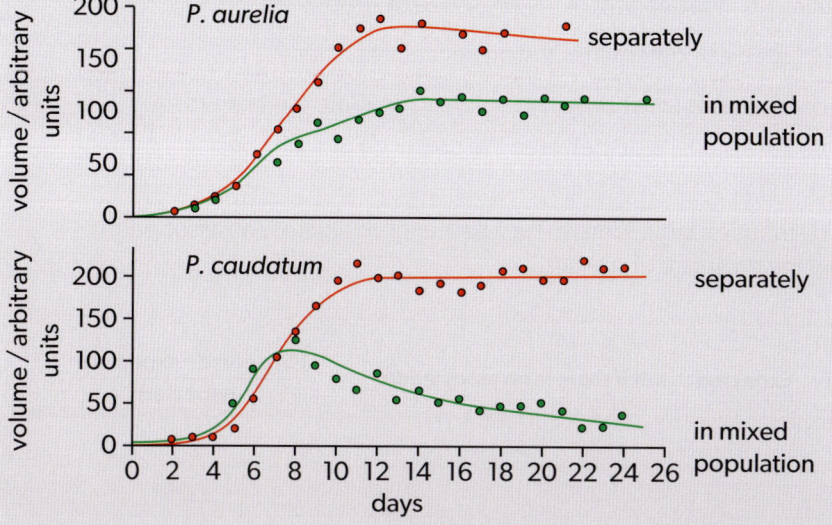

Explain the results shown in this experiment.

Solution

Two species cannot survive indefinitely in the same habitat if their niches are identical. *Paramecia* compete for food and space. *P. aurelia* is better suited than *P. caudatum*, which starts to disappear after 6–8 days in the mixed population. This is an example of interspecific competitive exclusion.

You studied fundamental and realized niches in subtopic B4.2.

Source: GAUSE, G. F. (1936). The Struggle for Existence. Soil Science, 41(2), 159. https://doi.org/10.1097/00010694-193602000-00018

Nature of science

Hypotheses can be tested by both experiments and observations. Experiments do not always take into consideration all the possible variables but are easier to monitor.

Chi-squared test

The chi-squared (χ^2) test can be used to determine whether there is a significant difference between the expected frequencies (E) and the observed frequencies (O) in one or more categories. For example, if you want to see if the distributions of two species are independent, you can use a chi-squared test. You should base your sampling on random numbers and in each quadrat record the presence or absence of the chosen species. The null hypothesis (also called H_0) is that the presence of one species is random in relation to the presence of the other species. The alternative hypothesis (also called H_a) is that the presence of one species is associated with or dependent on the presence or absence of the other species. The formula used to calculate the chi-squared test is:

$$\chi^2 = \sum \frac{(O-E)^2}{E}$$

The alternative hypothesis is accepted (and the null hypothesis is rejected) if the difference between the observed results and expected results is statistically significant (with a $p < 0.05$, where p is the probability). This means that if the calculated chi-squared result is higher than a tabulated chi-squared (critical value), it supports the association between the two species.

163

Interaction and interdependence

Example 6

On a field trip, a student used a quadrat to count the presence of two species (1 and 2) in an ecosystem. The student performed a chi-squared test to determine whether there is association between the presence of one species and the presence of the second species.

H_0 (null hypothesis) = The presence of species 1 is independent of the presence of species 2.

H_a (alternative hypothesis) = The presence of species 1 determines the presence of species 2.

a) The results for the observed numbers of organisms are shown in the table. Complete the table to:
   i) identify the expected results if there is no association between the presence of each species
   ii) calculate the chi-squared results.
b) Using the results and the chi-squared distribution table, determine whether or not there is an association between the presence of species 1 and the presence of species 2.

Solution

Species	Number of organisms observed	Expected result if there is no association	$\dfrac{(\text{Observed}-\text{expected})^2}{\text{expected}}$
1 only	3	2.5	0.1
2 only	1	2.5	2.25
1 and 2	4	2.5	2.25
none	2	2.5	0.1
total	10	10	4.7

	Probability			
Degrees of freedom	0.99	0.95	0.05	0.01
1	0.000	0.004	3.841	6.635
2	0.020	0.103	5.991	9.210
3	0.115	0.352	7.815	11.345
4	0.297	0.711	9.488	13.277
5	0.554	1.145	11.070	15.086

b) The degrees of freedom in this study is n − 1, therefore 3 (as there are 4 possible scenarios of the presence or not of species 1 and 2). The critical value for $p = 0.05$ is 7.815 (with a significance level of 5%). The calculated chi-squared is 4.7. Because the calculated value is smaller than the critical value, the alternative hypothesis is rejected and the null hypothesis accepted. This means there is no association between the presence of the two species.

Density-dependent, top-down and bottom-up control of populations

- **Density-independent** population limiting factors are food, pollutants, abiotic environment and climate.
- **Density-dependent** population limiting factors are predation, competition and disease.

C4.1 Populations and communities

Example 7

The interaction web shows top-down and bottom-up effects in a seagrass study system.

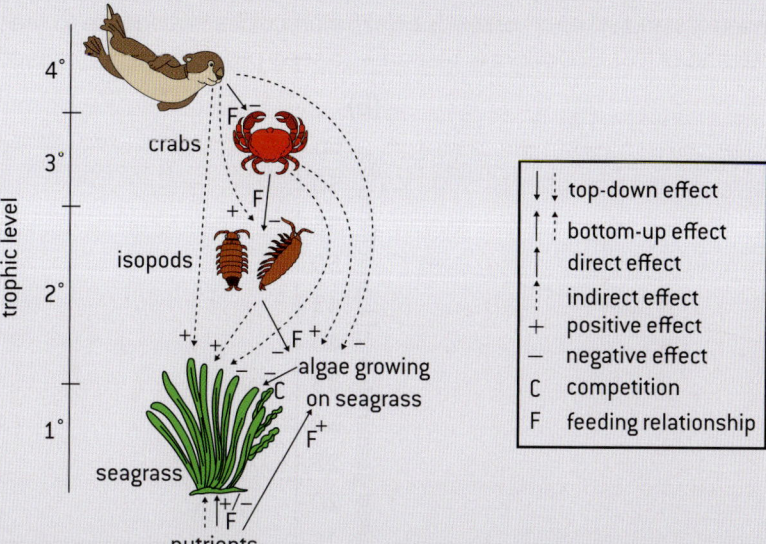

Source: Hughes, B. B., et al. (2013), Recovery of a top predator mediates negative eutrophic effects on seagrass. Proceedings of the National Academy of Sciences, 110(38), 15313–15318. www.doi.org/10.1073/pnas.1302805110

a) Identify which of the types of interaction shown in the key occur between crabs and seagrass.
b) Describe how the sea otter indirectly affects seagrass.
c) Explain how nutrients can have a positive or negative bottom-up effect on seagrass.

Solution

a) Top-down indirect negative effect. Algal epiphytes compete with seagrass for nutrients. The isopods modulate the negative algal epiphyte effects on seagrass by grazing on them. Predator crab populations predate on isopods, diminishing their grazing effects on algal epiphytes, therefore increasing algal competition with seagrass.

b) Sea otters have a positive indirect effect as sea otters feed on crabs that feed on isopods that feed on algae. Fewer crabs means more isopods so fewer algae. Fewer algae means more seagrass or that there is less competition between algae and seagrass.

c) The positive bottom-up effect is due to nutrient enrichment. Nutrient enrichment increases abundance of plants and their growth rates. The negative bottom-up effect is that an excess of nutrients causes a bloom in algae growth and therefore competition. Algae cover the sea grass and prevent them from receiving sunlight. Excess of algae can cause seagrass to die by eutrophication.

Allelopathy and secretion of antibiotics

Plants exert positive and negative effects on other organisms through chemical substances, described as allelochemicals—secondary metabolites that plants produce to obtain a competitive advantage. Density in inter and intraspecific plant–plant interactions affects the action modes of **allelopathy**. Likewise, some fungi secrete antibiotics—secondary metabolites that kill or inhibit bacterial growth.

- **Allelopathy** is when one plant reduces the growth of another organism through chemical mechanisms.

Example 8

A study was performed to determine the density-dependence of the intraspecific allelopathic effect among seeds of *Ipomoea murucoides* and *I. pauciflora* through the measure of seed germination. The graph shows the percentage of germination *in vitro* (mean ± standard error) of seeds of (a) *I. murucoides* in response to seed proportions of *I. pauciflora* : *I. murucoides* and of (b) *I. pauciflora* in response to seed proportions of *I. murucoides* : *I. pauciflora*.

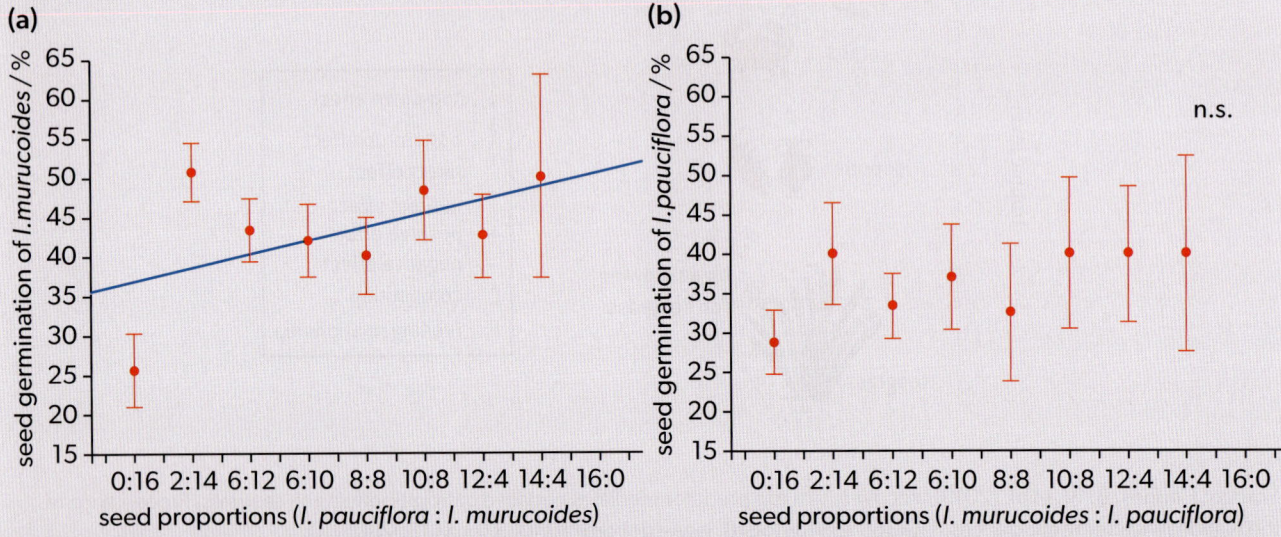

Source: Aguilar-Franco, Z. M., et al. (2019). Density-dependent effect of allelopathy on germination and seedling emergence in two *Ipomoea* species. *Revista Chilena de Historia Natural*, 92, 7. www.doi.org/10.1186/s40693-019-0087-z (CC BY 4.0)

a) State the germination percentage of *I. murucoides* when the proportion of seeds of both species was the same.
b) Compare and contrast the trends for seed germination at different seed proportions for *I. murucoides* and *I. pauciflora*.
c) Deduce whether the *Ipomoea* genus presents allelopathy.

Solution

a) 40%
b) In *I. murucoides* as the *I. pauciflora* proportion increases, the germination increases, while seed germination of *I. pauciflora* was not affected by the increment of seed proportion of *I. murucoides*. Both have a large range of results. *I. murucoides* reached lower germination rate in the absence of *I. pauciflora*.
c) Seed germination of *I. murucoides* was inversely proportional to the proportion of both species, which could be evidence of positive allelopathy by *I. pauciflora* and/or autotoxicity. There is no evidence for allelopathy in *I. pauciflora*.

- What are the benefits of models in studying biology?
- What factors can limit capacity in biological systems?

C4.2 Transfers of energy and matter

You should know:

- ecosystems are open systems in which both energy and matter can enter and exit.
- sunlight is the main source of energy in ecosystems.
- decomposers obtain their energy from dead matter.
- autotrophs convert carbon dioxide into carbohydrates and other carbon compounds.
- light is the external energy source in photoautotrophs and oxidation reactions are the energy source in chemoautotrophs.
- heterotrophs obtain their carbon compounds from other organisms.
- energy is released in both autotrophs and heterotrophs by oxidation of carbon compounds in cell respiration.
- most species occupy different trophic levels in multiple food chains.
- the percentage of ingested energy converted to biomass is dependent on the respiration rate.
- ecosystems are carbon sinks and carbon sources.
- carbon dioxide is produced by the combustion of biomass and fossilized organic matter.
- aerobic respiration depends on atmospheric oxygen produced by photosynthesis, and photosynthesis on atmospheric carbon dioxide produced by respiration.

You should be able to:

- construct and analyse food chains and webs.
- construct and compare pyramids of energy from different ecosystems.
- describe causes of energy loss between trophic levels
- explain why energy transfers are not 100% efficient.
- explain that at each successive stage in food chains there is less biomass, but the energy content per unit mass is not reduced.
- define primary production as the accumulation of carbon compounds in biomass by autotrophs.
- define secondary production as the accumulation of carbon compounds in biomass by heterotrophs.
- calculate primary and secondary production.
- construct a diagram of the carbon cycle.
- analyse the Keeling curve in terms of photosynthesis, respiration and combustion.
- explain that all chemical elements are recycled by living organisms in ecosystems.

Energy sources for ecosystems

An **ecosystem** is a community made up of living organisms and its interaction with the abiotic environment such as air, water and mineral soil. Ecosystems are open systems in which both energy and matter can enter and exit. Light is the main source of energy in most ecosystems, but there are exceptions such as ecosystems in caves where light cannot penetrate.

> - An **ecosystem** is a community and its interactions with the abiotic environment.

Food chains and food webs

Living organisms in an ecosystem can interconvert energy. Light energy can be converted into chemical energy, which can be converted into kinetic energy. Chemical energy can also be converted into heat energy. Heat energy cannot be converted into any other form of energy, therefore it is lost from the ecosystem.

Flow of chemical energy is through food chains. Organisms are classified into trophic levels. Producers are autotrophs that use light energy and transform it into chemical energy. Primary consumers are usually herbivores that feed on producers. Secondary consumers feed on primary consumers and are usually carnivores. Tertiary consumers feed on secondary consumers. Most species

Nature of science

Laws in science are generalized principles, or rules of thumb, formulated to describe patterns observed in nature. Laws are different from theories in that they do not offer explanations, but instead they describe phenomena. Like theories, laws can be used to make predictions.

Interaction and interdependence

have a varied diet so occupy different trophic levels in multiple food chains, therefore it is not very practical to show these relationships in a chain.
A food web is a more realistic model as it shows all the possible food chains in a community.

Sample student answer

The image shows a food web.

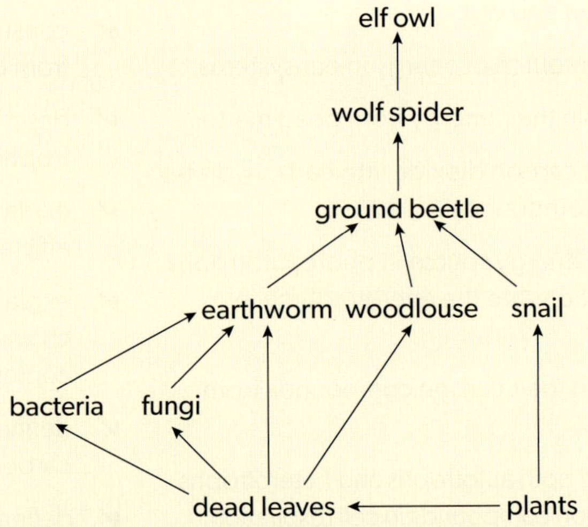

```
            elf owl
              ↑
          wolf spider
              ↑
          ground beetle
           ↑   ↑   ↑
     earthworm woodlouse snail
       ↑ ↑      ↑        ↑
  bacteria fungi
      ↑   ↑
      dead leaves  ←  plants
```

a) State the source of energy of this food web. [1]

This answer could have achieved 1/1 marks:

Sunlight

b) Identify a producer. [1]

This answer could have achieved 1/1 marks:

Plants and dead leaves.

▼ The answer is plants. The mark is still given even though the student mentioned dead leaves because they form part of the plant, but this answer should not have been here.

c) Identify a secondary consumer. [1]

This answer could have achieved 0/1 marks:

Woodlouse

▼ The woodlouse only feeds on dead leaves, so they are detritivores. The secondary consumers (carnivores) are ground beetles. Earthworms are decomposers.

d) Predict the consequences of the removal of all wolf spiders from this ecosystem. [3]

This answer could have achieved 3/3 marks:

As the ground beetles are not predated, they will increase in numbers, eating more earthworms, woodlice and snails. These would all decrease, so plants will be eaten less.

You studied the different types of nutrition in subtopic B4.2.

Pyramids of energy

Pyramids of energy model the energy flow through ecosystems. Energy is lost from one trophic level to the other. Most of the energy is lost as heat from respiration, both when ATP is produced and when it is used in cells. The percentage of ingested energy converted to biomass is dependent on the respiration rate. Energy can also be lost as indigestible material (for example, cellulose and fibre for humans) and as uneaten parts such as hair and teeth. Energy is also lost by **decomposers** and detritivores. There are restrictions on the number of trophic levels in ecosystems due to this energy loss. At each successive stage in food chains there are fewer organisms or smaller organisms. There is therefore less biomass, but the energy content per unit mass is not reduced. The loss of energy from one trophic level to the next can be shown with a pyramid of energy. The size of each bar should be proportional to the amount of energy in each trophic level.

tertiary consumers
secondary consumers
primary consumers
producer

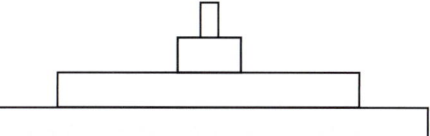

▲ Figure 3 Pyramid of energy

- **Decomposers** recycle elements needed by living organisms.
- **Primary production** is the accumulation of carbon compounds in biomass by autotrophs.
- **Secondary production** is the accumulation of carbon compounds in biomass by heterotrophs.
- A **carbon sink** (also known as a carbon reservoir) is a reserve of carbon.
- A **carbon flux** is the transfer of carbon from one carbon sink to another.

Primary and secondary production, and the carbon cycle

Primary production is the accumulation of carbon compounds in the biomass of autotrophs. It is measured in mass per unit area per unit time (for example, $g\,m^{-2}\,year^{-1}$). Biomass accumulates when autotrophs and heterotrophs grow or reproduce. Biomes vary in their capacity to accumulate biomass. **Secondary production** is the accumulation of carbon compounds in biomass by heterotrophs. Due to loss of biomass when carbon compounds are converted to carbon dioxide and water in cell respiration, secondary production in an ecosystem is lower than primary production. All elements used by living organisms, not just carbon, are recycled, especially by decomposers.

The continued availability of carbon in ecosystems depends on carbon cycling. The cycle shows the processes that transfer carbon from one **carbon sink** to another. **Carbon fluxes** should be measured in gigatonnes, which are $1 \times 10^{15}\,g$ (1,000,000,000,000,000 grams).

Nature of science

Monitoring of carbon dioxide levels was started by David Keeling at Mauna Loa Observatory in Hawaii. The continuous monitoring of the atmospheric greenhouse gases provides important information for policymakers in matters concerning global warming.

Example 9

The diagram shows part of the carbon cycle.

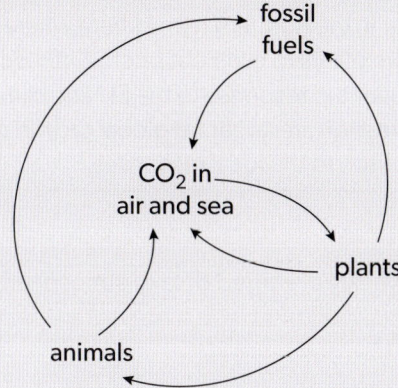

a) State one process by which carbon dioxide is transferred:
   (i) from fossil fuels to the air.
   (ii) from plants to fossil fuels.
   (iii) from air to plants.

b) The carbon flux from plants and animals to carbon dioxide in the air and sea is about 120 gigatonnes per year. Explain the process by which this carbon is transferred to the air.

c) On the cycle, draw a box showing the carbon sink of decomposers and the carbon flux due to decomposition leading to the carbon sink of decomposers.

d) Peat is a carbon sink. Describe the formation of peat.

Interaction and interdependence

Solution

(a) (i) Combustion
(ii) Fossilization
(iii) Photosynthesis

b) Carbon dioxide is produced by plants and animals by the process of respiration. Respiration involves many chemical reactions that convert food into energy. The reactions are catabolic, as they involve the breakdown of glucose to produce energy in the form of ATP. This process occurs in the mitochondria of cells. This carbon dioxide diffuses out of cells and passes into the atmosphere, or into the water in the case of aquatic plants. In aquatic habitats carbon dioxide dissolves, forming hydrogencarbonate ions.

c) A box (added anywhere in the cycle) labelled decomposers with arrows from animals and from plants to decomposers (usually in soil).

d) Waterlogged soils do not have oxygen, as all air gaps are full of water. In these soils, anaerobic respiration occurs, resulting in acid conditions. Saprotrophic bacteria and fungi are unable to function in acidic conditions, so full decomposition is inhibited, resulting in peat formation.

Example 10

The graph shows the weekly averaged data fCO_2 from the Mauna Loa station (the **Keeling curve**) in a comparison with globally (and weekly) averaged data gCO_2 from the four observatories in Barrow (Alaska), Mauna Loa (Hawaii), American Samoa and the South Pole (Antarctica) with data from 2010 to 2020.

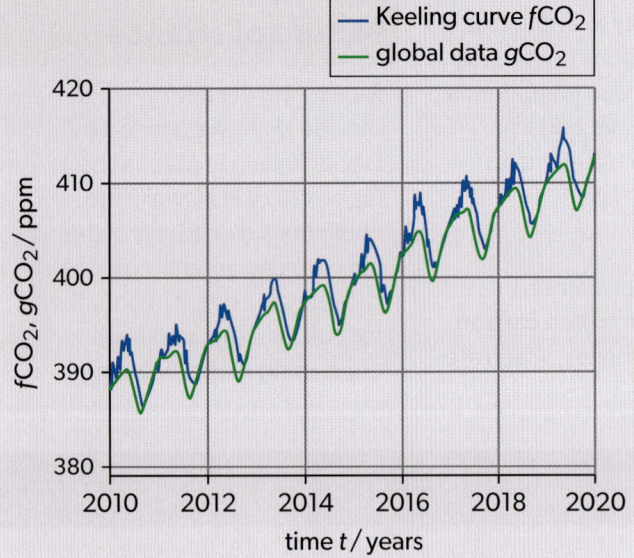

Source: Nordebo, S., et al. (2020). Estimating the short-time rate of change in the trend of the Keeling curve. *Scientific Reports*, **10**, 21222. www.doi.org/10.1038/s41598-020-77921-2 (CC BY 4.0)

a) Compare and contrast the trends in the Keeling curve and in global data.

b) Suggest reasons for these trends.

Solution

a) Both show fluctuating, rapidly increasing carbon dioxide levels. The timing of the annual variability for global data is very similar to the Keeling curve, although somewhat ahead. The increasing rate of the global trend seems to be slightly delayed with respect to the Keeling trend.

b) The increase in atmospheric carbon dioxide is due to the release of carbon dioxide into the atmosphere during combustion of biomass, peat, coal, oil and natural gas. If photosynthesis exceeds respiration there is a net uptake of carbon dioxide and if respiration exceeds photosynthesis there is a net release of carbon dioxide. Carbon sinks vary in date of formation. Combustion following lightning strikes sometimes happens naturally, but it is usually human activities that have greatly increased combustion rates. The fluctuations correspond to the seasonal change in uptake of carbon dioxide by plants. Levels decrease during the spring and summer in the northern hemisphere where there are more plants to carry out photosynthesis. Fluctuations in response to the El Niño climate phenomenon are larger for global growth rates in comparison to the local responses.

- A **Keeling curve** shows weekly averaged data of atmospheric carbon dioxide.

Example 11

The global carbon cycle involves sinks where carbon is stored, and fluxes where carbon is transferred. What are the largest sink and flux?

	Sink	Flux
A.	atmosphere	combustion
B.	oceans	respiration
C.	oceans	photosynthesis
D.	limestone	photosynthesis

Solution

The correct answer is **C**, as the oceans with their photosynthetic algae hold the greatest amount of carbon biomass. Photosynthesis is the largest flux of carbon, as it is taken in to form sugars and turned into biomass.

Practice problems

1. Describe how cooperation affects the chances of survival.

2. Bream (*Lepomis macrochirus*) are recent invaders of the environment where the sunfish (*Lepomis gibbosus*) lives. Suggest, with a reason, whether the bream and the sunfish show intraspecific competition.

3. Scientists investigated the antimicrobial effects of penicillic acid extracted from the fungus *Chaetomium elatum*. Ampicillin, gentamicin and fluconazole were used as a positive control for Gram-positive bacteria, Gram-negative bacteria and fungi, respectively. Discs with antimicrobials were placed on the surface of agar plates that had been previously inoculated with the different microorganisms. Following the incubation period, the diameter of the inhibition zone (including the disc diameter) was measured and recorded. This process was repeated three times in independent assays, and an average zone of inhibition was calculated from the three replicates.

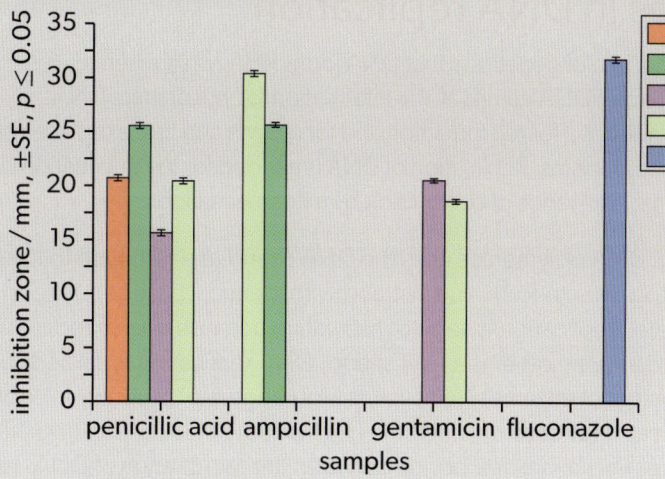

Source: Eshboev, F., et al. (2023). Antimicrobial and cytotoxic activities of the secondary metabolites of endophytic fungi isolated from the medicinal plant *Hyssopus officinalis*. Antibiotics, **12**, (7) 1201. https://doi.org/10.3390/antibiotics12071201 (CC BY 4.0)

a) State the inhibition zone of penicillic acid on *Bacillus subtilis*.

b) Identify the organism not affected by penicillic acid.

c) Suggest one reason fluconazole only affected one organism.

d) Comment on the efficacy of penicillic acid as an antimicrobial.

4. Explain how carbon compounds serve as sources of energy to organisms in different trophic levels of a food chain.

- What are the direct and indirect consequences of rising carbon dioxide levels in the atmosphere?
- How does the transformation of energy from one form to another make biological processes possible?

D Continuity and change

D.1.1 DNA replication

You should know:
- ✔ DNA replication produces exact copies of DNA with identical base sequences.
- ✔ DNA replication is semi-conservative and depends on complementary base pairing.
- ✔ PCR can be used to amplify small amounts of DNA.
- ✔ gel electrophoresis separates proteins or fragments of DNA according to size.

Additional higher level:
- ✔ DNA polymerases add the 5' of a DNA nucleotide to the 3' end of a strand of nucleotides.
- ✔ the proofreading role of DNA polymerase III.

You should be able to:
- ✔ explain the roles of helicase and DNA polymerase in DNA replication.
- ✔ explain how Taq DNA polymerase produces multiple copies of DNA by PCR.
- ✔ analyse DNA profiles.

Additional higher level:
- ✔ compare DNA replication on the leading and the lagging strand.
- ✔ explain DNA replication (only in prokaryotes), describing the functions of DNA primase, DNA polymerases I and III, and DNA ligase.

Enzymes in DNA replication

The central principle of genetics is that DNA forms RNA, which then makes proteins. Therefore, DNA carries all the information for inheritance. DNA **replication** is required for reproduction and for growth and tissue replacement in multicellular organisms. Replication of DNA is needed for the duplication of the genetic material in advance of cell division by mitosis or meiosis.

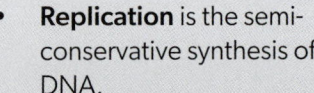

- **Replication** is the semi-conservative synthesis of DNA.
- **Helicase** breaks hydrogen bonds between strands.

In replication, DNA polymerase uses the DNA strand as a template to link nucleotides together to form a new strand with identical base sequences as the template. Replication is semi-conservative because one strand is an original strand and the other is a new strand. DNA replication is carried out by a complex system of enzymes that allow a high degree of accuracy in copying base sequences. **Helicase** unwinds the DNA double helix, separating the strands by breaking hydrogen bonds between the two strands of DNA. Next, DNA polymerase starts replication by adding nucleotides, thus synthesizing the complementary strand. If in the original strand there is an adenine, the base that is joined to the new strand is a thymine.

You will study transcription and gene expression in more detail in subtopic D1.2.

Example 1

Which statement is related to DNA replication?
A. It always requires Taq DNA polymerase to occur.
B. DNA polymerase is an enzyme that separates the DNA strands.
C. New nucleotide bases attach to the original sugar–phosphate backbone.
D. The new DNA contains one original strand and one new strand.

Solution

The correct answer is **D**, as replication is semi-conservative. Taq DNA polymerase is a bacterial enzyme that is heat resistant, which makes it ideal for use in polymerase chain reaction (PCR). Helicase is the enzyme that separates the DNA strands. The original strand acts as a template and nucleotides attach by complementary base pairing by hydrogen bonds—they do not attach to the sugar–phosphate backbone.

D.1.1 DNA replication

In 1958, Meselson and Stahl designed an experiment to test the semi-conservative replication of DNA. They grew bacteria in a medium with heavy nitrogen ^{15}N. After a few generations they transferred the bacteria to a ^{14}N medium, where they obtained bacteria from several generations. They separated their DNA according to size using density gradient centrifugation.

> **Assessment tip**
> To identify the DNA sequence, you must find the complementary base: for A it is T, and for C it is G. Remember that in DNA you have T instead of U.

◀ **Figure 1** Semi-conservative replication of DNA

Source: Meselson, M. & Stahl, F. W. (1958). The replication of DNA in *Escherichia coli*. *Proceedings of the National Academy of Sciences*, **44** (7), 671–682. www.doi.org/10.1073/pnas.44.7.671

PCR and gel electrophoresis

Polymerase chain reaction (PCR) is a technique by which small amounts of DNA can be amplified into large quantities. The DNA fragments obtained by this method can be separated by size in gel electrophoresis. Because DNA has a negative charge, the fragments move towards the positive electrode. The smaller fragments migrate further than the larger ones. A marker with known sizes of DNA fragments is used to compare sizes.

- **Polymerase chain reaction (PCR)** uses Taq polymerase to amplify DNA.

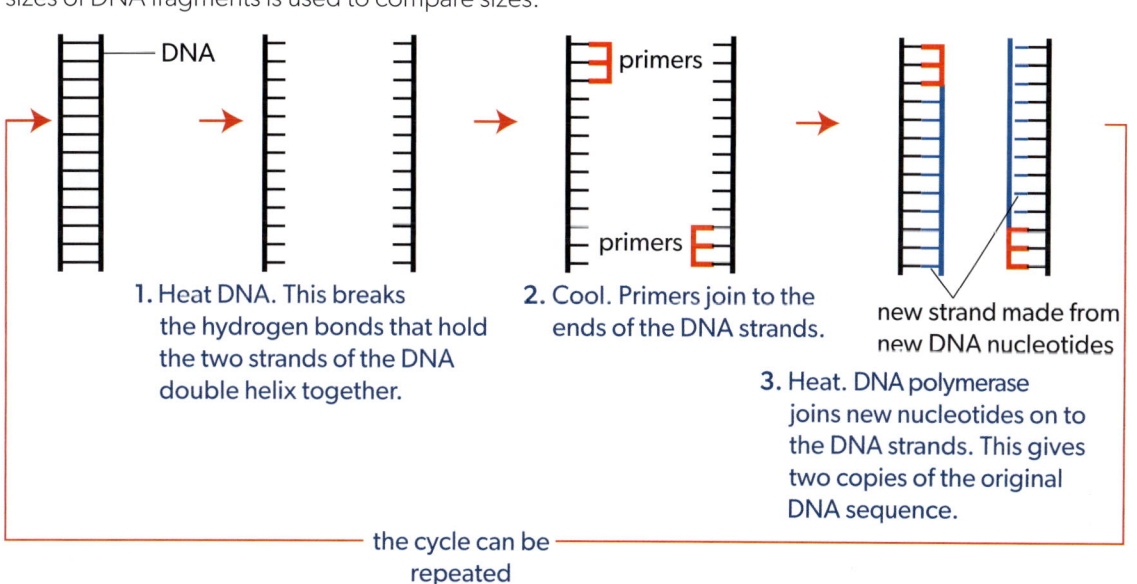

1. Heat DNA. This breaks the hydrogen bonds that hold the two strands of the DNA double helix together.
2. Cool. Primers join to the ends of the DNA strands.
3. Heat. DNA polymerase joins new nucleotides on to the DNA strands. This gives two copies of the original DNA sequence.

the cycle can be repeated

◀ **Figure 2** Polymerase chain reaction (PCR) cycle

173

Continuity and change

Sample student answer

PCR was performed to amplify a small amount of DNA. Eight tubes were prepared as shown in the table.

	Mix of nucleotides, salts, buffer and polymerase	DNA	Primers
Control	✓	✓	✗
Control	✓	✗	✓
Six tubes	✓	✓	✓

The tubes were placed in a thermal cycler with the temperatures shown in the diagram and run for 25 cycles.

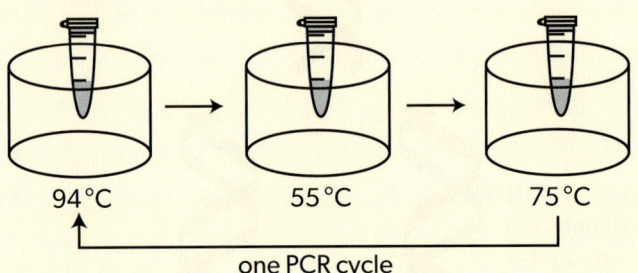

The image shows the result of gel electrophoresis on the eight samples.

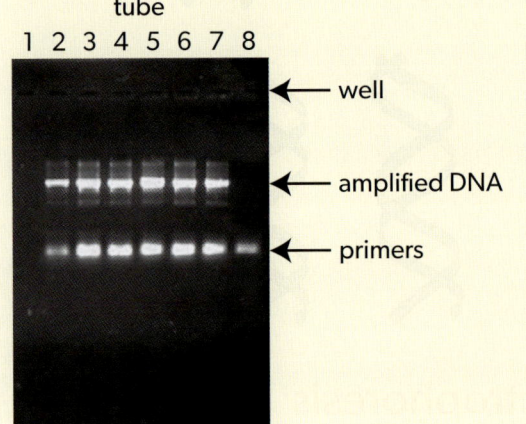

▼ The answer is 8. Tube 1 has no primers, so even if it had contained DNA, no bands would have been seen.

a) State the number of the tube used as a control without DNA. [1]
   This answer could have achieved 0/1 marks:

   Tube 1

▼ A short description of at least two temperatures is expected. Heat (94°C) separates the DNA strands by breaking hydrogen bonds. Although the terminology "denature" is used for this process, the answer has not specified at which temperature this occurs. Cooling (55°C) allows complementary primers to bind to the template by annealing. Heating again (72°C) is optimum for DNA Taq polymerase to add nucleotides, extending the DNA.

b) Explain the reason for changing the temperature during each cycle. [2]
   This answer could have achieved 0/2 marks:

   To denature DNA.

c) Predict the result that would be obtained if fewer cycles were used in this PCR process. [1]
   This answer could have achieved 1/1 marks:

   Bands of DNA much lighter.

▲ This answer is correct, because fewer cycles will lead to less amplification of DNA so the bands will be lighter or non-existent.

Example 2

Which of the three DNA profiles indicate(s) that the alleged father could be the biological father of the child?

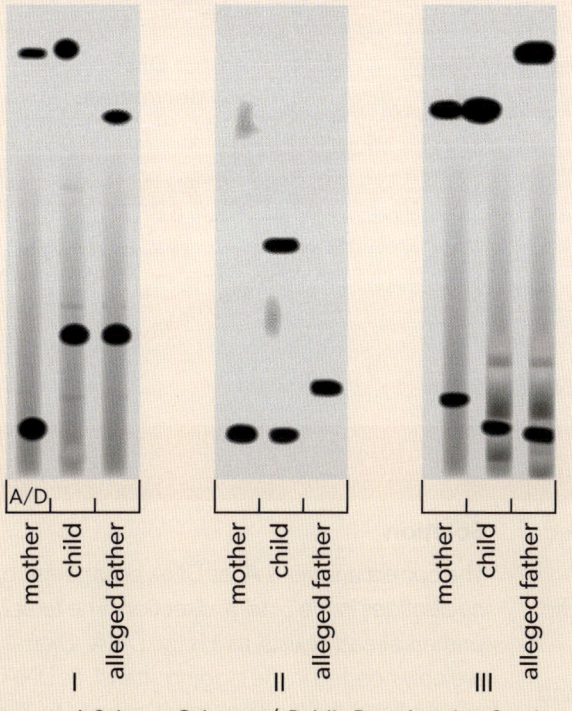

Source: A Science Odyssey / Public Broadcasting Services

A. I and II only
B. II and III only
C. I and III only
D. III only

Solution

The correct answer is **C**, as there is coincidence between some of the bands in I and III. In II, there are no common bands between the child and the alleged father.

Directionality of DNA polymerases

In replication, **DNA primase** adds an RNA primer, which is a short length of RNA. Next, **DNA polymerase III** starts replication by adding nucleotides at the primer. It synthesizes the complementary strand in a 5′ to 3′ direction. **DNA polymerase I** removes the primer by replacing RNA with DNA. **DNA ligase** seals the nicks linking sections of replicated DNA, or Okazaki fragments, in the lagging strand. Once replication is completed, DNA polymerase III proofreads for mistakes. It removes any nucleotide from the 3′ terminal with a mismatched base, followed by replacement with a correctly matched nucleotide.

Nature of science

DNA profiling is performed in paternity tests. To increase its reliability, the number of markers used is increased as this reduces the probability of a false match.

- **DNA primase** adds RNA primer at the 5′ end.
- **DNA polymerase III** adds complementary nucleotides at the primer in the 5′–3′ direction and proofreads the new DNA.
- **DNA polymerase I** removes the primer.
- **DNA ligase** seals the nicks.

- How is genetic continuity ensured between generations?
- What biological mechanisms rely on directionality?

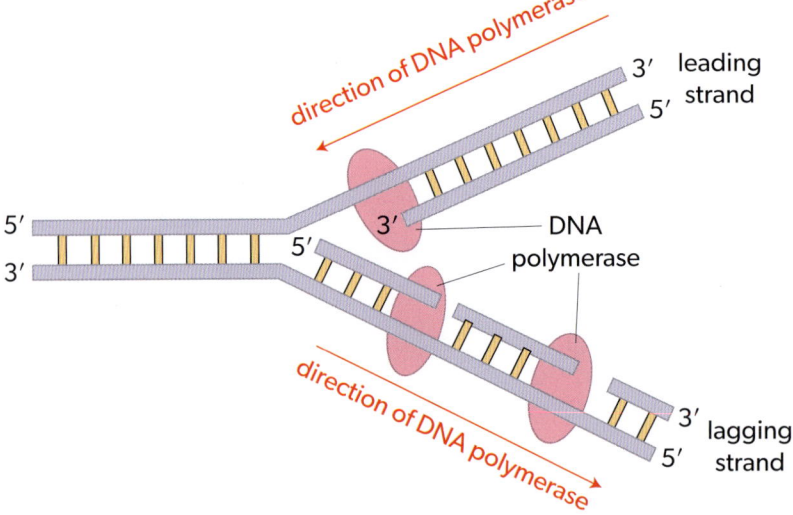

▲ Figure 3 DNA replication

Example 3

What is the reason for Okazaki fragments being formed during DNA replication?

A. To enable replication of the 3' → 5' (lagging) strand
B. To form the template for the RNA primers
C. To initiate replication on the 5' → 3' (leading) strand
D. To help the DNA helicase unwinding the DNA helix

Solution

The correct answer is **A**, as DNA polymerase only adds nucleotides in the 5' to 3' direction. The template for the primers already exists, as it is the DNA. Okazaki fragments are only needed in the lagging strand, as DNA polymerase can replicate the leading strand in the 5' to 3' direction. Helicase does not need help to unwind the DNA.

D1.2 Protein synthesis

You should know:

✔ mRNA is copied from the DNA base sequences by RNA polymerase in transcription.
✔ the DNA sequence must be conserved throughout the lifetime of a cell.
✔ transcription is required for the expression of genes.
✔ mRNA is translated into the amino acid sequence of polypeptides according to the genetic code.
✔ translation depends on complementary base pairing between codons on mRNA and anticodons on tRNA.
✔ the genetic code is universal and degenerate.

Additional higher level:

✔ transcription and translation occur in a 5' to 3' direction.
✔ transcription factors bind to the promoter to initiate transcription.
✔ regulators of gene expression, introns, telomeres, and genes for rRNA and tRNA are non-coding sequences in DNA that do not code for polypeptides.
✔ many proteins are modified into their functional state.
✔ amino acids are recycled by proteasomes.

You should be able to:

✔ describe the role of hydrogen bonding and complementary base pairing in transcription.
✔ explain the roles of mRNA, ribosomes and tRNA in translation.
✔ deduce the sequence of amino acids coded by an mRNA strand.
✔ describe stepwise movement of the ribosome along mRNA in elongation of the polypeptide.
✔ describe a point mutation that affects protein structure.

Additional higher level:

✔ explain post-transcriptional changes to RNA and how they affect gene expression.
✔ explain alternative splicing.
✔ describe initiation of translation at the ribosome.

D1.2 Protein synthesis

Transcription and stability of DNA templates

Transcription is the synthesis of RNA using a DNA template, whereas translation is the synthesis of proteins from mRNA. DNA contains the blueprint for the synthesis of mRNA, which will determine the primary structure of the protein. The two DNA strands are separated due to the breaking of the hydrogen bonds between complementary nucleotides. Complementary base pairing between DNA and mRNA is the same as in the DNA strands except for adenine (A) on the DNA pairs with uracil (U) on the mRNA strand. RNA polymerase adds the free 5′ end of the RNA nucleotide to the 3′ end of the growing mRNA molecule. The temporary hydrogen bonds between DNA and RNA are broken, liberating the mRNA. The mRNA either remains in the nucleus or leaves via a nuclear pore.

DNA is a stable molecule that can resist different environmental conditions. The sequence is maintained unchanged. In a few somatic cells that do not divide after differentiation, for example, in heart muscle cells, the sequences must be conserved throughout the lifetime of the cell for the cell to perform its function.

- **Transcription** is the synthesis of mRNA copied from the DNA base sequences by RNA polymerase.

Transcription is required for the expression of genes

Transcription is a process required for the expression of genes. These gene products are usually proteins, but can also be tRNA or small RNAs that can act as **gene expression** regulators. Most genes are turned off, and are therefore not being transcribed at any one time. This means that gene expression is regulated, and some genes are expressed only at certain times in certain cells of different tissues. Although the genome of an organism is the same in all its cells, only some genes are expressed in different tissues.

- **Gene expression** is the process by which a gene product is made using information in the DNA.

Example 4

Why is insulin produced only in the β cells of the pancreas?

A. Only the DNA of β cells of the pancreas contains the gene for insulin.
B. Insulin gene expression is repressed in all other cells.
C. The insulin gene is expressed in all cells, but its mRNA only survives in the β cells of the pancreas.
D. The β cells of the pancreas receive all the mRNA for insulin from the rest of the cells.

Solution

The correct answer is **B**, as the gene for insulin is only turned on in the β cells of the pancreas. The DNA of all cells contain the gene for insulin, but its expression is turned off. The mRNA is labile and easily degraded, so it usually remains inside the cell it is produced in.

Steps of translation

When the mRNA enters the ribosome, **translation** will occur. The mRNA determines the amino acid sequence of polypeptides according to the genetic code. Once transcribed, the mRNA binds to the small subunit of the ribosome and two tRNAs bind simultaneously to the large subunit. Translation depends on complementary base pairing between codons on mRNA and anticodons on tRNA. Codons of three bases on mRNA correspond to one amino acid in

- **Translation** is the synthesis of polypeptides from mRNA on ribosomes.

a polypeptide. The stepwise movement of ribosomes along the mRNA allows the entrance of one new complementary tRNA at a time. Peptide bonds are formed between the amino acids in the newly attached tRNA and the growing polypeptide chain.

> ## Example 5
>
> The information needed to make polypeptides is carried in mRNA from the nucleus to the ribosomes of eukaryotic cells. This information is decoded during translation. The diagram below represents the process of translation.
>
>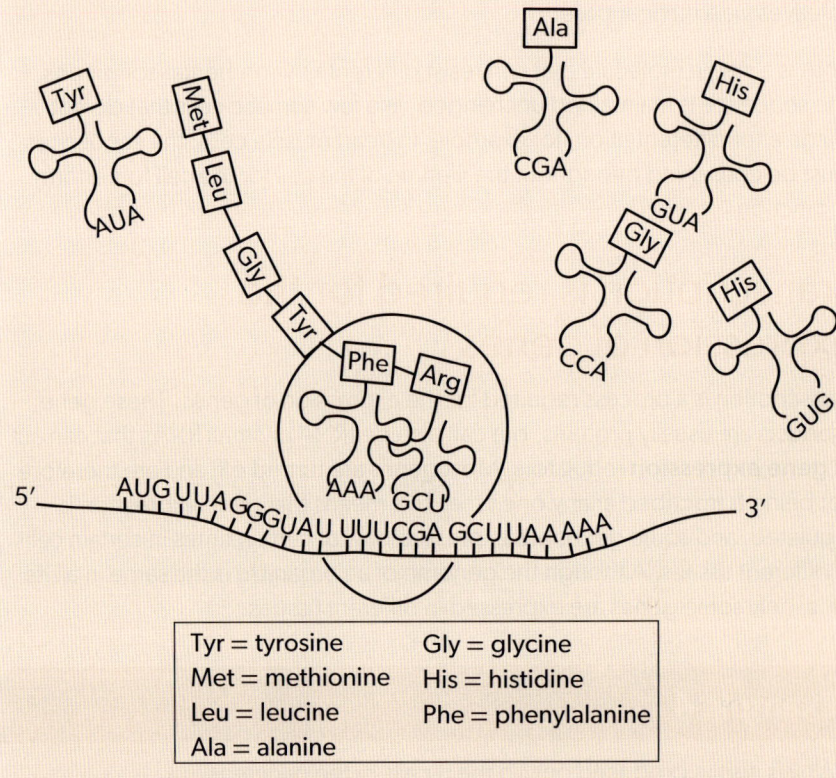
>
> Tyr = tyrosine   Gly = glycine
> Met = methionine   His = histidine
> Leu = leucine   Phe = phenylalanine
> Ala = alanine
>
> a) Annotate the diagram to show the direction in which the ribosome moves during translation.
>
> b) State the name of the next amino acid which will attach to the polypeptide.
>
> ### Solution
>
> a) From left to right (5' → 3')
>
> b) Alanine (Ala)

- The **genetic code** is universal because it can be found in most organisms and degenerate because more than one codon codes for one amino acid.

The **genetic code** is universal, which means all organisms have the same code. For example, human insulin can be produced in bacteria as genes can be transferred between species. The genetic code is also degenerate. This is because more than one codon codes for one amino acid. Most polypeptides start with the amino acid methionine, so the start codon of mRNA is usually AUG. Transcription starts at TAC on the antisense DNA (or ATG in the DNA coding sequence) as this is where the polymerase starts transcription.

D1.2 Protein synthesis

Example 6

The first, second and third positions of the three bases of the codons on messenger RNA (mRNA) are shown in this genetic code.

		Second base								
		U		C		A		G		
First base	U	UUU UUC	Phe	UCU UCC UCA UCG	Ser	UAU UAC	Tyr	UGU UGC	Cys	U C
		UUA UUG	Leu			UAA UAG	Stop Stop	UGA UGG	Stop Trp	A G
	C	CUU CUC CUA CUG	Leu	CCU CCC CCA CCG	Pro	CAU CAC	His	CGU CGC CGA CGG	Arg	U C A G
						CAA CAG	Gln			
	A	AUU AUC AUA	Ile	ACU ACC ACA ACG	Thr	AAU AAC	Asn	AGU AGC	Ser	U C A G
		AUG	Met			AAA AAG	Lys	AGA AGG	Arg	
	G	GUU GUC GUA GUG	Val	GCU GCC GCA GCG	Ala	GAU GAC	Asp	GGU GGC GGA GGG	Gly	U C A G
						GAA GAG	Glu			

A sequence of nucleotides of mRNA is shown:

AUGUUUCCACUAAAUGUAAGCAGA

a) (i) Identify the DNA sequence that has been transcribed to produce this mRNA.
   (ii) Deduce the peptide sequence translated from this mRNA sequence.
b) (i) Explain what would have happened if instead of AGA you had UGA in the last codon of this portion of mRNA.
   (ii) Suggest a reason how the change in b) (i) could have happened.

Solution

a) (i) TAC AAA GGT GAT TTA CAT TCG TCT

To identify the DNA sequence, you must find the complementary bases: for A it is T, and for C it is G, keeping in mind that in DNA you have T instead of U.

(ii) Met Phe Pro Leu Asn Val Ser Arg

To deduce the peptide sequence, you first separate the sequence of mRNA every three nucleotides. You then look in the table for A as the first letter, U for the second letter and G for the third. This is done for each codon.

b) (i) UGA is a stop codon, so the translation would be stopped here. There is no tRNA capable of joining this codon, so the peptide leaves the ribosome. The peptide produced would be shorter.

(ii) A point mutation replacing T with A on the DNA could have caused the change of A to U in the mRNA.

Mutations that change protein structure

Some **mutations** cause a change in the structure of the protein produced. In sickle cell anaemia, one specific base substitution causes glutamic acid to be substituted by valine as the sixth amino acid in the haemoglobin polypeptide. As a result, the structure of the haemoglobin protein is altered and it does not carry oxygen efficiently in erythrocytes. The shape of affected erythrocytes is also altered (they are "sickle-shaped").

- A **mutation** is a change to the structure of a gene caused by the alteration of single base units in the DNA. A mutation may be inherited by subsequent generations or may affect only the individual in which it occurs.

Directionality of transcription and translation and initiation of transcription

Transcription and translation always occur in the 5' to 3' direction. Initiation of transcription is in the promoter that is upstream of the 3' end of the gene in the DNA template. The first nucleotide of the transcribed mRNA is at the 5' end, forming the codon that first enters the ribosome to be translated (usually AUG on mRNA, which codes for methionine).

Example 7

The diagram shows the process of transcription.

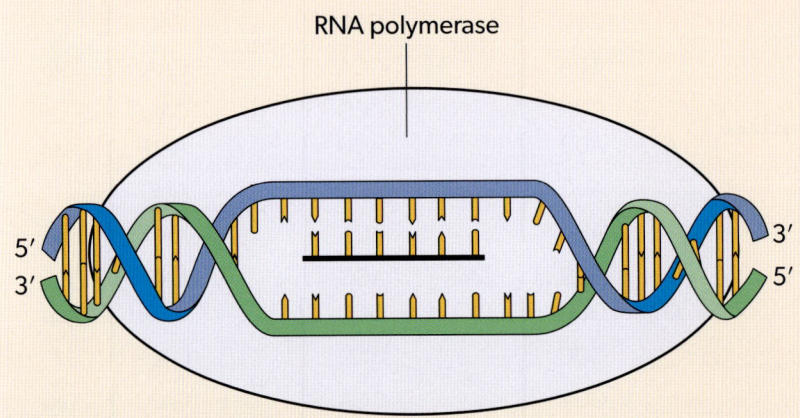

a) Label the sense and antisense strands.
b) Draw an arrow on the diagram to show where the next nucleotide will be added to the growing mRNA strand.

Solution

a) The antisense strand is the blue strand on top (the one that is copied by RNA) and the sense strand is the green strand on the bottom (the one that has a very similar sequence to the RNA transcribed, with exception of the change from T in DNA to U in RNA).

b) A clearly drawn arrow pointing at the free 3' end of the RNA strand—this is on the left of the growing mRNA. This is because the RNA grows in the 5' to 3' direction and therefore copies the antisense DNA strand in the 3' to 5' direction.

Non-coding sequences in DNA do not code for polypeptides

Non-coding regions of DNA do not code for proteins but have other important functions, such as regulators of gene expression, introns, telomeres, and genes for rRNAs and tRNAs in eukaryotes. Regulators of gene expression can be areas of the DNA such as the promoter, where the RNA polymerase joins to start transcription. Many promoters are regulated by transcription factors. Activator proteins will start transcription and repressor proteins will downregulate transcription. **Introns** are sections found in eukaryotic DNA that are transcribed to mRNA but are edited out before translation. Telomeres are the extremes of chromosomes and protect the DNA from damage. There are 170 to 570 genes coding for tRNAs in eukaryotes. The genes for rRNA are highly conserved.

- **Introns** are sections of mRNA that are edited out before translation.

D1.2 Protein synthesis

Example 8

Which are examples of non-coding regions of DNA?
I. Telomeres
II. Promoter
III. Exons

A. I only
B. I and II only
C. I and III only
D. I, II and III

Solution

The correct answer is **B**, as exons are the coding regions.

Post-transcriptional modification in eukaryotic cells

In eukaryotes the mRNA undergoes **post-transcription modifications** such as polyadenylation of the 3' end (polyA tail), capping of the 5' end (methyl-7-guanosine cap) and splicing. The addition of caps helps to stabilize the mRNA molecule and avoid digestion by nuclease enzymes. The removal of introns by special proteins and enzymes and the splicing together of exons occurs before the mature mRNA leaves the nucleus. If alternative sequences are used as introns, different peptides can be obtained from the same gene (alternative splicing). Alternative splicing reduces the amount of DNA required to produce polypeptides.

- **Post-transcription modifications** in mRNA include splicing of exons and addition of 3' polyA chains and 5' caps.

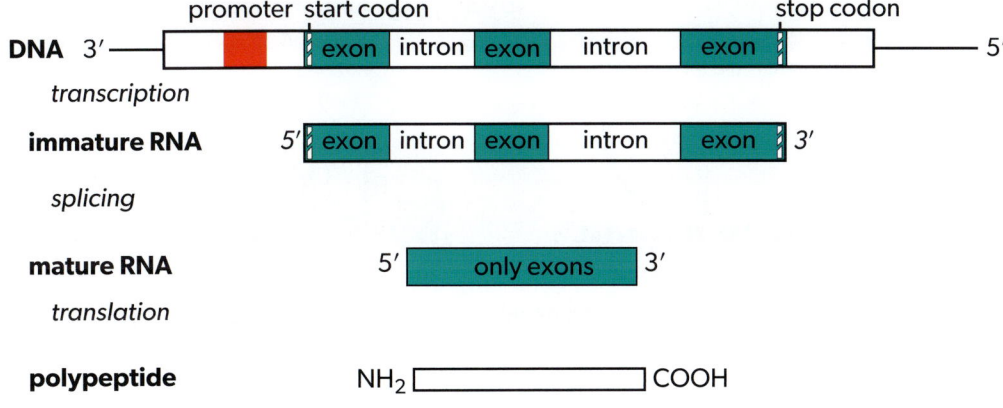

▲ Figure 4 Transcription and translation in protein synthesis

Initiation of translation

During initiation, the amino acids in the cytoplasm need to be activated before joining their corresponding tRNA. A tRNA-activating enzyme (aminoacyl tRNA synthetase) uses ATP to form a reactive molecule, the AMP-amino acid. The enzyme joins the activated amino acid to its corresponding tRNA, liberating the AMP.

Sample student answer

Explain how information in the mRNA can be translated into a polypeptide. [8]

This answer could have achieved 8/8 marks:

> Information transferred from DNA to mRNA is translated into an amino acid sequence. Synthesis of the polypeptide involves a repeated cycle of events. Initiation of translation involves assembly of the components that carry out the process. Firstly the tRNA is

▲ This is a complete answer, including all the parts of mRNA translation. Although the diagram is not necessary, in this case it includes important and clear information.

activated by phosphorylation in the cytoplasm. A tRNA-activating enzyme attaches a specific amino acid to the 3' end of a determined tRNA, using ATP for energy. It recognizes tRNA by its shape or chemical properties. To begin the process of translation, an mRNA molecule binds to the small ribosomal subunit at an mRNA binding site. An initiator tRNA molecule carrying methionine binds at the start codon (AUG). The large subunit of the ribosomes then joins the small subunit. It has three sites, the A (aminoacyl) site, P (peptidyl) site and E (exit) site. The initiator tRNA is in the P site of the ribosome. The next codon signals another tRNA to bind at the A site. A peptide bond is formed between the amino acids in the P and A sites. The ribosome translocates three bases along the mRNA (in the 5'-3' direction), moving the tRNA in the P site to the E site. This tRNA is now free in the cytoplasm to be activated again. The tRNA with the appropriate anticodon will bind to the next codon and occupy the vacant A site. This will continue until a stop codon (UAG, UGA or UAA) is reached. The disassembly of the components follows termination of translation.

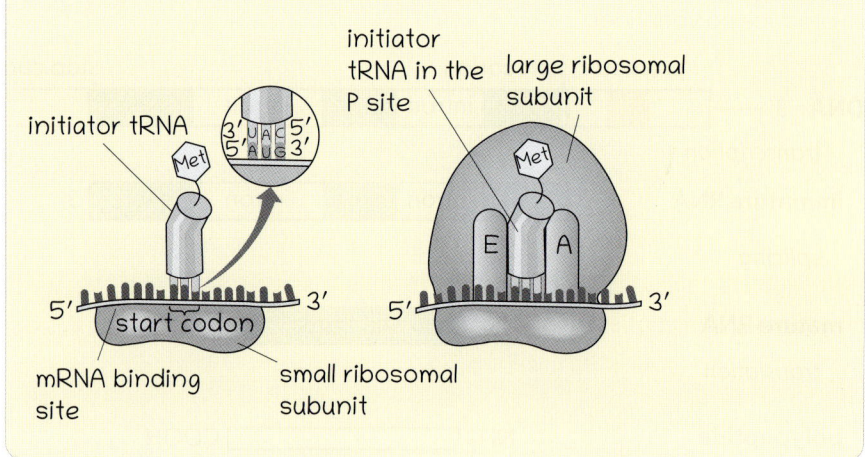

Modification of polypeptides into their functional state

Insulin is a peptide hormone that regulates glucose levels in blood. It is synthesized as an inactive precursor molecule, pre-proinsulin. The signal peptide (shown in red in **Figure 5**) is removed to produce the also inactive 110 amino acid-long protein called proinsulin. The protein forms disulfide bonds and has part of its structure removed to become an active molecule. A single protein of human insulin is composed of 51 amino acids. The proinsulin is an inactive form with long-term stability, which serves to keep the highly reactive insulin protected, yet readily available. The active molecule is a much faster-reacting hormone because diffusion rate is inversely related to particle size.

Recycling of amino acids by proteasomes

Proteasomes are protein complexes, found in the nucleus and cytoplasm of eukaryotes, which oversee protein degradation. Recycling of amino acids through protein enzymatic degradation enables the cell to maintain basal metabolic activities such as gene expression and cell cycle regulation. A functional proteome requires constant protein breakdown and synthesis.

- **Proteasomes** are protein complexes involved in the degradation of proteins to amino acids.

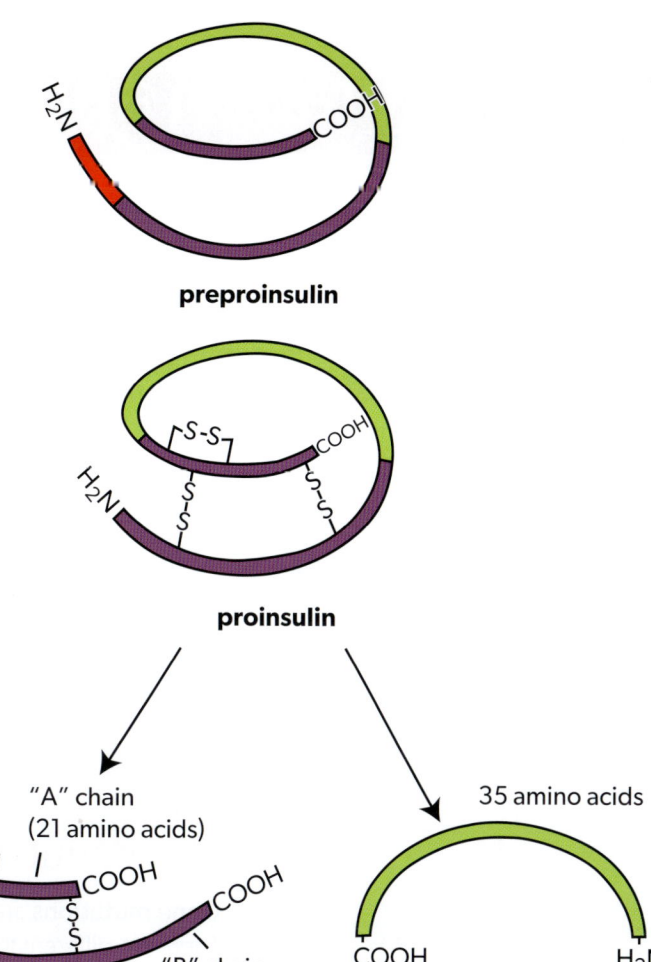

▶ Figure 5 Two-stage modification of pre-proinsulin to insulin

Example 9

The degradation of casein protein by the proteasome complexes 26S and 20S was studied in test tubes in the presence or absence of ATP.

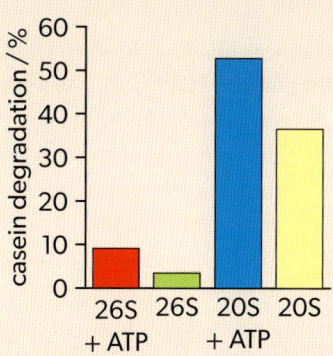

Source: Peters, J. M., et al. (1994). Distinct 19 S and 20 S subcomplexes of the 26 S proteasome and their distribution in the nucleus and the cytoplasm. *Journal of Biological Chemistry*, 269(10), 7709–7718. www.doi.org/10.1016/s0021-9258(17)37345-3 (CC BY 4.0)

a) Calculate the difference in percentage degradation of casein between 26S with and without ATP.

b) Deduce, with a reason, whether casein degradation in the proteasome is ATP dependent.

c) Discuss which proteasome is most likely involved in casein degradation in cells.

Solution

a) 9 − 4 = 5%

b) The 26S and 20S degraded casein in an ATP-dependent way as the percentage of casein degraded increased in the presence of ATP in both cases.

c) Both show casein degradation; 20S showed more casein degradation than 26S, with and without ATP. But cells require constant breakdown and synthesis of proteins, so the most degradation is not always convenient. Also, the conditions in cells might not be the same as in the test tube.

- How does the diversity of proteins produced contribute to the functioning of a cell?
- What biological processes depend on hydrogen bonding?

Continuity and change

D1.3 Mutation and gene editing

You should know:
- mutations are random structural changes to genes.
- gene mutation can be caused by mutagens and by errors in DNA replication or repair.
- some bases have higher possibility of mutation although mutations are random.
- mutations in germ cells are inherited whereas mutations in somatic cells may cause cancer.
- gene mutations are the source for genetic variation, therefore essential for evolution by natural selection.

Additional higher level:
- gene knockout technology can be used to study the function of a gene.

You should be able to:
- distinguish between substitutions, insertions and deletions.
- explain the consequences of base substitutions.
- explain the consequences of insertions and deletions.

Additional higher level:
- describe an example of CRISPR sequences and the enzyme Cas9 in gene editing.
- discuss the hypotheses to account for conserved or highly conserved sequences in genes.

Gene mutation

- **Gene mutations** are structural changes to genes at the molecular level.
- **Single-nucleotide polymorphisms (SNPs)** are caused by base substitution mutations. Because of the degeneracy of the genetic code, they may or may not change a single amino acid in a polypeptide.

Gene mutations are structural changes to genes at the molecular level. There are different types of gene mutations. A base substitution is when one nucleotide is replaced by another in the DNA sequence. A change in the DNA determines a change in the mRNA codon. The change in mRNA could cause a different tRNA to join by its anticodon, determining a different primary sequence of the protein. If the new codon is a stop codon, the protein will be much shorter. Because the genetic code is degenerate, if the change in nucleotide determines for a codon that codes for the same amino acid, no change will occur to the protein (neutral mutation). **Single-nucleotide polymorphisms (SNPs)** are a result of base substitution mutations. These SNPs determine variability in a population.

An insertion is when one or more nucleotides are added to DNA. This will cause a frameshift in the DNA reading frame. The codons will be displaced, so the final peptide produced will be completely different from the original peptide. In many cases, a new stop codon is formed, and the peptide is shorter. In the rare case where the insertion is of three nucleotides, the protein will be similar, but with an extra amino acid added. In the case of a deletion, a section of the DNA is eliminated. If the section where the insertion or deletion occurred is not a coding section and the number of nucleotides is a multiple of three so there is no frameshift, then the insertion or deletion does not affect the peptide produced.

Causes of gene mutation

- **Mutagens** damage the DNA molecule.

Mutations can arise from exposure to radiation such as ultraviolet (UV) rays, the presence of chemical **mutagens**, carcinogens, papilloma virus or cigarette smoke. The most common chemical mutagens are substances that greatly resemble nucleotides (base analogues) and are incorporated into the DNA during replication. Examples of chemical mutagens are mustard gas and ethidium bromide. There are many highly reactive oxygen species generated by normal cellular processes that are also mutagenic. Mutagens are also carcinogens if the damage they produce leads to cancer. Exposure to radiation such as X-rays, gamma rays or UV light can also cause mutagenesis. UV light induces cross-linking between DNA molecules that inhibits replication and transcription. Mutagenesis can also be caused by spontaneous mutations that result from errors in DNA replication, recombination and repair. If the error is in replication, there is usually a point mutation due to the addition of a wrong amino acid to the sequence of the peptide. The error can also be in the rearrangement of chromosomes during meiosis and mitosis and can cause a change in the number of chromosomes. Some of these changes are silent as they affect non-coding or regulating regions, or because of the degeneracy of the genetic code they form an alternative codon for the same amino acid. Some carcinogens, such as anabolic steroids, increase the chances that repair mechanisms fail, therefore causing cancer.

Example 10

The ABO blood group gene locus is located on human chromosome 9, and encodes for an enzyme that mediates the expression of A and B antigens on erythrocytes. The A and B genes differ in a few single-base substitutions. The protein sequence of the allele for blood group O differs in more amino acids from the sequences for blood groups A and B. The table shows part of a protein sequence alignment of the alleles of blood groups (an asterisk means the sequence is the same and a dash means the amino acid is not present).

Allele	Amino acid
blood group A	P K V L T P W K – D V L V V T P W L A P I V W E G T F
blood group B	* * * * * * * C R * * * * * * * * * * * * * * * * *
blood group O	* * * * * * * C R * * * * * P L G W L P L S G R A H S

a) Identify the number of amino acids that are different in this sequence between blood group A and blood group B.

b) Suggest whether this data shows that human diversity is due to SNPs.

Solution

a) 2

b) There are more SNPs between the sequence for blood group O and both A and B. There are few differences between A and B. Therefore, the SNPs make a difference in the blood group, confirming that SNPs can contribute to human diversity.

Example 11

Scientists estimated thyroid cancer risks related to the type of diagnostic X-ray procedure in 75,494 people, 251 of whom were diagnosed with thyroid cancer. The risk estimates were determined as hazard ratios compared to the norm. A ratio of 2 means that the patients are two times more likely to have thyroid cancer than the norm.

X-ray procedure	Hazard ratio
cervical spine radiograph	0.95
skull radiograph	0.99
other head and neck radiograph	1.02
angiogram	1.04
dental radiograph (per 10 radiographs)	1.11
mammogram	0.99
chest radiograph (per 5 radiographs)	0.92
upper gastrointestinal series	0.98
barium swallow	0.94
lumbar/thoracic spine radiograph	0.99

Source: Neta, G., et al. (2013). A prospective study of medical diagnostic radiography and risk of thyroid cancer. *American Journal of Epidemiology*, 177(8), 800–809.

a) State the procedure that has the highest risk of producing thyroid cancer.

b) Comment on the evidence of thyroid cancer risk associated with diagnostic X-rays.

Solution

a) Dental radiographs

b) The data shows no evidence of the relationship between X-ray procedures and thyroid cancer, except for dental X-rays where there is 11% more risk than the norm and 10 radiographs need to be taken. More tests should be performed, and more patients studied.

Randomness and consequences of mutation in germ cells and somatic cells

Mutations can occur anywhere in the base sequences of a genome, although some bases have a higher probability of mutating than others. There is no natural mechanism for making a deliberate change to a particular base with the purpose of changing a trait. The driving force for the generation of new mutations is in the replication of DNA when germ cells are created. Germ cells are especially good at preventing or repairing DNA damage but, in the unlikely event that this mutation in the germ cell is not repaired, it will be inherited by the next generation. In contrast, if the mutation is in a somatic cell, it will not be inherited by the offspring, but it may lead to cancer. Mutations are present in most or all malignant cells of a tumour and have probably been selected because they confer a proliferative advantage.

> **Nature of science**
>
> Commercial genetic tests can yield information about ancestry and potential future health and disease risk using only a saliva sample. The results are usually sent back to the consumer without much explanation or expert interpretation.

Mutation as a source of genetic variation

Most mutations either have no effect on the individual or are detrimental to health and survival. There are very few truly beneficial mutations. Nevertheless, mutations are the fuel of evolution, and are therefore beneficial to the adaptation of species to changes in their environment. Mutations function as facilitators of evolution, providing variation and enabling rapid evolution of new forms.

Example 12

DNA from different cancer tumours was sequenced and compared to normal cell DNA to detect transitions (change from purine to purine or pyrimidine to pyrimidine) and transversions (change from purine to pyrimidine or pyrimidine to purine). The graph shows the mutations observed only in a tumour due to random events (blue bars) and mutations that appear due to reproduction of a mutant cell (yellow bars).

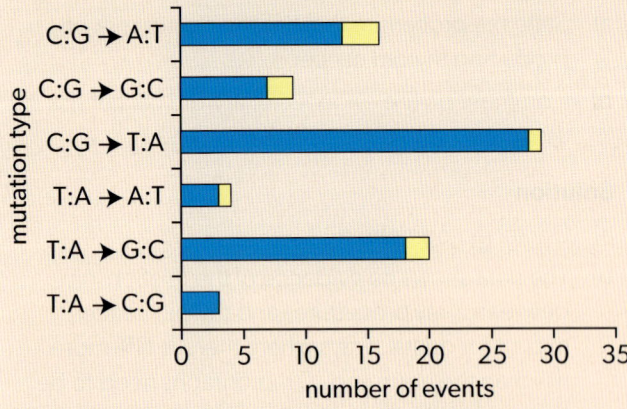

Source: Bielas, J. H., et al. (2006). Human cancers express a mutator phenotype. *Proceedings of the National Academy of Sciences,*, **103**(48), 18238–18242. www.doi.org/10.1073/pnas.0607057103 (CC BY 4.0)

a) State the most common change in base.
b) Identify the most common transversion.
c) (i) Compare and contrast random mutant events and mutations due to reproduction of mutated cells.
   (ii) Suggest one reason for the difference in number of events caused by random mutation and by mutations that appear due to reproduction of a mutant cell.

Solution

a) The transition CG to TA.
b) The transversion TA to GC.
c) (i) Random events happen much more often than mutations due to reproduction of a mutated cell. Both occur for most changes, except for TA to CG that does not appear in reproduction of a mutated cell.
   (ii) Mutations appear randomly in tumours, they are not just replicated. Cancer cells have increased rates of mutagenesis.

Gene knockout technology

Gene knockout (KO) is a technique for identifying the functions of coding and non-coding genomic regions by preventing the expression of this gene. Gene function can be studied using model organisms with similar sequences by using KO technology. By causing a specific gene to be inactive in an organism and observing any differences in behaviour or physiology, researchers can infer the probable function of the gene. A library of KO organisms is available for some species to use as models in research.

CRISPR sequences and Cas9 in gene editing

CRISPR–Cas9 is a complex formed by a short guide RNA (sgRNA) sequence that will target one selected region or gene on the DNA. This system can design an RNA complementary to a given DNA. Cas9 enzyme will cut the DNA at the targeted sequence, which will be repaired either by deleting some sequence or adding some sequence. A sequence of viral or prokaryotic DNA (PAM) prevents nucleases from destroying the target sequence. Several blood disorders, including leukaemia, lymphoma and thalassemia have been treated in tissues using this technology. In 2023, CRISPR–Cas9 technology was approved to treat sickle cell disease.

Nature of science

CRISPR–Cas9 is powerful in terms of the potential to start a biotechnological revolution in the field of crop development and human pathology. However, in the wrong hands, it can lead to abuse and misuse in multiple ways, including manipulation of germline genetics. Scientists across the world are subject to different regulatory systems. Therefore, an international effort is needed to harmonize regulation of the application of genome editing technologies such as CRISPR.

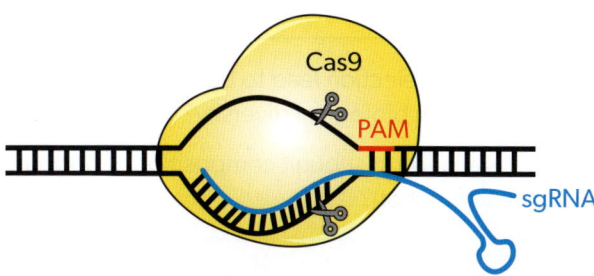

▲ Figure 6 Mechanism of action of CRISPR–Cas9

Source: Choi, E. & Koo, T. (2021). CRISPR technologies for the treatment of Duchenne muscular dystrophy. *Molecular Therapy*, **29**(11), 3179–3191. www.doi.org/10.1016/j.ymthe.2021.04.002 (CC BY-NC-ND)

- The **CRISPR–Cas9** system allows the editing of nucleotides in DNA.
- **Conserved DNA sequences** (or an amino acid sequence in a protein) have remained unchanged throughout evolution.

Conserved sequences

Conserved sequences are DNA molecules (or an amino acid sequence in a protein) that have remained unchanged throughout evolution. Conserved sequences are identical or similar across a species or a group of species. Highly conserved sequences are identical or similar over long periods of evolution. An example is the cytochrome c gene and its corresponding protein, which are highly conserved in vertebrates. This protein is involved in aerobic respiration, and is therefore essential for organisms. Another example is haemoglobin, which is essential for the transport of oxygen in vertebrates. In addition to protein-coding sequences, the human genome contains a significant amount of regulatory DNA, which contains highly conserved sequences. In bacteria and archaea, the 16S and 23S ribosomal genes are the most conserved genomic regions.

One hypothesis for the mechanism that maintains conserved sequences is the functional requirements for the gene products. Because the protein is essential, if there is a mutation in the gene, the organism does not survive, and the mutation is not inherited. Another hypothesis is that these genes have slower rates of mutation and that is why they do not change much throughout evolution.

Continuity and change

Practice problems

1. Explain semi-conservative replication in terms of the type of nitrogen in the DNA molecules (^{15}N and ^{14}N) in Meselson and Stahl's experiment.
2. Mutations are the ultimate source of genetic variation and are essential to evolution.
   a) State one type of environmental factor that may increase the mutation rate of a gene.
   b) Identify one type of gene mutation.
3. Using one example, discuss the CRISPR–Cas9 system and the ethical issues that must be addressed before its implementation.

- How can natural selection lead to both a reduction in variation and an increase in biological diversity?
- How does variation in subunit composition of polymers contribute to function?

D2.1 Cell and nuclear division

You should know:

- two genetically identical daughter nuclei are produced from the division of the nucleus during mitosis—cell division.
- oogenesis in humans and budding in yeast are examples of unequal cytokinesis.
- mitosis maintains the chromosome number and genome of cells, whereas meiosis halves the chromosome number and generates genetic diversity.
- DNA replication must occur prior to mitosis and meiosis.
- in both mitosis and meiosis, chromosomes condense by supercoiling with the help of histones.
- meiosis is a reduction division in which one diploid nucleus divides to produce four haploid nuclei.

Additional higher level:

- cell proliferation through mitosis is needed for growth, cell replacement and tissue repair.
- cyclins are involved in the control of the cell cycle.
- mutagens, oncogenes and metastasis are involved in the development of primary and secondary tumours.

You should be able to:

- describe cytokinesis as splitting of cytoplasm in a parent cell between daughter cells.
- outline the stages of mitosis—prophase, metaphase, anaphase and telophase.
- identify the phases of mitosis in diagrams or micrographs.
- outline the two rounds of segregation in meiosis.
- explain how non-disjunction can cause Down syndrome.
- outline how genetic variation is promoted by fusion of gametes from different parents, random orientation of chromosomes and crossing over.

Additional higher level:

- outline the phases of the cell cycle—interphase, mitosis and cytokinesis.
- describe interphase as an active phase of cell growth.
- compare tumours in terms of rates of cell division, growth and ability to metastasize.
- determine the mitotic index from a micrograph.

Cell division and cytokinesis

Generation of new cells in living organisms is by cell division called **mitosis**, where a parent cell (mother cell) divides to produce two daughter cells. **Cytokinesis** is the splitting of the cytoplasm of the parent cell, which divides into two daughter cells. Cytokinesis in animals is produced by cell strangling by a ring of contractile actin and myosin proteins which pinch the cell membrane together. In plant cells, the splitting is by formation of a plate by vesicles that assemble sections of membrane and cell wall. Division of cytoplasm is usually—but not in all cases—equal, and both daughter cells must receive at least one mitochondrion and any other organelle that can only be made by dividing a pre-existing structure. In humans during oogenesis, and in yeast during budding, cytokinesis is unequal.

- **Mitosis** is the generation of new cells in living organisms by cell division.
- **Cytokinesis** is a process that occurs along with telophase. The cytoplasm of the parental cell divides into two daughter cells. Cytokinesis in animals is produced by cell strangling while in plants it is by formation of a plate.

Overview of mitosis and meiosis

Nuclear division is needed before cell division to avoid production of **anucleate** cells. Mitosis maintains the chromosome number and genome of cells, whereas meiosis halves the chromosome number and generates genetic diversity. In both meiosis and mitosis, DNA replication is needed to maintain the amount of DNA in the new cells. After replication, each chromosome consists of two elongated DNA molecules (chromatids) held together until anaphase.

- **Anucleate** cells are cells without a nucleus.

Condensation and movement of chromosomes

During mitosis and meiosis, chromatin condenses into chromosomes using histone proteins. The chromosomes attach to **spindle fibres** made of microtubules. These microtubules move the chromosomes or **chromatids** to the poles using microtubule motor proteins that utilize energy derived from ATP hydrolysis to produce force and movement.

- **Spindle fibres** are made of microtubule proteins that attach and move chromosomes and chromatids during cell division.
- A **chromatid** is a DNA molecule that is the copy of a newly formed chromosome.

Phases of mitosis

Mitosis is a process that produces two genetically identical daughter cells. The stages of mitosis are **prophase**, **metaphase**, **anaphase** and **telophase**.

You studied the condensation of chromosomes into nucleosomes in subtopic A1.2.

- **Prophase** is the stage where chromatin condenses and associates with histones forming chromosomes, the nuclear membrane disappears, and spindle fibres are formed. Chromosomes attach to spindle fibres.
- **Metaphase** is the stage where chromosomes are aligned in the equator of the cell.
- **Anaphase** is the stage where sister chromatids (V shape, with the vertex pointing to the poles) are separated to the opposite poles of the cell.
- **Telophase** is the last stage, where a nuclear membrane forms around each set of chromosomes that begin to uncoil.

Continuity and change

Example 1

The image shows a micrograph.

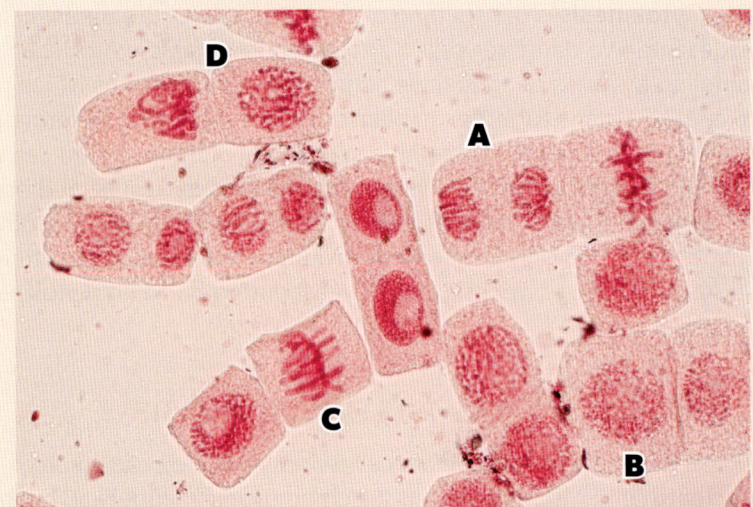

Which cell is in telophase?

Solution

The correct answer is **A**, as the chromatids have migrated to the poles and a nuclear membrane is being formed in the new cells. **B** and **D** are in prophase and cell **C** is in metaphase.

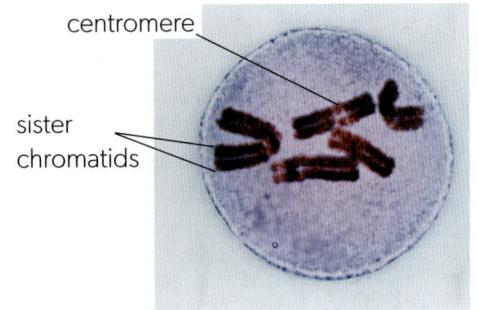

▲ Figure 1 Micrograph of a cell in mitosis

- The **centromere** is the structure of the chromosome (DNA) that holds together both chromatids. It is also the point of attachment to the spindle fibres.
- **Sister chromatids** are two DNA molecules formed by DNA replication prior to cell division until the splitting of the centromere at the start of anaphase.

Features of meiosis

Meiosis produces gametes. Gametes are cells that contain the **haploid** number of chromosomes, allowing two gametes (one from each parent) to fuse and create an embryo with the **diploid** number of chromosomes. Meiosis is divided into two processes, meiosis I and meiosis II. The halving of chromosome number occurs in meiosis I, where each chromosome of a homologous pair migrates to the opposite pole. DNA is replicated before meiosis so that all chromosomes consist of two sister chromatids. In metaphase I, the **homologous chromosomes** pair up and **crossing over** occurs. The homologous chromosomes are randomly orientated prior to separation in anaphase I. Meiosis II is similar to a mitotic cell division, as each sister chromatid migrates to an opposite pole. Overall, one diploid nucleus divides by meiosis to produce four haploid nuclei at the end of meiosis II. Alleles segregate during meiosis, allowing new combinations to be formed by the fusion of gametes. On rare occasions there is an error in the distribution of chromosomes, called non-disjunction, with an outcome of cells with a different number of chromosomes. An example is Down syndrome, where the cells have an extra chromosome 21.

- **Haploid** nuclei have one chromosome of each pair.
- **Diploid** nuclei have pairs of homologous chromosomes.
- **Homologous chromosomes** are one set of maternal and one set of paternal chromosomes that have the same genes in the same loci. They pair up during meiosis.
- **Crossing over** is the process in which DNA is exchanged between non-sister chromatids of homologous pairs.

Meiosis as a source of variation

Genetic variation is promoted by crossing over, random orientation of chromosomes and fusion of gametes from different parents. Meiosis leads to an independent assortment of chromosomes and a unique composition of alleles in daughter cells. The independent assortment of genes is due to the random orientation of pairs of homologous chromosomes in meiosis I.

Crossing over is the exchange of DNA material between non-sister homologous chromatids occurring during prophase I. A single strand break in the DNA is a cut between the non-sister chromatids of homologous chromosomes. Chiasmata (or **chiasma**) formation between non-sister chromatids can result in an exchange of alleles. This exchange of alleles can produce recombinants. Crossing over produces new combinations of alleles on the chromosomes of the haploid cells. When crossing over occurs, homologous chromosomes form structures with strange shapes called tetrads (named after the presence of four chromatids) or bivalents (named after two chromosomes). The chromatids pull away from each other but are held together at the chiasmata.

- The **chiasma** is the point where genetic material is exchanged in crossing over.

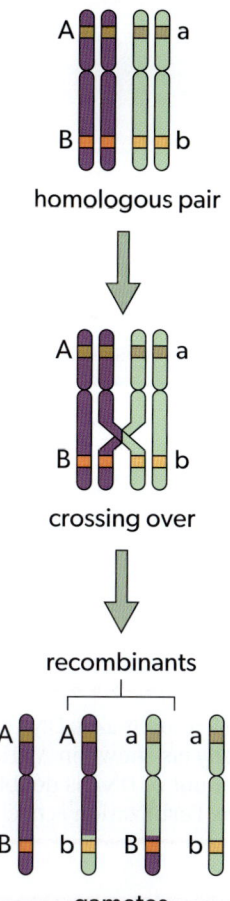

▲ Figure 2 Formation of recombinants by crossing over during prophase I

Sample student answer

The diagram shows the changes in the amount of DNA per cell in the anther of a plant.

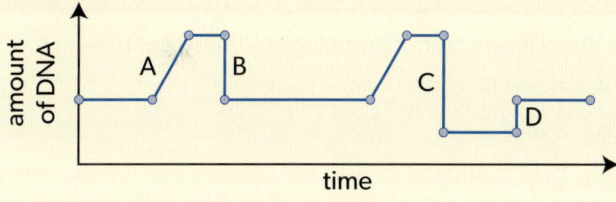

a) State the process occurring at A. [1]

   This answer could have achieved 1/1 marks:

   DNA replication.

b) Explain the reason the mass of DNA decreases at B. [2]

   This answer could have achieved 2/2 marks:

   At B the cell is going through mitosis. The sister chromatids of each chromosome have migrated to the opposite pole in anaphase. In telophase the nuclear membrane reappears and the cell has divided into two cells in cytokinesis.

▲ The answer identifies the process and stages of DNA reduction.

Continuity and change

c) Suggest what is happening at C. [2]

This answer could have achieved 2/2 marks:

At C the process of meiosis is occurring. The amount of DNA decreases to half during meiosis I, as each homologous chromosome migrates to a different pole in anaphase I, producing two different cells with half the number of chromosomes in cytokinesis. In the second division, in meiosis II, each sister chromatid of each chromosome migrates to the opposite pole in anaphase II, and new cell nuclear membranes form in telophase II. In cytokinesis four cells are formed. These cells have half the number of chromosomes and 1/4 the amount of DNA compared with the original cell.

▲ This answer is complete.

d) State the name of the process occurring at D. [1]

This answer could have achieved 0/1 marks:

Replication.

▼ Replication makes DNA gradually (as shown in A). Here the amount of DNA is doubled all at once. Fertilization is occurring.

Example 2

The diagram shows six chromosomes of a diploid cell in early metaphase I (although not shown as bivalents).

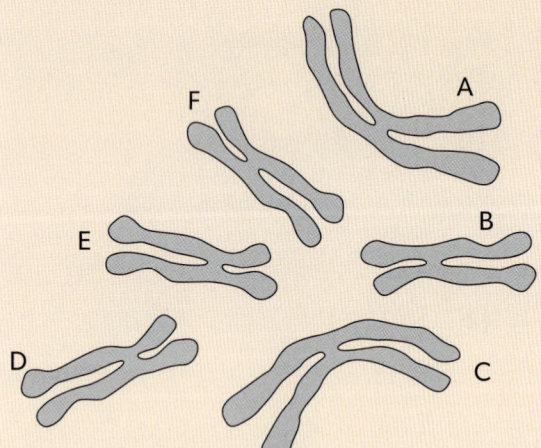

Which is one pair of homologous chromosomes?

A. A and B
B. A and C
C. B, D, E and F
D. There is no pair of homologous chromosomes.

Solution

The correct answer is **B**, as the homologous chromosomes have the centromere in the same place and the chromatids are of the same length. **C** is incorrect, as although the chromosomes all look the same, they are not a pair—there are four not two chromosomes.

The cell cycle

- **Interphase** is the stage of cell division before mitosis and meiosis. Cells grow, forming organelles (G_1 stage), DNA is duplicated (S stage) and synthesis of proteins that are involved in nuclear division occurs (G_2 stage).

Cell division is necessary for growth, cell replacement and tissue repair. Proliferation is needed for growth within plant meristems and early-stage animal embryos. Skin is an example of cell proliferation during routine cell replacement and during wound healing. The cell cycle includes several phases. The sequence of events includes **interphase**, followed by mitosis and then cytokinesis. Interphase is a metabolically active period which has three stages. In the G_1 stage, there is an increase in the number of organelles (such as mitochondria and ribosomes) and growth. In the S stage, there is DNA replication and in G_2 there is synthesis of proteins.

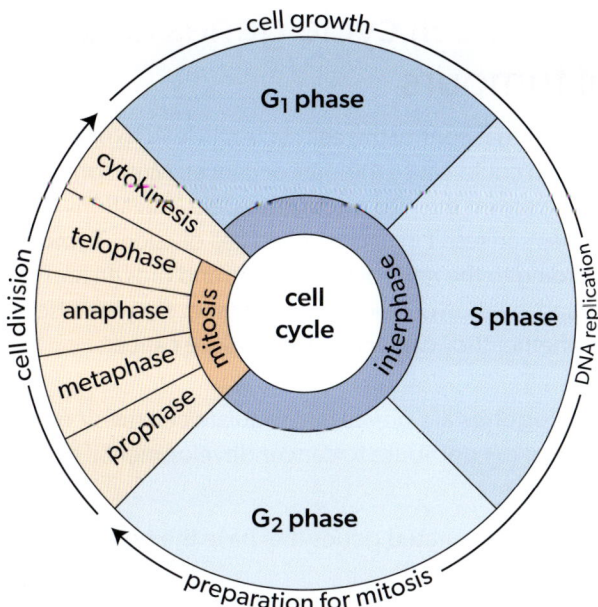

▲ Figure 3 Phases of the cell cycle

Control of the cell cycle using cyclins

Cyclins are synthesized soon after an egg is fertilized and increase in levels during interphase. However, the levels decrease quickly in the middle of mitosis. The fact that the amount of cyclins drops periodically at different cell division stages is important for cell cycle control. The concentration of different cyclins increases and decreases during the cell cycle and a threshold level of a specific cyclin is required to pass each checkpoint in the cycle.

Example 3

Cyclins are proteins that bind to cyclin-dependent kinase enzymes to activate them. These kinases attach phosphate groups to other proteins in the cell, also activating them to perform a specific function. The concentrations of cyclins rise and fall in cells at certain times.

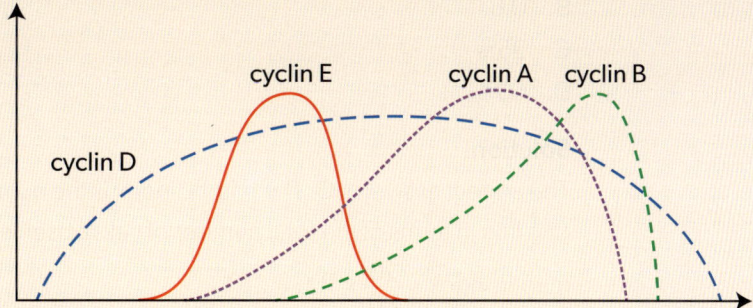

Source: https://en.wikipedia.org

What times are these?

A. Day and night
B. Seasons of the year
C. Prophase of mitosis only
D. Developmental stages in the life cycle

Solution

The correct answer is **D**, as cyclins are important in regulating the cell cycle. The levels of cyclins do not change in relation to the time of day or seasons of the year, or only during prophase.

Continuity and change

Mutations in cell cycle genes and types of tumours

Mutations in genes that control the cell cycle can lead to uncontrolled cell proliferation which could cause tumours or cancer. Mutations in **proto-oncogenes** can convert them to **oncogenes**, and mutations in tumour suppressor genes can result in uncontrolled cell division. **Tumours** can be classified according to the rates of cell division and growth and in the capacity for metastasis and invasion of neighbouring tissue. They are also classified according to whether they can or cannot produce cancer.

- **Proto-oncogenes** are genes that regulate cell division which have the potential to contribute to cancer development if their expression is altered.
- **Oncogenes** are mutated genes that have the potential to cause cancer.
- A **benign tumour** does not invade nearby tissue or spread to other parts of the body while a **malignant tumour** invades tissues or spreads through the bloodstream or lymphatic system.
- A **primary tumour** is where the cancer starts. Cells from a primary tumour may spread to other parts of the body and form a **secondary tumour**.
- **Mitotic index** is the number of cells in mitosis divided by the total number of cells seen under the microscope. It can be shown as a percentage too.

Example 4

A micrograph shows a total of 80 cells, 20 of which are in mitosis. What is the **mitotic index** in this tissue?

A. 0.4
B. 0.6
C. 25%
D. 60%

Solution

The correct answer is **C**. The mitotic index is the number of cells dividing in terms of the total number of cells observed.

mitotic index = $\frac{20}{80} \times 100 = 25\%$

- What processes support the growth of organisms?
- How does the variation produced by sexual reproduction contribute to evolution?

D2.2 Gene expression

You should know:
Additional higher level:

✓ gene expression affects phenotype.

✓ gene expression is regulated by proteins that bind to specific base sequences in DNA.

✓ mRNA is degraded by nucleases—a method of regulating translation.

✓ epigenesis, such as methylation, develops patterns of differentiation.

✓ most epigenetic tags are removed in eggs and sperm.

You should be able to:
Additional higher level:

✓ distinguish between the genome, transcriptome and proteome of individual cells.

✓ explain the effect of methylation of cytosine in DNA or histones on gene expression.

✓ describe heritable changes to gene expression.

✓ give examples of how the environment of a cell and of an organism have an impact on gene expression.

✓ investigate the effects of the environment on gene expression using monozygotic twin studies.

✓ describe external factors affecting gene expression.

Regulation of gene expression

Gene expression is the mechanism by which information in genes has effects on the phenotype. The most common stages of gene expression are transcription, translation and the function of a protein product, such as an enzyme. The promoter is essential in the **regulation of gene expression**. It is a non-transcribed sequence of DNA usually found upstream of the gene. If the RNA polymerase cannot join the promoter on the DNA, the transcription into mRNA will not occur. **Transcription factors** and methylated cytosines in DNA both have major roles in regulating gene expression. Transcription factors regulate gene expression by enhancing or repressing transcription of a gene. Some transcription factors bind to promoters to prevent RNA polymerase binding and so block gene expression. Other transcription factors bind to enhancers to promote gene expression.

Control of mRNA degradation

The rate of mRNA degradation is as important as the rate of synthesis in regulating the concentration of mRNA. The average mRNA half-life in mammalian cells is 24 hours, with short-lived messages lasting 20 minutes and long lived having four days before being broken down by nucleases. The 3' poly(A) tail and the 5' cap serve to block decay pathways. Once the tail has been shortened, the mRNA is then subject to decay by the enzymes in a 3' → 5' direction and the 5' decapping in the 5' → 3' direction.

Epigenetic changes affect phenotype not genotype

Epigenetic factors can modify gene expression without changing the DNA sequence. Epigenetic changes modulate the expression of a genotype into a particular phenotype. The **genome** includes all the DNA in a cell, regardless of whether it is expressed or not. The **transcriptome** is all the mRNA produced, and includes all genes that are transcribed at that time in the cell. The **proteome** includes all the proteins that can be expressed in a cell. No cell expresses all its genes. The pattern of gene expression in a cell determines how it differentiates.

- **Regulators of gene expression** are non-coding regions of DNA that increase or decrease gene expression.

- **Transcription factors** are proteins that regulate gene expression by joining a promoter, blocking expression, or joining an enhancer to promote expression. They can also modify histones or recruit activators or repressors.

- **Epigenesis** is the development of patterns of differentiation in the cells of a multicellular organism.

- A **genome** is the entire set of DNA instructions found in a cell, including DNA in mitochondria and in chloroplasts in plants.

- A **transcriptome** is the full range of mRNA expressed in an individual cell.

- A **proteome** is the complete set of proteins expressed by an individual cell.

Methylation

Epigenetic tags are changes in DNA that can repress or activate genes without changing the DNA sequence. Methylation can occur in DNA or in the histones forming the nucleosome. In both cases, these epigenetic tags alter DNA accessibility and chromatin structure, thereby regulating patterns of gene expression. Methylation of DNA is the addition of a methyl group ($-CH_3$) to a nucleotide, usually a cytosine found in a CpG island. CpG islands are areas of DNA in or close to the promoters where there is a cytosine nucleotide followed by a guanine nucleotide in a linear sequence. Normally the cytosines of these islands are not methylated, but if they are, this affects transcription inversely, repressing expression of genes downstream. Methylation of specific amino acids in histones causes changes in the nucleosome.

▲ **Figure 4** DNA methylation

Example 5

The diagram shows the methylation pattern of the DNA promoters of homologous chromosomes in two homozygous organisms.

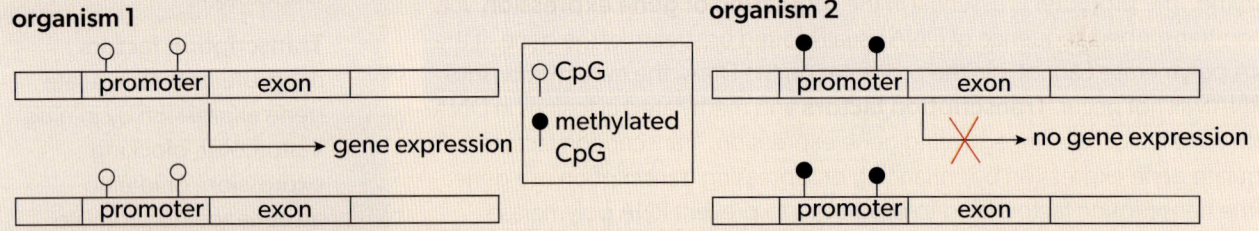

If **organism 2** were a heterozygote instead of homozygous, what would be the result of CpG methylation?

A. The gene would be expressed at the same level as in the first organism.
B. There would be no gene expression.
C. Only half the gene would be expressed, producing a shorter protein.
D. Only the methylated strand would be expressed.

Solution

The correct answer is **A**, as if one chromosome has the promoter unmethylated, then that strand can be transcribed to RNA. In this case, methylation is inhibiting transcription, therefore the strand that is not methylated can be transcribed. Usually only one chromosome is used as a template for transcription, therefore the amount of gene expression will not change.

Epigenetic inheritance and effects of the environment on gene expression

If epigenetic tags remain in place during mitosis or meiosis they will be inherited by offspring, without changes in the nucleotide sequence of DNA. Most epigenetic changes are either neutral or deleterious. Some can have the potential to be adaptive, and in some cases might even respond to environmental challenges, with major implications for heredity, breeding and evolution. Air pollution affects gene expression by changes in DNA or histone methylation and chromatin remodelling. These processes can influence health outcomes during the life course and even across generations.

- **Imprinted genes** maintain their epigenetic tag determining parent-of-origin-dependent growth in hybrids.

Most epigenetic tags are removed from eggs and sperm

Soon after egg and sperm meet, most of the epigenetic tags that activate and silence genes are removed to allow for sexual reproduction and the adoption of the specialized, hypomethylated germ cell and embryo. However, in mammals, some genes keep their epigenetic tags (**imprinted genes**). Because imprinted genes have only one copy active at a time, they are under greater selective pressure than normal genes, evolving more rapidly. Therefore, imprinted inheritance depends on the phenotype of the parent that contributed the allele, rather than which sex inherits it.

Lions and tigers can produce hybrid offspring. A liger is a hybrid from a female tiger (*Panthera tigris*) and a male lion (*Panthera leo*). The tigon is a hybrid of a male tiger and female lion. Ligers are larger than tigons due to genomic imprinting— the unequal expression of genes is due to differences in the methylation patterns of their parents. In ligers, the single imprinted gene is only expressed from the maternal allele, as the male allele is methylated (–CH$_3$) and therefore silenced.

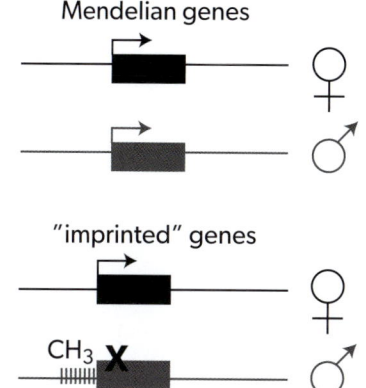

▲ **Figure 5** Inheritance of imprinted genes

Source: Vrana, P. B. (2007). Genomic imprinting as a mechanism of reproductive isolation in mammals. *Journal of Mammalogy*, 88(1), 5–23. www.doi.org/10.1644/06-mamm-s-013r1.1

Example 6

Epigenetic differences were recorded in twins. Methylation of histones and of CpG islands in DNA were recorded at the ages of 4 years and 40 years.

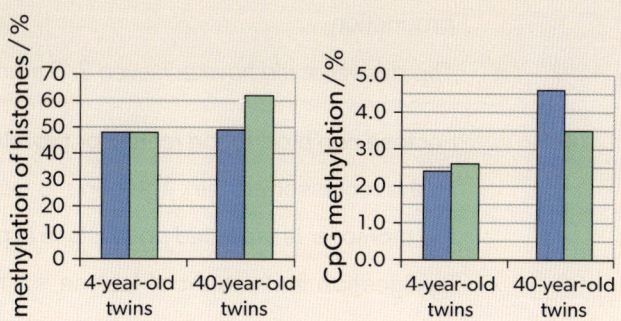

a) (i) Describe the histone methylation patterns of the twins.
   (ii) Explain how the methylation of histones affects gene expression.
b) Suggest a reason for the differences in CpG methylation patterns between the twins.

Solution

a) (i) At 4 years, both twins have 47% methylation of histones, but at 40 years old, one twin has nearly 50% while the other has 62%.
   (ii) Methylation of histones changes the structure of the nucleosome. Therefore, if the condensed chromatin is transformed into a more relaxed structure there are greater levels of gene transcription. If the methylation condenses the chromatin even more, transcription is repressed.
b) Methylation patterns depend on environmental factors such as diseases, hormones, drugs and lifestyle, such as feeding and exercise. When the twins were small, they probably lived together and were exposed to similar environmental factors. As they grew, these factors changed.

External factors impacting gene expression

There are external factors that can impact the pattern of gene expression. For example, hormones in mammals and lactose or tryptophan in bacteria can regulate the expression of mRNA. Synthesis of the amino acid tryptophan is regulated by negative feedback.

When there is a lot of tryptophan, it joins a repressor protein that at the same time joins a DNA sequence close to the promoter in the gene it regulates, inactivating the synthesis of tryptophan.

> **Sample student answer**
>
> The image shows the regulation of the gene responsible for producing lactase. 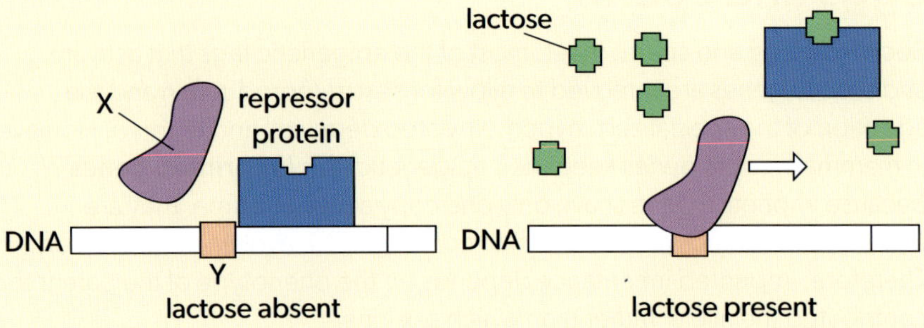
>
> [Diagram showing DNA with repressor protein X binding at site Y when lactose is absent, and lactose binding to the repressor protein when lactose is present]
>
> a) Identify:
>
>   (i) X, the enzyme which copies a DNA sequence. [1]
>
>   *This answer could have achieved 0/1 marks:*
>
>   Polymerase
>
> ▼ *Given that there are different polymerases, the answer had to refer to RNA polymerase.*
>
>   (ii) Y, non-coding DNA at the start of a gene. [1]
>
>   *This answer could have achieved 1/1 marks:*
>
>   Promoter
>
> b) Explain the role of lactose in the expression of the gene for lactase production. [3]
>
>   *This answer could have achieved 3/3 marks:*
>
>   Lactose can bind to the repressor protein and prevents inhibition by the repressor protein. The gene to make lactase is turned on as the promoter is activated and lactase is produced. Lactase reduces by digestion the amount of lactose, so the repressor protein binds again to the DNA and stops the manufacture of lactase.
>
> c) State **one** reason that identical twins may show different methylation patterns as they grow older. [1]
>
>   *This answer could have achieved 1/1 marks:*
>
>   Disease
>
> ▲ *Other causes are a different diet or a different environment, for example a polluted one.*

- What mechanisms are there for inhibition in biological systems?
- In what ways does the environment stimulate diversification?

The mRNAs encoding hormone receptors are commonly regulated by their own hormones to create autoregulatory feedback loops. Oestradiol has distinct negative and positive effects on the stabilities of different mRNAs in each tissue. Oestradiol is secreted by the ovaries and placenta in case of pregnancy. It travels through blood to the target cell where it crosses the cell membrane and binds to an oestradiol receptor in the cytoplasm. This receptor enters the nucleus, binding to the DNA and changing the transcription, therefore the expression of genes.

D2.3 Water potential

You should know:
- water is a universal solvent.
- cells that lack a cell wall swell and burst in hypotonic solutions and shrink and crenate in hypertonic solutions.
- cells with a cell wall develop turgor pressure in hypotonic solution and plasmolyse in hypertonic medium.
- isotonic solutions are used in intravenous fluids and to bathe organs for transplant.

Additional higher level:
- water moves from a higher to a lower water potential.
- in cells with walls, water potential is solute potential plus pressure potential.

You should be able to:
- explain that water moves from less concentrated to more concentrated solutions.
- distinguish between hypotonic, isotonic and hypertonic solutions.
- explain and predict the movement of water by osmosis.
- analyse data to deduce isotonic solutions.

Additional higher level:
- define water potential as the potential energy of water per unit volume.
- explain the changes in plant tissues bathed in hypotonic and hypertonic solutions.

Water is a solvent

Molecules interact with their solvent according to their polarity by solvation. In the case of water, this solvation is called hydration. Hydrogen bonds are formed between solute and water molecules, and both positive and negative ions are attracted to the different polarity of water.

> You learned about the hydrogen bonds between water molecules in subtopic A1.1.

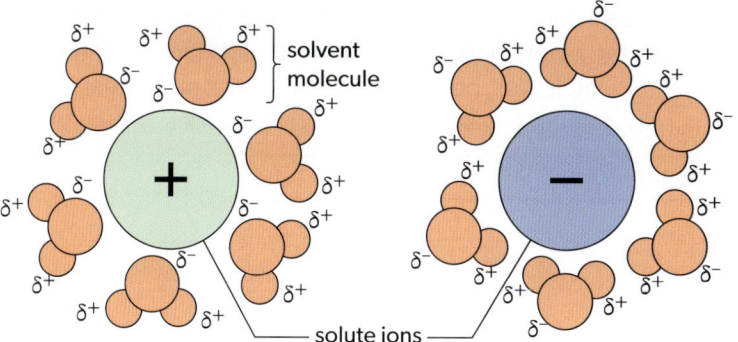

▲ Figure 6 Solvation

Water movement by osmosis

Osmosis is the net movement of water molecules across a semi-permeable membrane in a direction that equalizes the solute concentrations on either side. Water therefore moves from a less concentrated (**hypotonic**) solution to a more concentrated (**hypertonic**) solution until it reaches an equilibrium. In an **isotonic** environment there is dynamic equilibrium rather than no movement of water.

- A **hypotonic** solution has less solute than the solution compared.
- A **hypertonic** solution has more solute than the solution compared.
- An **isotonic** solution has equal solute concentration to the solution compared.

Example 7

In an experiment, potato strips with a mass of 2 g were placed in different sucrose solutions, with three replicates per solution. After 30 minutes, they were dried with a paper towel, and the mass was measured. The mean and standard deviation are shown in the table.

Concentration of sucrose solution / mol dm^{-3}	Mean mass / g	Standard deviation
0.00	2.80	0.3
0.20	2.20	0.2
0.40	1.82	0.2
0.60	1.72	0.1
0.80	1.46	0.2
1.00	1.40	

a) Identify the hypotonic solutions.
b) Predict the range in which the isotonic solution could be found.
c) Describe what happens when the potato strips are placed in the 0.80 mol dm^{-3} solution.
d) Explain the effect of the 0.20 mol dm^{-3} solution on the cells of the potato strips.
e) The potato strips at 1.00 mol dm^{-3} had a mass of 1.40, 1.30 and 1.50 g. Calculate the standard deviation (SD) and standard error (SE) for this concentration.

$$SD = \sqrt{\frac{\Sigma(x-\bar{x})^2}{n-1}} \qquad SE = \frac{SD}{\sqrt{n}}$$

Solution

a) 0.00 and 0.20 mol dm^{-3}
b) Between 0.20 and 0.40 mol dm^{-3} (excluding these values).
c) Water leaves the potato cells. The cells become flaccid, as the cell membrane separates from the cell wall.
d) The solute concentration of the cells of the potato tissues decreases as water enters the potato by osmosis, until an equilibrium is reached. Water still enters and leaves, but there is no further net movement of water. The cells are turgid, making the potato strips larger.

e) $SD = \sqrt{\frac{(1.40-1.40)^2 + (1.50-1.40)^2 + (1.30-1.40)^2}{3-1}} = 0.1$

$SE = \frac{0.1}{\sqrt{3}} = 0.058$

Effects of water movement on cells with and without a cell wall

- **Crenation** is the curved shape a cell without a wall has due to the loss of water when placed in a hypertonic solution.
- **Plasmolysis** is the loss of water by a cell with a cell wall when placed in a hypertonic solution.

Cells that do not have a cell wall swell and burst in a hypotonic medium and shrink and **crenate** in a hypertonic medium. Freshwater unicellular organisms have contractile vacuoles to remove excess water from the cell. Tissue fluid in multicellular organisms needs to be maintained as isotonic to the cells to prevent harmful changes. Cells that have a cell wall swell and become turgent (but do not burst) in a hypotonic medium and shrink and plasmolyse in a hypertonic medium. When **plasmolysis** occurs, the membrane separates from the cell wall.

Example 8

In an experiment on osmosis, red blood cells were immersed in a salt solution for 2 hours. The drawings show the appearance of these cells before and after immersion in the salt solution.

before immersion

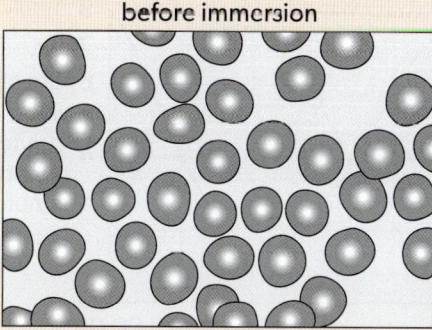

after immersion

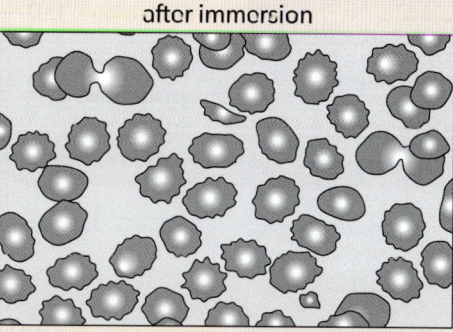

What explains the observed changes?

A. The salt solution was hypertonic and entered the red blood cells.
B. The salt solution was hypotonic and disrupted the membranes of the red blood cells.
C. The salt solution was hypertonic and water moved into it from the red blood cells.
D. The salt solution was hypotonic and mineral salts were lost from the red blood cells.

Solution

The correct answer is **C**, as the cells shrink and suffer crenation because water leaves the red blood cells towards the more concentrated (hypertonic) solution.

Medical applications of isotonic solutions

There are medical applications of isotonic solutions. Intravenous fluids are given as part of medical treatment, for example, in cases of cholera. Organs ready for transplantation are kept in an isotonic solution to avoid tissue damage.

Water potential

It is impossible to measure the absolute quantity of the potential energy of water, so values relative to pure water at atmospheric pressure and 20°C are used. The units are usually kilopascals (kPa). Water moves from a higher **water potential** to a lower water potential because it moves in favour of a potential energy gradient. Water potential is the sum of the **solute potential** plus the **pressure potential** in cells with cell walls, and the solute potential only in cells without cell walls. The water potential of pure water is zero. The solute potentials are negative and the higher the solutes in the solution, the lower the solute potential. Pressure potentials are generally positive inside cells, although negative pressure potentials occur in xylem vessels where sap is being transported under tension.

$$\text{water potential } (\psi_w) = \text{solute potential } (\psi_s) + \text{pressure potential } (\psi_p)$$

The transpiration pull of water from the soil to the leaves causes water to be under negative pressure, decreasing the water potential below zero. The effect of osmotic concentration slightly decreases the leaf water potential too.

- **Water potential** (ψ_w) is the potential energy of water per unit volume.
- **Solute potential** (ψ_s) is the pressure applied on a solution to avoid water from entering a cell across a semi-permeable membrane.
- **Pressure potential** (ψ_p) or turgor pressure is the pressure on the cell wall due to the inflow of water.

 You learned about water transport in plants in subtopic B3.2.

Continuity and change

Examiner tip

When reading a graph with two axes, make sure you read the correct scale for the value you are asked for.

Sample student answer

In an experiment, the pressure inside the xylem and the rate of water flow were measured at different times of the day in common oak trees (*Quercus robur*). The graph shows the results obtained during one day.

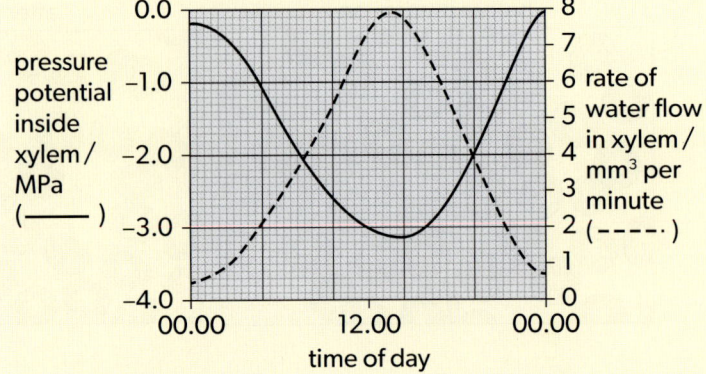

a) State the rate of water flow in the xylem at 12.00 hours. [1]

This answer could have achieved 0/1 marks:

> 7.6

▼ Although the value is correct, the units (mm^3 per minute) are missing.

b) Calculate the change in pressure potential in the xylem from 00.00 hours to 12.00 hours. [1]

This answer could have achieved 1/1 marks:

> Pressure in xylem at midnight = −0.2 MPa
> Pressure in xylem at noon = −3.0 MPa
> Difference = −2.8 MPa

▲ When asked for a change it is important to keep the negative sign or mention it is less or decreases.

c) Explain the changes in water flow at different times of the day. [3]

This answer could have achieved 3/3 marks:

> At midnight the stomata are closed and there is little transpiration, therefore little water flow in the xylem. As the sun comes out, the leaves of the oak start doing photosynthesis. This requires gaseous exchange, so the stomata open. Water is lost by transpiration, causing the transpiration pull. Water flows from an area of high pressure (soil) to an area of low pressure (leaf air gaps). The cohesion and adhesion of water molecules allow for this transport.

d) Suggest a reason for the change in pressure potential in the xylem. [1]

This answer could have achieved 0/1 marks:

> Xylem supplies water to tissues.

▼ The answer does not explain the change in pressure. Pressure potential in the xylem is the pressure exerted by the rigid cell wall that limits intake of water. It causes turgor pressure, making the water move up the xylem tissue. At night the pressure in the xylem is high, as the roots are loading minerals. These minerals attract water into the xylem, increasing the pressure. During the day, while transpiration is rapid, the xylem solution becomes diluted by the entering water. Water is lost by transpiration, decreasing the pressure in the xylem.

Practice problems

1. The following events occur in mitosis.

   I. Supercoiling of chromosomes

   II. Movement of sister chromatids to opposite poles

   III. Attachment of spindle microtubules to centromeres

202

What is the correct sequence of events?

A. I, II, III   C. I, III, II
B. II, I, III   D. III, II, I

2. In which stage of meiosis is the number of chromosomes halved?

   A. Prophase I
   B. Anaphase I
   C. Metaphase II
   D. Telophase II

3. In an experiment, potato strips of the same length were placed in different sugar concentration solutions. After 20 minutes, the potatoes were dried, and the length was recorded.

   What change would be seen in potatoes placed in a hypotonic solution?

   A. They were longer and turgent.
   B. They were shorter and flaccid.
   C. They were the same length but flaccid.
   D. They were longer and plasmolysed.

- What variables influence the direction of movement of materials in tissues?
- What are the implications of solubility differences between chemical substances for living organisms?

D3.1 Reproduction

You should know:
- the menstrual cycle is controlled by negative and positive feedback mechanisms involving ovarian and pituitary hormones.
- the process of fertilization in humans.
- how flowering plants carry out sexual reproduction.
- methods of promoting cross-pollination in flowering plants.
- self-incompatibility mechanisms increase genetic variation within species.

Additional higher level:
- the developmental changes of puberty are controlled by gonadotropin-releasing hormone and steroid sex hormones.
- fertilization involves mechanisms that prevent polyspermy.
- implantation of the blastocyst in the endometrium is essential for the continuation of pregnancy.
- the placenta facilitates the exchange of materials between the mother and foetus.
- hormone replacement therapy has a small risk of coronary heart disease (CHD).

You should be able to:
- distinguish between sexual and asexual reproduction.
- explain the role of meiosis and fusion of gametes in the sexual life cycle.
- distinguish between male and female sexes in reproduction.
- draw and annotate diagrams of male-typical and female-typical reproductive systems.
- explain the roles of follicle-stimulating hormone (FSH), luteinizing hormone (LH), oestradiol and progesterone in the menstrual cycle.
- outline the use of hormones in in vitro fertilization (IVF) treatment.
- draw half-views of insect-pollinated flowers.
- distinguish between pollination and seed dispersal.

Additional higher level:
- explain the formation of gametes in spermatogenesis and oogenesis.
- describe the use of monoclonal antibodies in pregnancy tests to detect human chorionic gonadotropin (hCG) during early pregnancy.
- describe how hormones regulate pregnancy and childbirth.

Continuity and change

You learned about meiosis in subtopic D2.1 and will learn about homeostasis in subtopic D3.3.

You learned about the adaptations of sperm and egg cells in subtopic B2.3.

Comparing sexual and asexual reproduction

Reproduction can be asexual or sexual. Asexual reproduction does not require another organism and produces offspring that are genetically identical to the parent. Asexual reproduction occurs in individuals that are adapted to an existing environment, whereas sexual reproduction produces offspring with new gene combinations. This variation gives the organisms an advantage for adaptation to a changed environment.

Meiosis and fusion of gametes in the sexual life cycle

Sexual reproduction involves the development and fusion of haploid gametes. Gametes therefore need to undergo a reduction in the number of chromosomes through meiosis. Fusion of gametes or fertilization produces new combinations of alleles.

Example 1

Sperm and egg cells are adapted for sexual reproduction.
a) Which part of a sperm provides energy for movement?
b) Distinguish between the structure of the male and female gametes in humans.
c) Explain why humans produce more male gametes than female gametes.

Solution
a) Mitochondria (middle section).
b) The male gamete is smaller, with less food reserves than the egg. Only the sperm has a tail and an acrosome in the head. In the ovum the mitochondria are scattered in the cytoplasm while in the sperm they are clustered in the neck. Only the ovum is surrounded by follicle cells.
c) During sexual reproduction, the male gamete travels to the female gamete, so more male gametes are needed to ensure fertilization occurs.

- **Testes** are the male reproductive organs that produce sperm and testosterone.
- The **scrotum** holds the testes.
- The **epididymis** is a duct where sperm are stored until ejaculation.
- **Sperm ducts** transfer the sperm from the epididymis to the urethra during ejaculation.

Anatomy of the human male and female reproductive systems

Example 2

The diagram shows a section through the male reproductive system. Which structure represents the prostate gland?

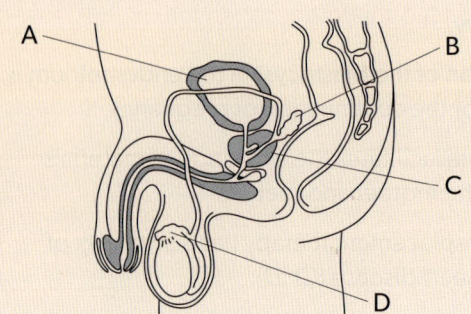

Solution

C shows the prostate gland. **A** shows the bladder, which is part of the urinary system and where urine is stored. **B** shows the seminal vesicles that produce seminal fluid, and **D** shows the epididymis where sperm are stored.

Example 3

The diagram shows the female reproductive system.

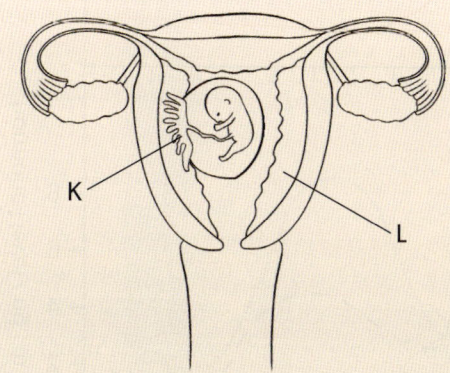

Which structures do K and L identify?

	K	L
A.	endometrium	uterine wall
B.	placenta	endometrium
C.	amnion	placenta
D.	foetus	uterine wall

Solution

The correct answer is **B**, as the uterine wall is outside the endometrium and the amnion surrounding the foetus is in the centre of the diagram.

- **Seminal vesicles** and the **prostate gland** secrete fluid that makes semen.
- The **penis** is the organ that penetrates the vagina during sexual intercourse.
- **Ovaries** are the organs where eggs, oestradiol and progesterone are produced.
- **Oviducts** are the canals that receive the eggs at ovulation and are the site of fertilization.
- The **uterus** is the organ where the embryo is implanted after fertilization.
- The **cervix** is the lower part of the uterus. Its muscles dilate to provide a birth canal.
- The **vagina** is the duct joining the vulva with the uterus, allowing sexual intercourse.
- The **vulva** protects the internal parts of the female reproductive system.

Hormonal control of the menstrual cycle

The menstrual cycle is controlled by negative and positive feedback mechanisms involving ovarian and pituitary hormones. **Follicle stimulating hormone (FSH)** stimulates the development of follicles and secretion of **oestradiol** by the follicle wall. Oestradiol stimulates the repair and thickening of the endometrium after menstruation. High levels of oestradiol inhibit the secretion of FSH (negative feedback) and stimulate **luteinizing hormone (LH)** secretion. LH stimulates the completion of meiosis and allows ovulation. LH also promotes the development of the wall of the follicle after ovulation into the corpus luteum, which secretes oestradiol (positive feedback) and **progesterone**. Progesterone promotes the thickening and maintenance of the endometrium. It also inhibits FSH and LH secretion by the pituitary gland (negative feedback).

- **FSH** stimulates the development of follicles and the secretion of oestradiol by the follicle wall.
- **Oestradiol** stimulates the repair and thickening of the endometrium.
- **LH** stimulates the development of follicles, leading to ovulation.
- **Progesterone** promotes the thickening and maintenance of the endometrium.

Continuity and change

Example 4

The graph shows the levels of hormones during the menstrual cycle.

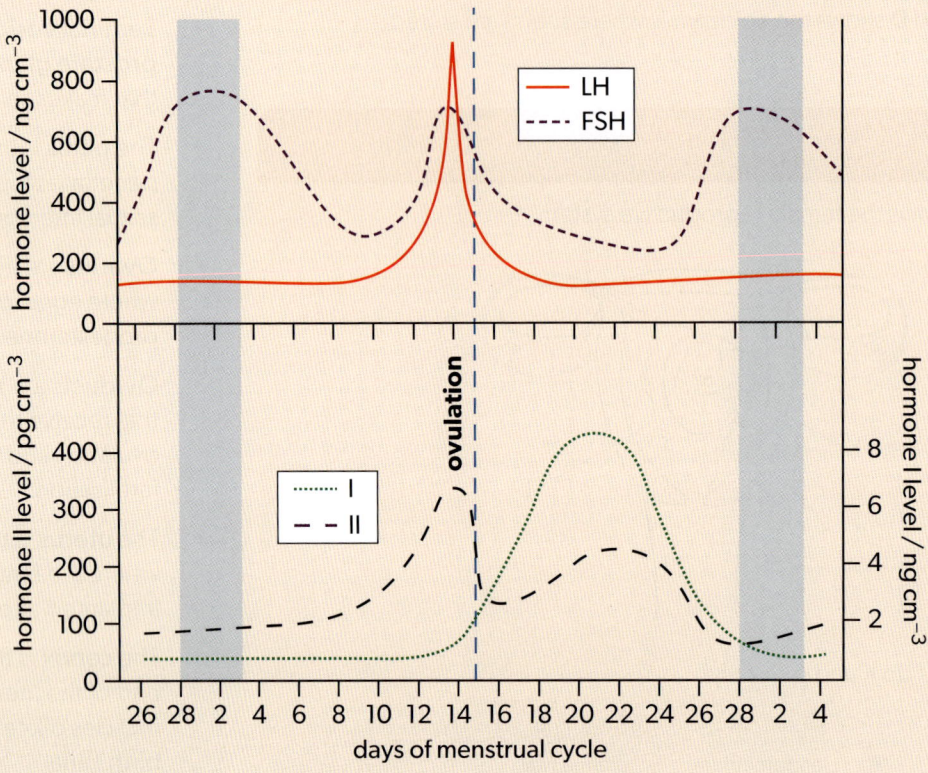

a) Identify hormones I and II.
b) Outline the roles of FSH in the menstrual cycle.
c) FSH is secreted by the pituitary gland. During pregnancy, FSH secretion is inhibited. Suggest how FSH secretion could be inhibited during pregnancy.

Solution

a) I is progesterone and II is oestradiol.
b) FSH is a peptide hormone that stimulates the development of the follicle, which contains an oocyte. It also stimulates the production of oestradiol by the follicle.
c) High levels of oestradiol inhibit the secretion of FSH by negative feedback.

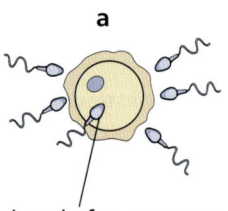

▲ Figure 1 Fertilization in humans

Fertilization in humans

The process of fertilization in humans occurs in the oviduct (fallopian tube) when the sperm's cell membrane fuses with an egg cell membrane. The sperm nucleus enters the egg, with destruction of the tail and mitochondria. The nuclear membranes of sperm and egg nuclei dissolve. The newly formed cell, still with condensed chromosomes, undergoes mitosis and produces two diploid nuclei.

In vitro fertilization

In vitro fertilization (IVF) is used in cases of infertility. FSH and LH are used to stimulate development of follicles, causing ovulation of many ova (superovulation). Human chorionic gonadotropin (hCG) is used to mature the ova. Fertilization is performed outside the body and fertilized embryos are then implanted in the woman. Progesterone is used to ensure that the uterus lining is maintained. Scientists are aware that the drugs women take in fertility treatment pose potential risks to health. Nevertheless, this method is widely used throughout the world in treating couples with fertility problems.

Sexual reproduction in flowering plants

Flowering plants reproduce by sexual reproduction. The flowers are the reproductive structures produced by the shoot apical meristem that contain the gametes. Female gametes or ovules are found in the ovary. Male gametes or pollen are found in the anther. Success in plant reproduction depends on **pollination**, **fertilization** and **seed dispersal**. Plants can pollinate themselves in self-pollination or pollinate with plants of different genetic composition through cross-pollination. **Hermaphroditic** plants can undergo self-pollination and are not dependent on pollinators. Most flowering plants use mutualistic relationships with pollinators for sexual reproduction while some plants are wind pollinated.

Plants have different methods to promote cross-pollination, therefore avoiding self-pollination. For example, they can have different maturation times for pollen and stigma, female flowers separated from the male flowers in the same plant by enough distance that these do not self-pollinate, or directly separate male and female plants. Self-pollination leads to inbreeding, which decreases genetic diversity and vigour. Some plants have self-incompatibility mechanisms to increase genetic variation within a species. Genetic mechanisms in many plant species ensure male and female gametes fusing during fertilization are from different plants.

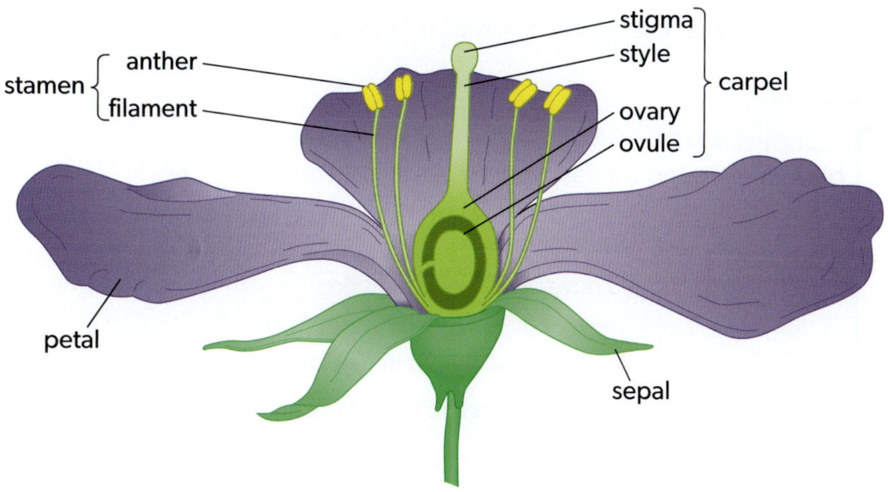

- **Pollination** is the transfer of pollen from the anther to the stigma.
- **Fertilization** is the union of the female and male gametes.
- **Seed dispersal** is the transport of the seed away from the parent plant.
- **Hermaphroditic** plants have flowers of both sexes.

▲ **Figure 2** Features of an insect-pollinated flower

Sample student answer

To prevent transfer of pollen from an anther of one plant to the stigma of the same plant (self-pollination), the sunflower (*Helianthus spp.*) anther sheds its pollen before the stigma is mature enough to receive it. Early in the morning the anther is exposed by elongation of the filaments. The anthers open at this time to release their pollen (anthesis). The stigma appears above the anthers by late afternoon, and by the following morning it is fully receptive.

To see how the filament (F) and the style (S) are affected by light, their lengths were measured at time intervals starting 12 hours before anthesis (−12). Some plants were grown in continuous white light (L24) and some plants grown under cycles of 16 hours white light followed by 8 hours dark (L16/D8). The results are shown in the graph.

Continuity and change

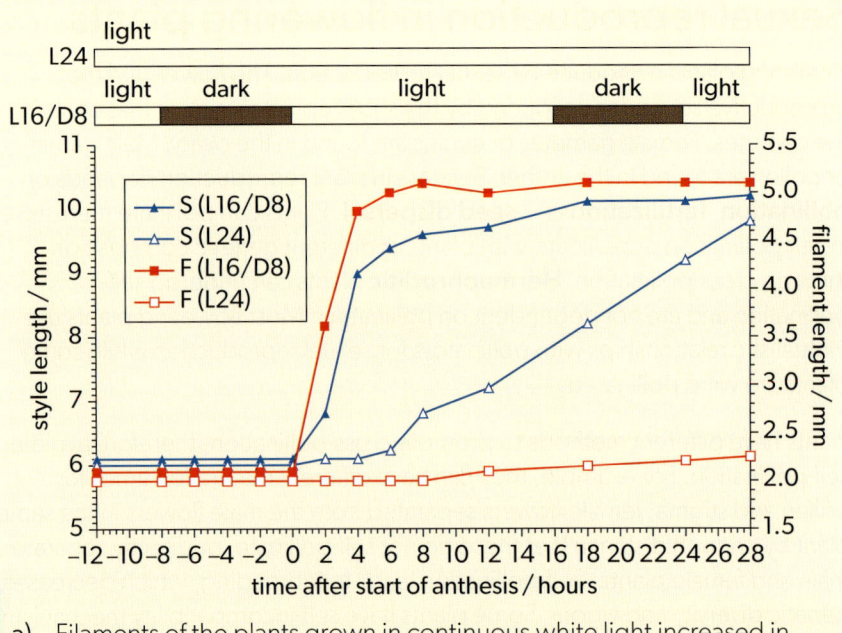

▶ Source: Lobello, G., Fambrini, M., Baraldi, R., Lercari, B., & Pugliesi, C. (2000). Hormonal influence on photocontrol of the protandry in the genus Helianthus. Journal of Experimental Botany, 51(349), 1403–1412. https://doi.org/10.1093/jexbot/51.349.1403

a) Filaments of the plants grown in continuous white light increased in length by 0.25 mm in the 28 hours after anthesis. Calculate how much the filaments of the plants grown in alternating white light and dark increased during the same period. [1]

This answer could have achieved 1/1 marks:

• 3 mm

▲ Any value between 2.7 and 3.1 mm scored a mark.

b) Compare and contrast the increase in the length of the style in the plants grown in continuous white light with those grown in alternating white light and dark. [2]

This answer could have achieved 1/2 marks:

• Cyclic light makes style grow almost immediately while with continuous light it takes longer to start to grow.

▲ Marks for saying that cyclic light makes the style grow almost immediately whereas with continuous light it takes longer to start to grow. L16/D8 starts growing in the first hour whereas the L24 style starts growing after 6 hours—growth is more gradual in L24. With continuous light the style grows less (to 9.8 mm) whereas with cyclic light it grows to 10.2 mm.

▼ The answer did not give the similarity, which was that there is little difference after 28 hours. In both cases growth only starts with anthesis.

c) Explain the disadvantages to a plant of self-pollination. [2]

This answer could have achieved 2/2 marks:

• Self-pollination reduces variation and increases chances of disease.

▲ The answer is correct, as in self-pollination there is no new combination of alleles, so no variation for natural selection. The offspring are more susceptible to infectious diseases and more prone to genetic disease. In inbreeding, it is more likely to be homozygous for a disease.

Examiner tip

When you compare and contrast, you give an account of similarities and differences between two (or more) items or situations, referring to both (all) of them throughout.

Dispersal and germination of seeds

During germination, the seed reserves of starch, protein and lipid are mobilized or degraded. The seed reserve present in the highest amount is the one that is most heavily used during germination. Some plants require oils such as sorghum, others such as lentils require proteins and others such as rice require starch.

Hormone changes at puberty

At puberty, sexual maturation occurs. Increased release of gonadotropin-releasing hormone (GnRH) by the hypothalamus in childhood triggers the onset of increased LH and FSH release. The production of steroid sex hormones (oestradiol, progesterone and testosterone) stimulates the growth and development of secondary sexual characteristics.

Spermatogenesis and oogenesis in humans

Spermatogenesis and oogenesis both involve mitosis, cell growth, two divisions of meiosis and differentiation. Spermatogenesis is the development of the male gametes (haploid sperms) and occurs in the seminiferous tubules of the testes. Oogenesis is the formation of the female gamete (a haploid ovum) from a diploid germ cell and it occurs in the ovaries.

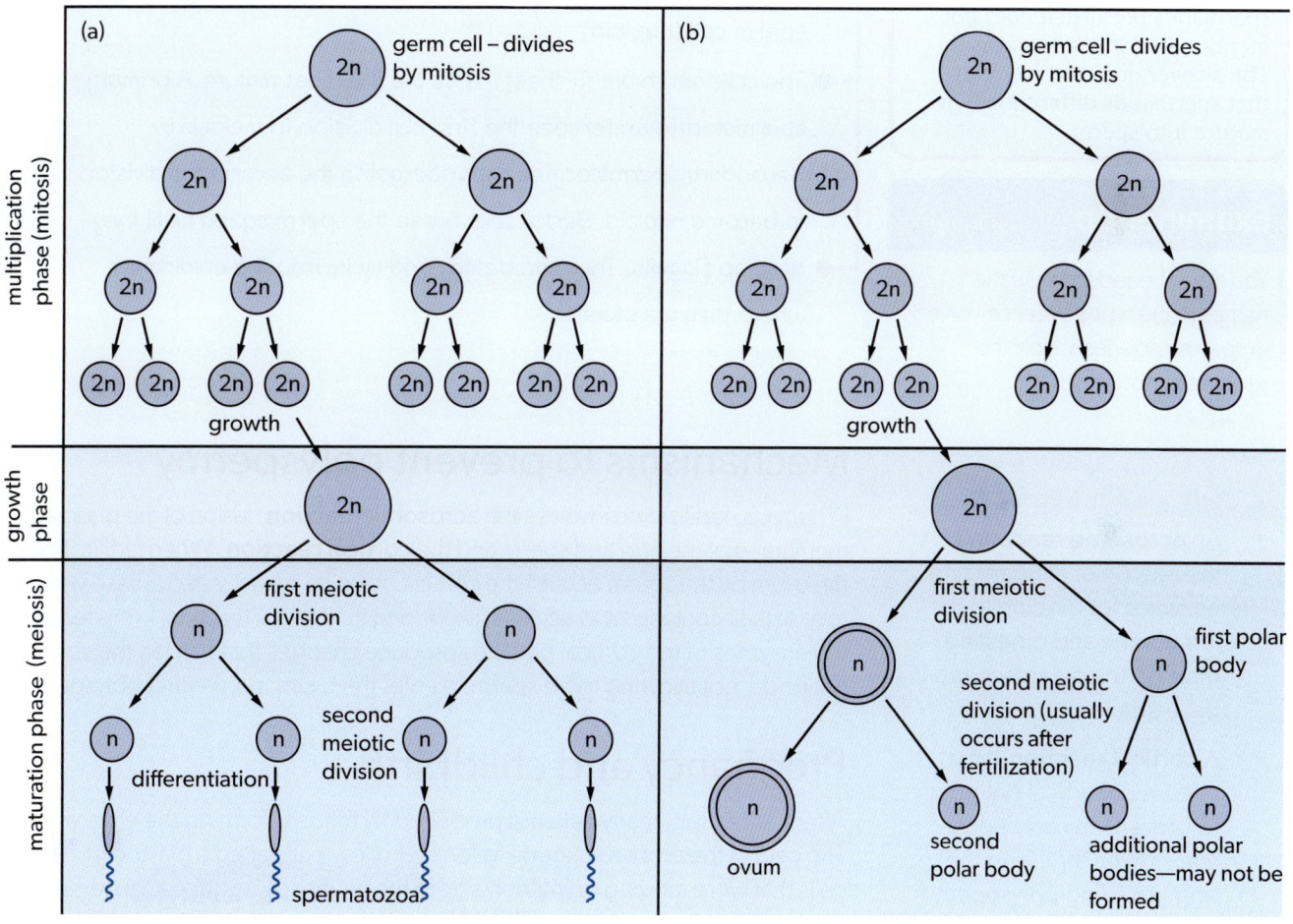

▲ Figure 3 (a) Spermatogenesis and (b) oogenesis

The germ cells in the ovary of the foetus divide by mitosis (multiplication phase). Around the fourth month of development, the cells start dividing by meiosis. Just before or shortly after birth, the oogonia differentiate (growth phase). These cells enter prophase of meiosis I and primary follicles are formed. Each primary follicle consists of one ovum and follicle cells around it. At the start of each menstrual cycle, a primary follicle is stimulated by FSH to mature into a secondary oocyte by completing meiosis I (maturation phase). Of the two cells produced, only one survives. The other cell turns into a smaller cell called a polar body, which degenerates. The surviving cell halts meiosis II in metaphase II. Upon fertilization, the cell will finish the second meiotic division to transform the ovum and another polar body. The result is one ovum which is much larger than the male sperm.

Continuity and change

▲ The answer scored a mark for primary spermatocytes dividing by meiosis I into secondary spermatocytes. The second mark is for Sertoli or nurse cells providing nourishment to these developing cells.

▼ The answer could have mentioned that spermatogonia (2n) are undifferentiated germ cells. The spermatogonia mature and divide by mitosis into primary spermatocytes. The answer did not score a mark for secondary spermatocytes dividing into spermatids because it does not mention that it is by meiosis II. The answer does not mention that spermatids differentiate or mature into sperm.

Examiner tip

You do not need to know the names of the types of cells involved in spermatogenesis, only the processes occurring.

- An **acrosome reaction** is the release of antigens and enzymes for binding to the membrane and digesting through the ovum's zona pellucida.

- A **cortical reaction** is the release of cortical granules to prevent any additional sperm from fertilizing the egg, avoiding polyspermy.

Sample student answer

Describe the processes occurring in the different cell types in the seminiferous tubules that are involved in the process of spermatogenesis. [4]
This answer could have achieved 2/4 marks:

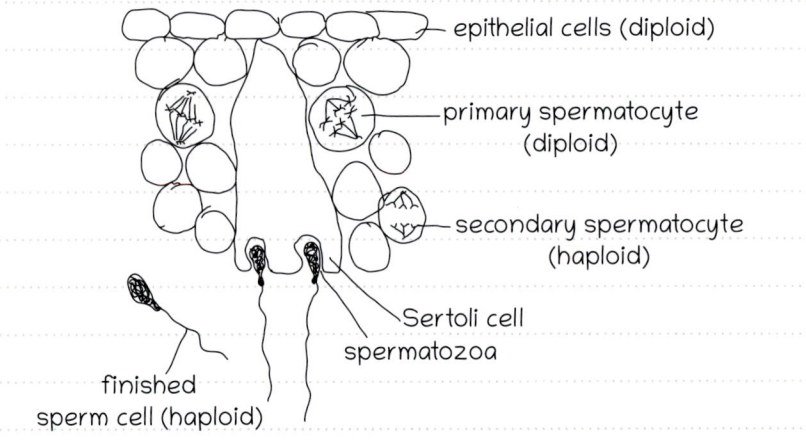

The cells which are furthest inside are the most mature. A primary spermatocyte undergoes the first cell division in meiosis I. Secondary spermatocytes are undergoing the second cell division to become haploid. Sertoli cells nurse the spermatozoa until they develop flagella. They can detach and swim into the epididymis, where they are stored.

Mechanisms to prevent polyspermy

In humans, fertilization involves the **acrosome reaction**, fusion of the plasma membrane of the egg and sperm and the **cortical reaction**. When fertilization of the ovum by the sperm occurs, the cortical granules are activated. These vesicles release their contents by exocytosis, producing the cortical reaction. In this reaction, the enzymes of the cortical granules produce changes that harden the zona pellucida, not allowing more sperm to enter the ovum, preventing polyspermy.

Pregnancy and childbirth

After fertilization, many cells are produced by mitosis, forming the embryo. The cells in the embryo migrate to form a hollow ball called a blastocyst. The cells that were feeding on nutrients from the ovum now require an external supply of food. Implantation of the blastocyst in the endometrium is essential for the continuation of pregnancy. After the third month of pregnancy, the placenta is fully formed. The placenta facilitates the exchange of materials between the mother and foetus due to the large surface area of the placental villi. The placenta allows the foetus to be retained in the uterus to a later stage of development than in mammals that do not develop a placenta.

The pregnancy is maintained by hormonal control. hCG stimulates the corpus luteum in the ovary to secrete progesterone during early pregnancy, which helps maintain the endometrium. Oestradiol and progesterone are secreted by the placenta once it has formed. Progesterone inhibits the secretion of oxytocin by the pituitary gland, thus inhibiting contractions of the uterus during pregnancy.

Birth is mediated by positive feedback involving oestradiol and oxytocin. Hormones produced by the foetus inhibit the secretion of progesterone.

Oxytocin induces the wall of the uterus to contract. The contractions are perceived by sensory neurons that send messages to the pituitary gland to produce more oxytocin through positive feedback. As the cervix dilates and the amniotic sac is broken, birth will take place as the baby is pushed out through the cervix and vagina.

Monoclonal antibodies have been used for pregnancy-detection kits. In these kits, the monoclonal antibodies used are anti-human chorionic gonadotropin (anti-hCG). hCG is a glycoprotein hormone that is synthesized by the embryo or developing placenta, and therefore appears only during pregnancy.

Hormone replacement therapy and the risk of coronary heart disease

Hormone replacement therapy (HRT) is a treatment for relief of menopausal symptoms. These symptoms, caused by lower levels of oestradiol at menopause, include hot flashes, sleep disturbances and vaginal dryness. The therapy consists of combining oestradiol plus progestogen. This treatment helps in the prevention of osteoporosis. Risks include increased risk of breast cancer, stroke and an increase in blood clots in the veins.

Nature of science

Women undergoing HRT had reduced incidence of coronary heart disease (CHD), so a causation relationship was established in early studies. Later, a small increase in the risk of CHD was detected in randomized controlled trials of HRT. This confirmed that the correlation between HRT and decreased incidence of CHD was not actually a cause-and-effect relationship. HRT patients had a higher socioeconomic status, and this status has a causal relationship with lower risk of CHD.

- How can interspecific relationships assist in the reproductive strategies of living organisms?
- What are the roles of barriers in living systems?

D3.2 Inheritance

You should know:

- ✔ inheritance depends on haploid gametes in parents and their fusion to form a diploid zygote.
- ✔ the effects of dominant and recessive alleles on phenotype.
- ✔ there are several alleles of a gene in the gene pool but an individual only inherits two.
- ✔ how phenylketonuria (PKU) is inherited.
- ✔ ABO blood groups as an example of multiple alleles.
- ✔ sex is determined in humans on the X and Y chromosomes.
- ✔ haemophilia is a sex-linked genetic disorder.
- ✔ continuous variation is due to polygenic inheritance and/or environmental factors.

Additional higher level:

- ✔ unlinked genes segregate independently as a result of meiosis.
- ✔ gene loci are said to be linked if on the same chromosome.
- ✔ chi-squared tests are used to determine whether the difference between an observed and an expected frequency distribution is statistically significant.

You should be able to:

- ✔ describe methods for carrying out genetic crosses in flowering plants.
- ✔ distinguish between genotype and phenotype.
- ✔ explain phenotypic plasticity by varying patterns of gene expression.
- ✔ distinguish between incomplete dominance and codominance.
- ✔ deduce patterns of inheritance of genetic disorders in pedigree charts.
- ✔ distinguish between discrete and continuous variables.
- ✔ construct box-and-whisker plots to represent data for a continuous variable.

Additional higher level:

- ✔ calculate the predicted genotypic and phenotypic ratio of offspring of dihybrid crosses involving unlinked autosomal genes.
- ✔ explore genes and their polypeptide products in databases.
- ✔ identify recombinants in crosses involving two linked genes.

Inheritance and genetic crosses

During reproduction, haploid gametes from one parent and haploid gametes from the other parent fuse to form a diploid zygote, which inherits the information from both parents. Therefore, a diploid cell has two copies of each autosomal gene. This pattern of inheritance is common to all eukaryotes with a sexual life cycle. In plants, the pollen contains the male gametes, and the female gametes are contained in the ovary, so pollination is required to carry out the cross. Some plants produce both male and female gametes on the same plant, allowing self-pollination and therefore self-fertilization. Genetic crosses are widely used to breed new varieties of crop or ornamental plants.

Example 5

Seeds in peas can be round (R) or wrinkled (r). In a cross between homozygous peas with round seeds and wrinkled seeds all the F1 generation had heterozygous round seeds.

a) Identify the dominant allele.
b) State the genotypes for the peas in the P generation.
c) Construct a Punnett grid to shows the inheritance of the F1 generation.
d) State the phenotype of peas in the F1 generation.
e) Predict the genotype ratios of peas in an F2 generation.

Solution
a) Round (R)
b) RR and rr
c) R: round seeds and r: wrinkled seeds

	R	R
r	Rr	Rr
r	Rr	Rr

d) Round seeds
e) 1 RR : 2 Rr : 1 rr

Assessment tip

It is a convention to use the same letter for alleles of the same gene, using a capital letter for the dominant characteristic.

- A **dominant** allele is one that produces its characteristic phenotype.
- **Alleles** are various specific forms of a gene.
- A **recessive** allele only produces its characteristic phenotype when homozygous.
- **Homozygous** has two identical alleles of a particular gene.
- **Heterozygous** has two different alleles of a particular gene.

Some traits in humans are due to **genotype** only and others are due to environment only. At the same time there are traits due to interaction between genotype and environment. Epidemiology datasets reveal that human body weight distributions and specifically body mass index (BMI) are environmentally related. Skin colour in humans is an example of **continuous variation** which is partly due to the environment and partly due to polygenic inheritance.

- **Genotype** is the combination of alleles inherited by an organism.
- **Continuous variation** is the combined effect of many genes and/or the effect of the environment on genes. It can be plotted as a normal distribution curve.

Phenotype effects

Dominant alleles prevail over **recessive** ones. The latter only appear if they are in homozygosity. When there is a dominant allele, whether in **homozygous** or **heterozygous** genotype, it will produce the same **phenotype**.

Phenotypic output is defined by DNA sequence, epigenetics and environmental variables (for example, nutritional sufficiency), and their interactions. Phenotypic plasticity is the capacity to develop traits suited to the environment experienced by an organism by varying patterns of gene expression. Phenotypic plasticity is not due to changes in genotype, and the changes in traits may be reversible during the lifetime of an individual. Examples are seasonal phenotypic plasticities in coat colours of Arctic animals and caste phenotypical distinctions in honeybees.

- **Phenotypes** are the observable traits of an organism resulting from genotype and environmental factors.

Phenylketonuria is due to a recessive allele

In humans, some diseases only appear in homozygosity because they are caused by a recessive allele. Phenylketonuria (PKU) is caused by a deficiency of an enzyme that converts phenylalanine into tyrosine, causing excess phenylalanine in blood. It is a recessive genetic condition, which means it only causes disease if the person has both chromosomes of the homologous pair for that autosomal gene mutated.

SNPs and multiple alleles

A gene pool is the sum of all the population's genetic material at a given time. Gene pools change over time through evolution. **Single-nucleotide polymorphisms (SNPs)** arise by point mutations. These mutations are either neutral or not harmful and therefore survive in the gene pool, giving rise to alternative alleles to that gene. In the gene pool, there is a great number of gene alleles, but as each chromosome can only carry one allele, an individual only inherits two of them.

- **Single-nucleotide polymorphisms** are caused by point mutations that give rise to different alleles containing alternative bases at a given nucleotide position within a locus.

ABO blood groups and codominance

In **codominance**, heterozygotes have a dual phenotype. Blood group is inherited by multiple alleles. A and B are codominant, while O is recessive. I_A is the allele for group A, I_B is the allele for group B, i is the allele for group O. $I_A I_A$ and $I_A i$ are genotypes for group A, $I_B I_B$ and $I_B i$ are genotypes for group B, ii is the genotype for group O and $I_A I_B$ is the genotype for group AB. If two people with the same blood group in homozygous state have a child, the child will have the same blood group as the parents. If the mother has blood group A and the father B and both are homozygous, all the children are blood group AB.

- **Codominant alleles** are those that are simultaneously expressed in the phenotype of the individual.

There are several crosses that can be done to show the inheritance of blood groups.

For example, if the mother is A heterozygous and the father O:

	I_A	i
i	I_Ai	ii
i	I_Ai	ii

there is a chance of 50% of any children being blood group A and 50% O.

In **incomplete dominance**, one characteristic is not dominant over the other, so both persist. In marvel of Peru plants (*Mirabilis jalapa*), when dark pink-flowered plants are crossed with white-flowered plants, light pink-flowered offspring are produced.

- **Incomplete dominance** is when offspring have an intermediate phenotype.

Sex determination and sex-linked disorders

The sex chromosome in sperm determines whether a zygote develops certain male-typical or female-typical physical characteristics. The female gamete always carries an X chromosome. If the sperm that fertilizes the ovum carries a Y chromosome a male is formed (XY), and if it carries an X chromosome a female (XX). Far more genes are carried by the X chromosome than the Y chromosome.

Continuity and change

- **Sex linkage** is the phenotypic expression of an allele related to the sex chromosomes.

Example 6

Haemophilia is a **sex-linked** recessive disorder inherited on the X chromosome. Draw a Punnett grid to shows the inheritance of haemophilia in a family where the father has haemophilia and the mother is a carrier.

Solution

(X^H shows the normal allele and X^h the haemophilia allele.) For females, there is a 50% chance of being a carrier ($X^H X^h$) and a 50% chance of having haemophilia ($X^h X^h$). For males, there is a 50% chance of having haemophilia ($X^h Y$) and a 50% chance of not having haemophilia ($X^H Y$).

	X^h	Y
X^H	$X^H X^h$	$X^H Y$
X^h	$X^h X^h$	$X^h Y$

Pedigree charts

The prohibition of marriage between close relatives in many societies is due to the higher probability of inheriting a genetic disease. This can be deduced from patterns of inheritance in pedigree charts.

Example 7

What type of inheritance is shown in the pedigree?

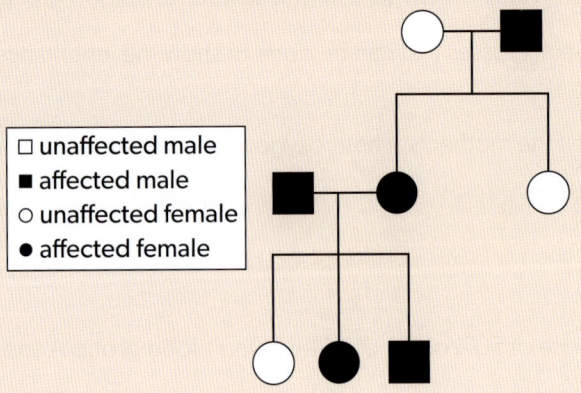

□ unaffected male
■ affected male
○ unaffected female
● affected female

A. Sex-linked recessive
B. Autosomal dominant
C. Multiple alleles
D. Codominant alleles

Solution

The correct answer is **B**, as it is showing an autosomal dominant inheritance. It is not sex-linked because if it was, the daughter in the F1 pairing would not be affected, as her mother is unaffected, or both daughters would be affected if the mother were a carrier. It is not showing multiple alleles or codominant alleles because the daughter in F2 is not affected.

Nature of science

Deductive reasoning involves progressing from general ideas to specific conclusions whereas inductive reasoning starts with specific observations to form general conclusions. A pattern of inheritance may be deduced from parts of a pedigree chart through inductive reasoning. The theory obtained may then allow genotypes of specific individuals in the pedigree to be deduced.

Continuous and discrete variation

Continuous variation is due to polygenic inheritance and/or environmental factors. For example, skin colour or height in humans is continuous. A continuous variable takes on an infinite number of possible values within a given range. In contrast, **discrete variables** are limited in number, countable and indivisible, for example ABO blood groups.

- **Discrete variation** is a discontinuous variation found in a number of distinct categories.

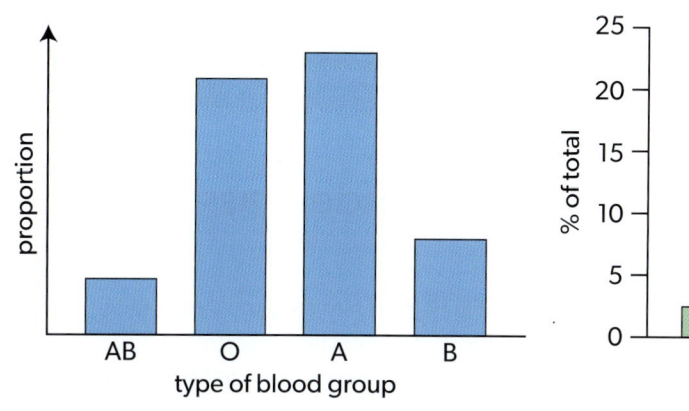
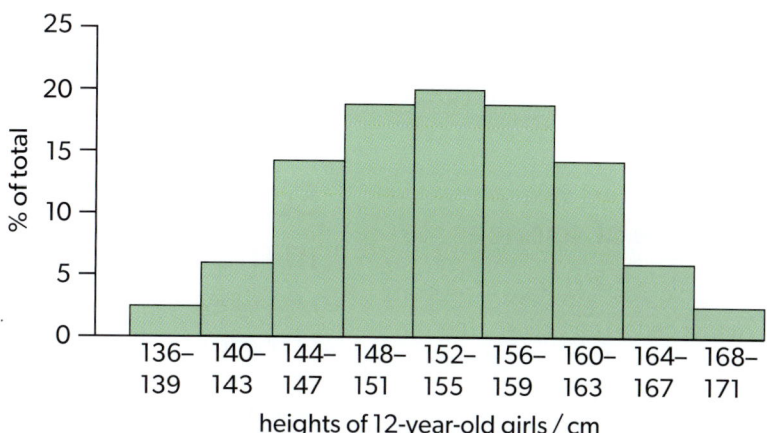

▲ Figure 4 Discrete variation (left) and continuous variation (right)

Assessment tip

The mean of a data set is found by adding all numbers and dividing by the number of values. The median is the middle value when a data set is ordered from least to greatest. The mode is the number that occurs most often in a data set.

Box-and-whisker plots

Box-and-whisker plots are used to represent data for a continuous variable.

Example 8

The box-and-whisker plots show the height of 16th–17th century Vilnius populations in Lithuania.

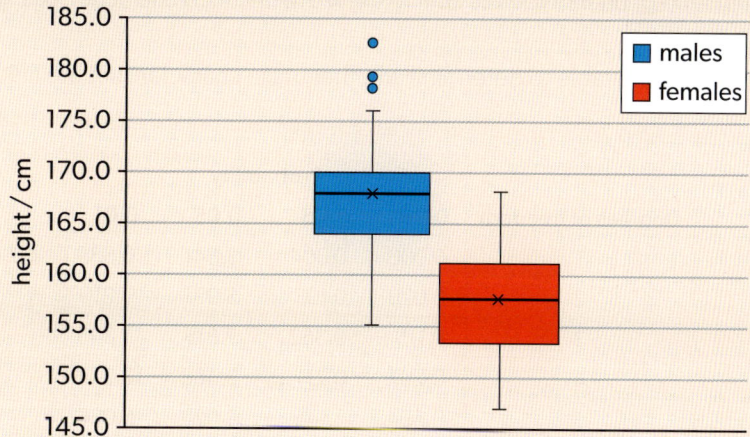

Source: Kozakaitė, J. & Miliauskienė, Z. (2019). Hidden, unwanted or simply forgotten? A bioarchaeological profile of the Subačius Street 41 population. *Archaeologia Lituana*, **20**, 116–138. www.doi.org/10.15388/archlit.2019.20.5 (CC BY 4.0)

a) State the mean for males.
b) Identify the group with most outliers.

Continuity and change

c) Calculate the percentage of the population of women whose height was less than 153 cm.
d) Explain what the range between 170 cm and 164 cm means for men's height.

Solution

a) 168 cm
b) Males
c) 25%, as the bar shows the range between the minimum and the lower quartile.
d) This shows the upper and lower quartiles. This means 50% of the men are between these values, 25% above the mean and 25% below the mean.

Nature of science

All biological laws have exceptions under certain conditions. The 9:3:3:1 and 1:1:1:1 ratios for dihybrid crosses are based on what has been called Mendel's second law. This law only applies if genes are on different chromosomes or are far apart enough on one chromosome for recombination rates to reach 50%.

Segregation and independent assortment in meiosis

Mendel's experiments with pea plants helped him to discover the laws of inheritance—the laws of segregation, independent assortment and dominance. Because each homologous chromosome migrates to a different pole during gamete formation, the alleles for each gene segregate from each other, so that each gamete carries only one allele for each gene (law of segregation). The homologous chromosomes migrate in a random way, therefore genes for different traits can segregate independently during the formation of gametes (law of independent assortment). Some alleles are dominant while others are recessive. An organism with at least one dominant allele will display the effect of the dominant allele (law of dominance).

Example 9

The seeds of a plant can be round-shaped (R) or wrinkled (r) and can have yellow (G) or green (g) colour. Round shape is dominant over wrinkled and yellow is dominant over green. In a cross between two plants with round and yellow seeds the following offspring were produced:

  947 round and yellow seeds
  284 round and green seeds
  322 wrinkled and yellow seeds
  109 wrinkled and green seeds.

a) Calculate the expected ratio of phenotypes assuming these are unlinked characteristics. Show your results in a Punnett grid.
b) Use the chi-squared formula and table to determine whether these characteristics are linked. Show your working.

	Probability			
Degrees of freedom	0.99	0.95	0.05	0.01
1	0.000	0.004	3.841	6.635
2	0.020	0.103	5.991	9.210
3	0.115	0.352	7.815	11.345
4	0.297	0.711	9.488	13.277
5	0.554	1.145	11.070	15.086

Solution

a) The Mendelian crossing for unlinked traits is as follows:

  R = allele for round seeds   r = allele for wrinkled seeds
  G = allele for yellow seeds   g = allele for green seeds

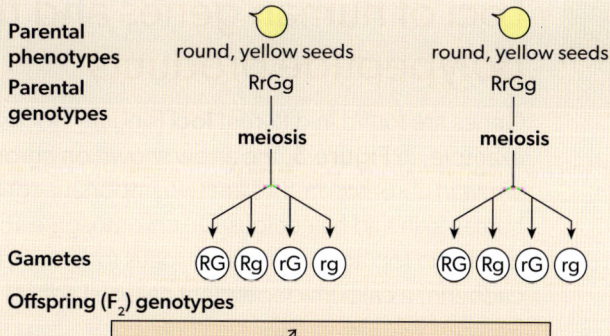

Offspring (F₂) phenotypes
9 round, yellow seeds; 3 round, green seeds; 3 wrinkled, yellow seeds; 1 wrinkled, green seed

b) There are two hypotheses: the null hypothesis (H_0) and the alternative hypothesis (H_1).

H_0 = The characteristics assort independently (are not linked).

H_1 = The characteristics do not assort independently (thus are linked).

The total number of seeds is 1,662.

To calculate the expected frequencies we use the data obtained in (a) for the expected ratios of round yellow : round green : wrinkled yellow : wrinkled green of 9 : 3 : 3 : 1.

First, add the expected frequencies: 9 + 3 + 3 + 1 = 16

The expected frequency for round yellow is 9/16 so it is calculated as: 1,662 × 9/16 = 935

This is done for all possible characteristics. The results are shown in the table:

Phenotypes	Observed (O)	Expected (E)	(O − E)	$\frac{(O-E)^2}{E}$
round yellow	947	935	−12	0.2
round wrinkled	284	312	28	2.5
wrinkled yellow	322	312	−10	0.3
wrinkled green	109	104	−5	0.2
Total	1,662			3.2

The value for chi-squared = **3.2**

The degrees of freedom for each characteristic is calculated as:

(number of rows −1) = (4 − 1) = 3

Then use the table to find the critical value at **p = 0.05** (significant at 5%) and 3 degrees of freedom. The value is **7.815**.

If your results give a value above the critical value in the table, you reject the null hypothesis and accept the alternative hypothesis. If it is below the critical value, you accept the null hypothesis.

In this example, the value for chi-squared is below the critical value (i.e. statistically insignificant), therefore the null hypothesis is accepted, confirming that the shape of the seed and the colour of the seed are not linked.

Continuity and change

Nature of science

Statistical testing often involves using a sample to represent a population, such as the F2 generation in dihybrid crosses.

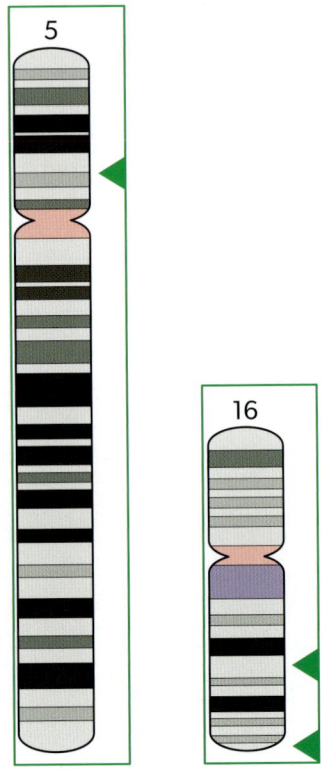

▲ **Figure 5** Loci on chromosomes 5 and 16

Loci of human genes and their polypeptide products

Genes are found in different **loci** (singular: locus) on chromosomes. For example, in **Figure 5**, the arrow shown on chromosome 5 and the bottom arrow in chromosome 16 are two important components in determining human skin and hair colour. On chromosome 16, although near the other locus, the top arrow shows the locus for a completely different protein, cadherin, a calcium-dependent cell–cell adhesion protein.

- The **locus** is the place on the chromosome where a gene is located.

There are many datasets to study genes in chromosomes. One of the most frequently used is the National Center for Biotechnology Information (NCBI).

Autosomal gene linkage

Genes may be linked or unlinked and are inherited accordingly. Gene loci are said to be linked if they are on the same chromosome. Unlinked genes segregate independently as a result of meiosis. Linkage can be autosomal or sex linked. In sex-linked inheritance the genes are usually located on the X chromosome.

Alleles are usually shown side-by-side in dihybrid crosses, for example AaBb. In representing crosses involving linkage, it is more common to show them as vertical pairs.

Example 10

The sketch shows two pairs of human chromosomes, 1 and 12. On chromosome 1, the gene for *CHD5* is represented by the alleles R and r, and the gene for *CLIC4* by S and s. On chromosome 12, the gene for *ZNF26* is represented by T and t.

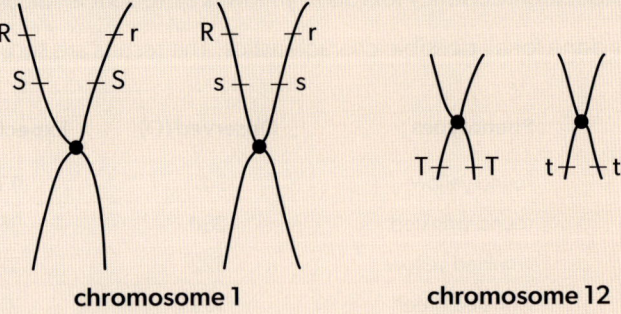

a) State two genes that are linked.
b) Suggest, with a reason, the process that resulted in this combination of alleles R and r on the homologous pair of chromosome 1.
c) Draw the possible genotypes of gametes formed with these chromosomes using a Punnett grid.

Solution

a) The linked genes are on the same chromosome, therefore the answer is *CHD5* and *CLIC4* (R and S), as they are on chromosome 1.

b) The alleles on one chromosome should be the same, as one sister chromatid is formed by replication of the other. In this case through the process of crossing over, the R from one chromosome was exchanged with the r of one of the non-sister chromatids of the homologous pair.

c)

	RS	rS	Rs	rs
T	RST	rST	RsT	rsT
t	RSt	rSt	Rst	rst

- What are the principles of effective sampling in biological research?
- What biological processes involve doubling and halving?

D3.3 Homeostasis

You should know:

✔ homeostasis is the maintenance of the internal environment of an organism controlled by negative feedback loops.

✔ thermoregulation is controlled by negative feedback.

Additional higher level:

✔ the ultrastructure of the glomerulus and Bowman's capsule facilitate ultrafiltration.

✔ the proximal convoluted tubule selectively reabsorbs useful substances by active transport.

✔ the loop of Henle maintains hypertonic conditions in the medulla.

✔ blood supply to organs changes in response to activity.

You should be able to:

✔ explain how insulin and glucagon regulate blood glucose levels.

✔ outline physiological changes, risk factors, and methods of prevention and treatment of type 1 and type 2 diabetes.

✔ explain thermoregulation in humans.

Additional higher level:

✔ distinguish between the kidney's roles of excretion and osmoregulation.

✔ explain how antidiuretic hormone (ADH) controls reabsorption of water in the collecting duct.

Homeostasis and negative feedback

In **homeostasis**, even when there are fluctuations in the external environment, homeostatic variables are kept within the norm. Body temperature, blood pH, blood glucose concentration and blood osmotic concentration are homeostatic variables in humans. These are maintained using negative feedback loops, which return homeostatic variables to the set point from values above and below the set point.

- **Homeostasis** is the maintenance of the internal environment of an organism.

Blood glucose regulation and diabetes

Glucagon and insulin are secreted by the α and β cells of the pancreas, respectively, to control blood glucose concentration. These hormones travel through blood to different tissues.

Diabetes is a medical condition in which blood glucose levels are higher than normal, even after fasting. Glucose can be detected in urine. Type 1 diabetes is an autoimmune disease that causes the destruction of the β cells of the islets of Langerhans of the pancreas. This causes the body to lack insulin, therefore raising the level of glucose in blood. It is treated with insulin injections. Type 2 diabetes is caused by resistance to insulin. It can be caused by obesity. Type 2 diabetes can be treated by controlling diet to reduce sugar intake, and therefore weight, and increasing exercise.

Thermoregulation

Homeotherms are animals that maintain a regulated body temperature by thermoregulation through negative feedback control. Peripheral thermoreceptors are in the skin and sense surface temperatures. Variations in body temperature activate these thermoreceptors that inform the hypothalamus which, at the same time, activates heat regulation mechanisms to increase or decrease body temperature and return it to set point. Heat is lost from the skin to the external environment by radiation, conduction, convection and evaporation.

When hot, heat is lost by:

- increased sweating
- vasodilation of skin arterioles (blood vessels increase in diameter)
- reduced production of thyroid-stimulating hormone (TSH) by the pituitary gland
- decreased thyroxin, lowering metabolic rate
- behavioural changes such as reducing movements, adopting an open body position, removing clothing and reducing appetite.

When cold, heat is generated or maintained by:

- erection of hairs (goosebumps), leading to heat-trapping
- vasoconstriction of skin arterioles
- increased production of TSH by the pituitary gland
- increased thyroxin to increase metabolic rate and subsequent heat production
- skeletal muscle contraction and shivering, leading to increased heat production
- uncoupled respiration in brown adipose tissue
- behavioural changes such as increased movements, adopting a closed body position, adding clothing and increasing appetite.

Function of the kidneys

The **kidney** is involved in **osmoregulation** and **excretion**. Osmoregulation is the maintenance of the internal environment, whereas excretion is the elimination of substances from the body.

- **Kidneys** are the organs in humans used to carry out osmoregulation and removal of nitrogenous wastes.
- **Osmoregulation** is the regulation of osmotic concentration. The units for osmotic concentration are osmoles per litre (osmol l^{-1}).
- **Excretion** is the elimination of waste matter.
- The **Bowman's capsule** is where glomerular filtrate collects to be further processed.
- The **proximal convoluted tubule** selectively reabsorbs useful substances by active transport. It contains many mitochondria and microvilli for absorption.
- The **loop of Henle** maintains hypertonic conditions in the medulla. Animals adapted to dry habitats have long loops of Henle.
- The **distal convoluted tubule** regulates mineral ion levels and maintains pH.
- The **collecting duct** is the site where the amount of water reabsorption can be varied as a part of osmoregulation.
- The **glomerulus** is the high-pressure capillary causing ultrafiltration in the Bowman's capsule.

Example 11

The table shows solute concentrations in normal blood plasma and the fluid in one section of the nephron.

Solutes	Plasma	Fluid inside the nephron
Cl⁻ ions	110 mol dm^{-3}	110 mol dm^{-3}
glucose	5 mol dm^{-3}	5 mol dm^{-3}
urea	5 mol dm^{-3}	5 mol dm^{-3}
proteins	750 mg dm^{-3}	3–4 mg dm^{-3}

In which section of the nephron would you expect to find these concentrations?

Solution

The correct answer is **A**, in the **Bowman's capsule**. The blood has entered the nephron by ultrafiltration, leaving the plasma proteins in the blood plasma because they are too big to pass through the fenestrations. **B** shows the **proximal convoluted tubule**. The glucose and ions are lower here because of reabsorption. **C** is the **loop of Henle** and **D** the **distal convoluted tubule** flowing into the **collecting duct**. Water and mineral ions are reabsorbed here. The urea becomes more concentrated.

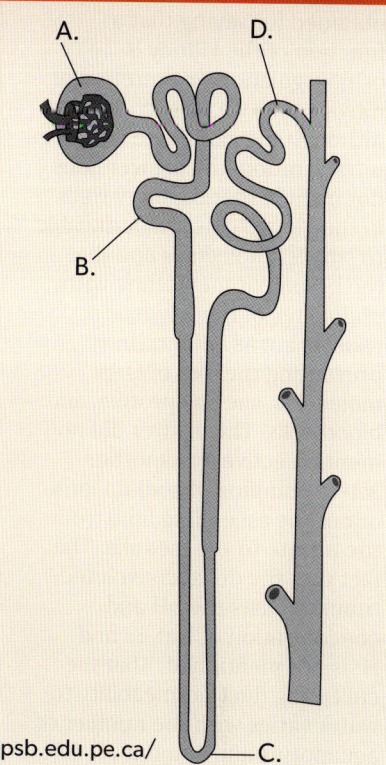

Source: www.psb.edu.pe.ca/

Sample student answer

Explain how the structure of the nephron and its associated blood vessels enable the kidney to carry out its functions. [8]

This answer could have achieved 7/8 marks:

> In the **glomerulus** of the nephron, the afferent arteriole is much larger than the efferent arteriole. This creates large amounts of pressure, forcing out the urea, glucose, water and mineral ions from the blood but leaving the proteins in the blood vessels. The rest enters the Bowman's capsule, which is lined by podocytes, which wrap capillaries. This is ultrafiltration. The filtrate moves to the proximal convoluted tubule for reabsorption. Here, the glucose and ions are reabsorbed by active transport. The cells have mitochondria to help with this. Water flows by osmosis. The large surface area increases the reabsorption in the proximal convoluted tubule. The fluid then moves to the loop of Henle.
> - Descending loop is permeable to water but not to ions.
> - Ascending loop is permeable to ions but not to water.
> This structure allows for a high concentration of salt ions in the medulla. The longer the loop of Henle, the more water is conserved. The fluid then moves to the distal convoluted tubule, where more is reabsorbed and to the collecting duct, where water is reabsorbed if ADH is present. Blood vessels surround the nephron to allow reabsorption to the bloodstream. Glucose is

▲ The answer scored one mark for high blood pressure in the glomerulus being due to a larger arteriole bringing blood than the arteriole taking away blood. One mark was given for the (selective) reabsorption of glucose and useful substances such as mineral ions in the proximal convoluted tubule. One mark for active transport needed in selective reabsorption. Another mark was given for water reabsorbed in the descending limb of loop of Henle and one for the ascending limb being impermeable to water. One mark was for the loop of Henle creating a high solute concentration in the medulla. One mark was for water reabsorbed in the collecting duct, which has microvilli to give a large surface area

▼ Marks could have been obtained for saying that a function of the kidney is osmoregulation or excretion of nitrogenous waste such as urea. Although the candidate mentions ultrafiltration, the answer does not include that capillary walls in the glomerulus are permeable to smaller molecules and that the basement membrane with podocytes of the Bowman's capsule act as a filter, thus preventing the loss of large molecules, such as protein, and blood cells. The answer did not mention active transport or active pumping of sodium ions out of the ascending limb from the filtrate to the medulla. The fact that the distal convoluted tubule adjusts the pH and concentration of Na+, K+ and H+ is not mentioned. That the collecting duct permeability to water varies with the number of aquaporins and amount of ADH or osmoregulation by varying the amount of water reabsorbed in the collecting duct would also have scored a mark.

reabsorbed into the blood capillaries as well as ions and water, but urea is removed as urine. The nephron is in the kidney and functions to filter the blood. Desert animals have long loops of Henle to conserve water.

ADH or **antidiuretic hormone** is a peptide hormone produced by the hypothalamus and stored in the pituitary gland. It controls the reabsorption of water by changing the permeability of the collecting ducts. When ADH is present, the collecting ducts become permeable to water and more is reabsorbed into blood, therefore the urine becomes more concentrated, and less urine is produced. Aquaporins are expressed in the kidney collecting ducts, where they move from intracellular sites to the plasma membrane under the control of ADH. The length of the loop of Henle positively correlates with the need for water conservation in animals, as does the amount of ADH produced.

- **ADH** is the antidiuretic hormone that controls water reabsorption by aquaporins.

You studied aquaporins in subtopic B2.1.

Changes in blood supply to organs in response to changes in activity

The chart shows the idealized distribution of blood flow at rest and during maximum exercise in a healthy untrained young male subject.

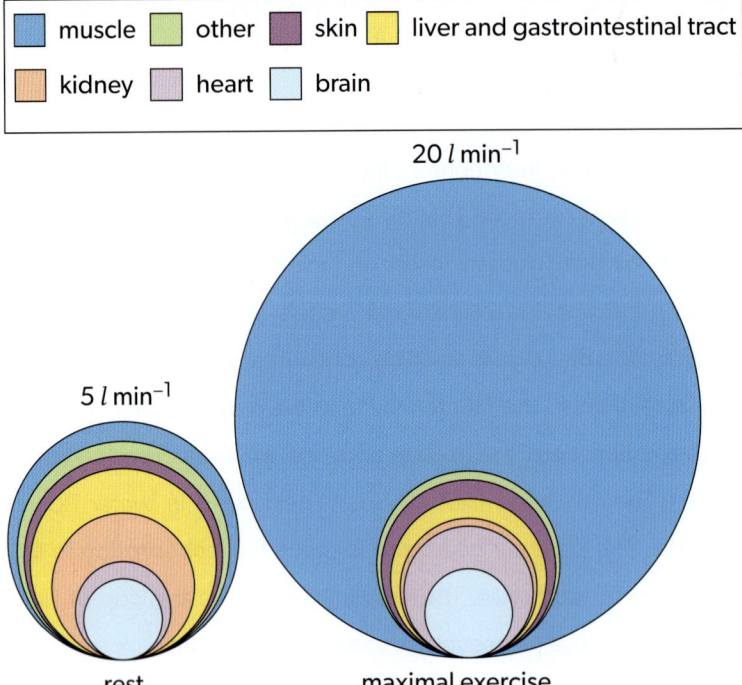

▲ Figure 6 Blood flow at rest and during maximum exercise

Source: Joyner, M. J. & Casey, D.P. (2015). Regulation of increased blood flow (hyperemia) to muscles during exercise: a hierarchy of competing physiological needs. *Physiological Reviews*, 95(2), 549–601. www.doi.org/10.1152/physrev.00035.2013

At rest, cardiac output is $5\,l\,min^{-1}$ and rises to $20\,l\,min^{-1}$ during maximum exercise. As cardiac output rises with exercise, brain blood flow remains constant (or increases slightly) while blood flow to the heart increases to meet the increased demands for myocardial blood flow that are primarily associated with exercise-induced increases in heart rate. Skeletal muscle blood flow increases dramatically, while blood flow to other tissues, especially the abdominal viscera and kidneys, is reduced. During heavy exercise, the vast increase in cardiac output is directed almost exclusively to contracting skeletal and cardiac muscles.

Example 12

The graph shows an increase in blood flow in the forearm immediately after a single brief contraction of the forearm muscles in a healthy person. The increase in forearm blood flow was obtained by subtracting the resting flow before each contraction from the flow measured just after contraction. Measurements were made in normal breathing air and with 99.6% oxygen.

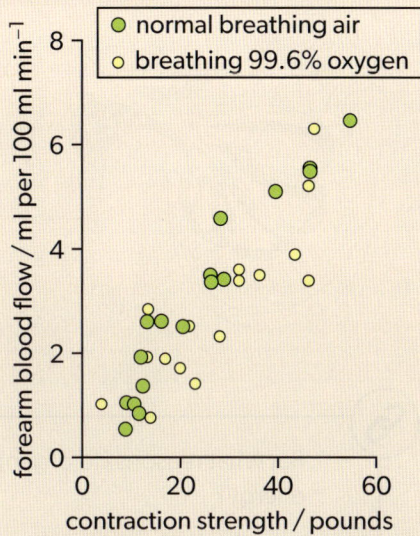

Source: Corcondilas, A., et al. (1964). Effect of a brief contraction of forearm muscles on forearm blood flow. *Journal of Applied Physiology*, **19**(1), 142–146. www.doi.org/10.1152/jappl.1964.19.1.142

a) Identify the relationship between contraction strength and forearm blood flow in normal breathing air.
b) Suggest the effect of increasing the oxygen concentration in breathing air on the forearm blood flow.

Solution

a) Positive relationship. The higher the contraction strength, the higher the blood flow.
b) No significant effect, as the distribution is similar to normal air.

Practice problems

1. A series of experiments was conducted with the seeds of cocklebur (*Xanthium pennsylvanicum*) to determine the conditions necessary for germination. Each lot of seeds was put on wet absorbent cotton in a jar and was subjected to certain conditions of atmospheric pressure at recorded room temperature for 10 days. The elongation of the shoot below the cotyledon (hypocotyl), followed by the geotropic response, was used as a criterion of germination.

Jar	Atmospheric pressure / mm	Oxygen pressure / mm	Temperature range / °C	Germinated seeds / %	Average length of hypocotyl / mm
A.	99	20	19–22	75	15
B.	90	19	21–22	80	23
C.	72	15	20–28	45	11
D.	72	15	20–22	30	6
E.	28	6	21–24	0	0

a) State the jar with the best conditions for germination in this experiment.
b) Describe the effect of oxygen pressure on germination and shoot elongation.
c) Suggest, with a reason, which jars could be used to see the effect of temperature on germination.
d) Evaluate the experimental design that allowed the researcher to obtain the results shown.

2. Describe an example of phenotypic plasticity.

3. Explain how unlinked genes segregate in meiosis.

4. The diagram shows a human kidney. Which label shows the renal artery?

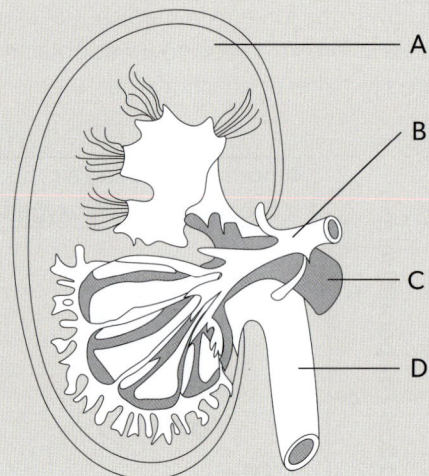

- For what reasons do organisms need to distribute materials and energy?
- What biological systems are sensitive to temperature changes?

D4.1 Natural selection

You should know:
✔ diversity of life has evolved and continues to evolve by natural selection.

✔ abiotic factors act as density-independent selection pressures.

✔ individuals that are better adapted tend to survive and produce more offspring while the less well adapted tend to die or produce fewer offspring.

✔ traits must be heritable for evolutionary change to occur.

✔ sexual selection is a selection pressure in animal species.

Additional higher level:

✔ a gene pool consists of all the genes and their different alleles present in a population.

✔ the differences between directional, disruptive and stabilizing selection.

✔ Hardy–Weinberg conditions must be maintained for genetic equilibrium in a population.

✔ artificial selection is carried out in crop plants and domesticated animals.

You should be able to:
✔ describe the roles of mutation and sexual reproduction in causing variation between individuals.

✔ describe overproduction of offspring and competition for resources as factors that promote natural selection.

✔ interpret data on John Endler's experiments with guppies.

Additional higher level:

✔ use databases to search for allele frequencies.

✔ explain evolution in terms of changing of allele frequencies.

✔ compare allele frequencies of geographically isolated populations.

✔ calculate allele and genotype frequencies with the Hardy–Weinberg equation.

Natural selection drives evolution

Evolution is cumulative change in allele frequency of the population over time. This change is due to **natural selection**. A population has **variations** among the individuals. These variations arise by meiosis or through sexual reproduction. Crossing over and independent assortment of homologous chromosomes during meiosis I produce different combinations of gametes. The combination of male and female gametes when fertilization occurs also produces variation. Although many mutations are neutral, mutations are also a source of variation. Certain variations give an advantage to some organisms over others in certain environments when subjected to different selection pressures. Populations (or species) produce more offspring than the environment can support. Individuals of the species compete for the same resources, which can limit the carrying capacity. The better-**adapted** organisms tend to survive and reproduce, and the less well-adapted organisms tend to die or produce fewer offspring. Individuals that reproduce pass on their heritable alleles to their offspring. Natural selection increases the frequency of heritable alleles of the better-adapted organisms. Characteristics acquired during an individual's life due to environmental factors are not encoded in the base sequence of genes and so are not heritable.

Different abiotic factors can act as selection pressures. **Density-independent factors**, such as high or low temperatures, may affect survival of individuals in a population. Selection favours competitive individuals that gain more resources to invest in reproduction and that have higher **fitness** than less competitive individuals. The outcome of competition will determine which individuals contribute genes to future generations and therefore determine how phenotypes evolve in response to competition. Fitness is determined by intraspecific competition where individuals compete for limited resources. Since intraspecific competition can create winners and losers, it is likely to lead to variation in the expression of resource-dependent traits, behaviours and fitness. If reproductive isolation occurs, new species can arise by natural selection. This often occurs across environmental gradients, generating hybrid zones. If the selection is very strong, the hybrid zones are smaller.

Example 1

Candelabrum coral (*Eunicea flexuosa*) lives in association with an algal symbiont. There are two lineages of coral, shallow and deep. Each is adapted to its own water depth, maintaining its own algal symbiont, even when both lineages meet at intermediate depths. To estimate variation in frequency of the two lineages, samples of candelabrum coral colonies were taken at Media Luna Reef, Puerto Rico. For each individual from a depth transect, five nuclear markers and one mitochondrial DNA marker were sequenced for the coral and one chloroplast marker for the algal symbiont.

The pie charts show the frequencies of shallow and deep lineages of candelabrum coral at the extremes of the depth gradient.

Nature of science

When Darwin's theory provided a convincing mechanism and replaced Lamarckism, there was a **paradigm shift** respecting evolution.

- A **paradigm shift** is a change in the worldview of concepts, models and practices in a determined subject.
- **Natural selection** is the mechanism driving evolutionary change, which operates continuously and over billions of years, resulting in the biodiversity of life on Earth.
- **Heritable variations** are differences in a species due to mutation, meiosis and sexual reproduction.
- **Adaptations** are characteristics that make an individual suited to its environment and way of life.
- **Density-independent factors** do not depend on the number of organisms.
- **Fitness** is the reproductive success in an environment. It determines the genetic pool of the next generation.

You studied negative feedback control of population size by density-dependent factors in subtopic C4.1.

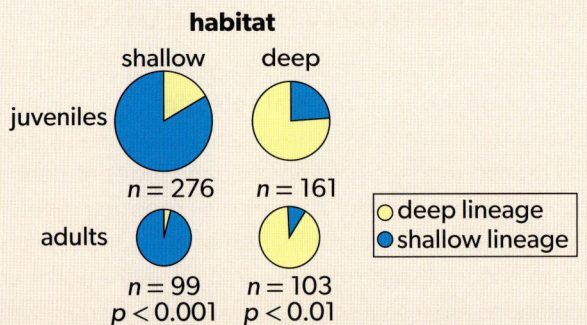

Source: Prada, C. & Hellberg, M. E. (2014). Strong natural selection on juveniles maintains a narrow adult hybrid zone in a broadcast spawner. *American Naturalist*, 184(6), 702–713. www.doi.org/10.1086/678403

a) Compare the distribution of shallow lineages in juveniles and adults.
b) Strong density-independent selection pressures provide a mechanism to explain the prevalence of depth-segregated lineages of the same species in the sea. Comment on this statement using the data provided.

Solution

a) Distributions of shallow lineages were more distinct in adults than juveniles, where most were shallow in shallow habitats.
b) To generate a small hybrid zone, like in this case, an environmental gradient strong enough to generate local adaptation between populations must be present. In sea habitats, temperature, salinity and light availability generate selection pressures.

Modelling sexual selection and natural selection

Example 2

The drawing shows a male and female Emperor bird of paradise (*Paradisaea guilielmi*) found in Papua New Guinea.

Explain how the differences between males and females shown in the diagram determine their sexual behaviour.

Solution

Birds of paradise present sexual dimorphism, which means the females and males are different. The plumage or feathers of the males are longer than in females. The elaborate feathers extending from the wings and tail in males are more colourful to attract or impress females for courtship. The more ostentatious birds will have more chances to mate, leaving more offspring.

Source: australianmuseum.net.au

Example 3

John Endler studied sexual and natural selection based on experimental control of selection pressures in guppies (*Poecilia reticulata*)—small fish commonly found in Trinidad. In a similar experiment, male guppies were divided into those with a lot of colour (bright) and those with little colour (dull). They were also exposed to a predator to see if they approached it (brave) or swam away from it (timid). Female guppies were allowed to select males to mate with, in the presence and absence of predators. The bar chart summarizes their choice of males.

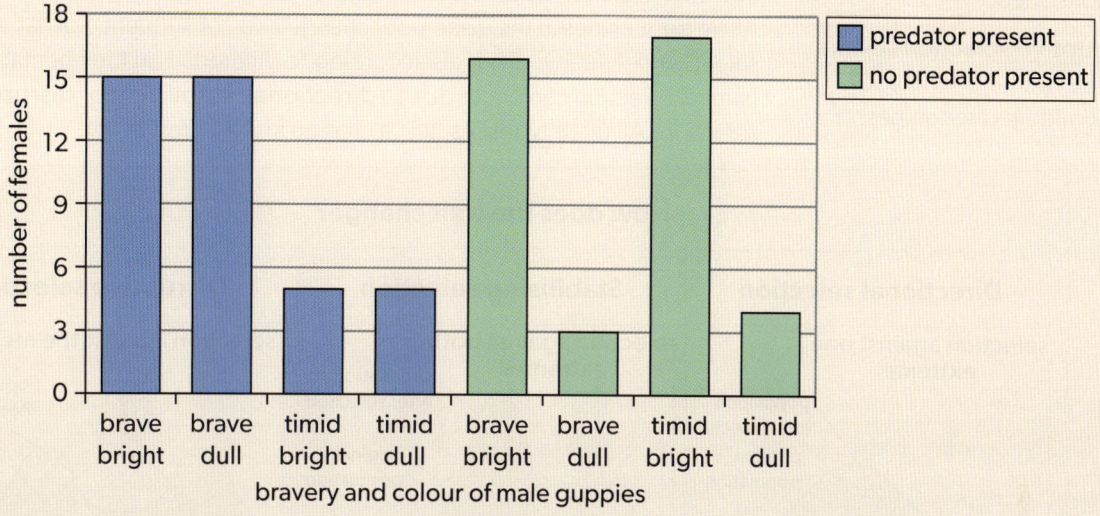

Source: Godin, J. G. & Dugatkin, L. A. (1996). Female mating preference for bold males in the guppy, *Poecilia reticulata*. *Proceedings of the National Academy of Sciences*, USA, 93, 10262–10267. www.doi.org/10.1073/pnas.93.19.10262

a) Compare mate selection by females in the presence and absence of a predator.
b) Suggest reasons for this pattern of mate selection.
c) Evaluate the hypothesis that bravery is more important than colour to females when selecting a mate.

Solution

a) With the predator present, females chose brave over timid, whereas without the predator present they chose bright over dull.
b) Brave males will defend females against predators and pass on genes for bravery to offspring. Bravery cannot be detected without a predator. Bright males are attractive to females.
c) In the presence of a predator, bravery is more important than colour, whereas in the absence of a predator, colour is more important than bravery. The difference between brave and timid is less than the difference between bright and dull, therefore bright is more important.

Gene pools

A gene pool is the sum of all the population's genes and their different alleles at a given time. Gene pools change over time through evolution. In different environments, there are different selection pressures as there are different habitats or niches to exploit. These changes cause natural selection. The allele frequencies change, causing the populations to diverge. **Neo-Darwinism** is a modified theory on Darwin's origin of species that integrates genetics with natural selection.

- **Neo-Darwinism** uses Darwin's theory of evolution and explains the origin of species on a genetic basis.

Continuity and change

Changes in allele frequencies

- **Directional selection** favours one extreme phenotype over the other.
- **Stabilizing selection** favours middle phenotypes.
- **Disruptive selection** favours both extreme phenotypes over middle phenotypes.

Allele frequencies change in different geographically isolated populations. There are databases where allele frequencies for different populations can be searched. For example, in the Allele Frequency Net Database the frequencies for immune gene (*HLA*) alleles can be found. The ALelle FREquency Database (ALFRED) stores frequencies of alleles at human autosomal polymorphic sites.

Evolution requires that allele frequencies change with time in populations. Allele frequencies in the gene pool change as a consequence of natural selection between individuals according to differences in their heritable traits. Natural selection can occur due to **directional**, **stabilizing** or **disruptive** selection. All three types result in a change in allele frequency.

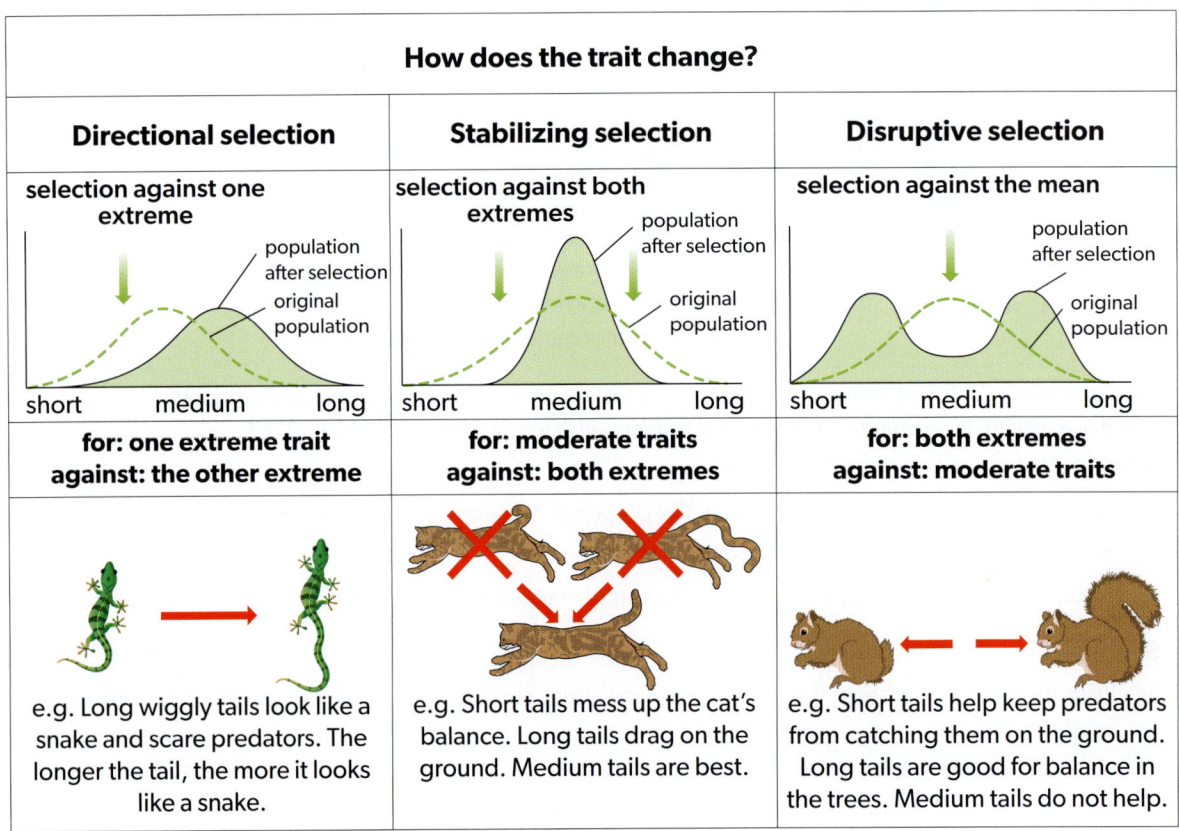

▲ Figure 1 Types of selection

Hardy–Weinberg equation and conditions

- The **Hardy–Weinberg equation** calculates the allele or genotype frequencies in a population as $p^2 + 2pq + q^2 = 1$.

The **Hardy–Weinberg equation** is used to calculate the allele or genotype frequencies in a population. The letters p and q denote the two allele frequencies so $p + q = 1$. Therefore, the genotype frequencies are predicted by the Hardy–Weinberg equation: $p^2 + 2pq + q^2 = 1$. If one of the genotype frequencies is known, the allele frequencies can be calculated using the same equations.

Hardy–Weinberg assumptions that must be maintained for a population to be in genetic equilibrium are:

- autosomal loci
- random mating
- absence of natural selection (survival rates do not vary between genotypes)
- very large population size
- no migration
- no gene flow
- no mutation.

Example 4

In peas (*Lathyrus oleraceus*), seeds can be yellow or green. The genotype frequency for homozygous green peas in a population is 0.16. What is the allele frequency for yellow seeds?

A. 0.84
B. 0.60
C. 0.16
D. 0.36

Solution

The correct answer is **B**, as if the genotype frequency p^2 is 0.16, the allele frequency for green is $p = 0.4$. The sum of the green and yellow allele frequencies according to the Hardy–Weinberg equation is 1, $p + q = 1$, so $q = 0.6$.

Artificial selection by deliberate choice of traits

Artificial selection by deliberate choice of traits is carried out in crop plants and domesticated animals by choosing individuals for breeding that have desirable traits. This choice will cause a directional selection towards one trait, leading to a change in the gene pool allele frequencies. Unintended consequences of human actions, such as the evolution of resistance in bacteria when an antibiotic is used, are not artificial selection but natural selection.

You learned about antibiotic resistance in subtopic C3.2.

- How do intraspecific interactions differ from interspecific interactions?
- What mechanisms minimize competition?

D4.2 Stability and change

You should know:

✔ natural ecosystems are stable.
✔ factors required for maintaining a stable ecosystem.
✔ deforestation of the Amazon rainforest could be a tipping point in ecosystem stability.
✔ community structure can be strongly affected by keystone species.
✔ harvesting sustainably depends on the rate of harvesting being lower than the rate of replacement.
✔ pollutants become concentrated in the tissues of organisms at higher trophic levels by biomagnification.

Additional higher level:

✔ succession can be triggered by changes in both an abiotic environment and in biotic factors.
✔ changes during primary succession.
✔ grazing by farm livestock and drainage of wetlands can arrest succession.

You should be able to:

✔ calculate the percentage change in deforestation.
✔ design sealed mesocosms to try to establish sustainability.
✔ describe factors that affect the sustainability of agriculture.
✔ analyse data illustrating the causes and consequences of biomagnification.
✔ describe how leaching causes eutrophication in aquatic and marine ecosystems.
✔ analyse data on microplastic and macroplastic pollution of the oceans.
✔ describe how rewilding can help to restore ecosystems.

Additional higher level:

✔ analyse data showing primary succession.
✔ describe an example of cyclical succession.
✔ explain how human influence can prevent succession.

Stability of ecosystems

Natural ecosystems are more resistant than the species compositions under environmental change. Only major disturbances can push ecosystems into a new ecological organization. There is evidence of ecosystem stability in forest, desert or other ecosystems that have shown continuity over long periods, some of which have persisted for millions of years.

To maintain the stability of an ecosystem the following are necessary:
- supply of energy
- recycling of nutrients
- genetic diversity
- climatic variables remaining within tolerance levels.

Tipping points

A large area of rainforest is required to generate atmospheric water vapour by transpiration, with consequent cooling, air flows and rainfall. Although there is uncertainty over the minimum area of rainforest that is sufficient to maintain these processes, the deforestation of Amazon rainforest affects ecosystem stability. Brazil is the country with the most tropical forest loss, with 43% of the global total. The 1.8 million hectares of primary forest loss in the year 2022 resulted in 1.2 gigatonnes of carbon dioxide emissions. Apart from the carbon impact, forest loss may lead to a tipping point where the ecosystem starts to turn into a savanna.

Example 5

The graph shows the fire and non-fire related primary forest loss in Brazil from 2002 to 2022. The three-year moving average may represent a more accurate picture of the data trends due to uncertainty in year-to-year comparisons.

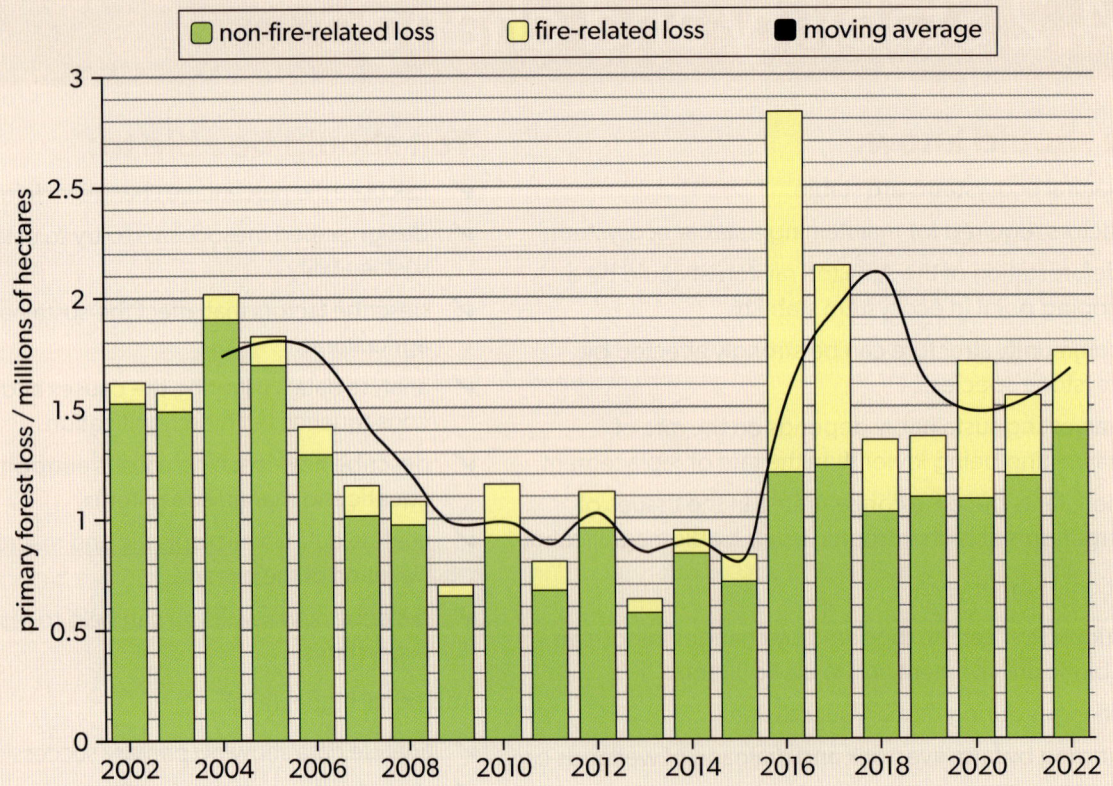

Source: Weisse, M., Goldman, E., & Carter, S. (2024, April 4). The Latest Analysis on Global Forests & Tree Cover Loss | Global Forest Review. Research.wri.org. https://research.wri.org/gfr/latest-analysis-deforestation-trends / Data from: Peter Potapov, Svetlana Turubanova and Sasha Tyukavina – University of Maryland's GLAD lab (CC BY 4.0)

a) State causes of primary forest loss other than fire (non-fire-related loss).
b) State the year with the most fires.
c) Calculate the percentage of non-fire- and of fire-related forest loss in 2022.
d) Describe the trend in primary forest loss in Brazil over the years 2002 to 2022.

Solution

a) Non-fire-related loss can occur from mechanical clearing for agriculture and logging, as well as natural causes such as wind damage and river meandering.
b) 2016
c) Total loss is 1.75 million hectares (100%); non-fire-related loss = 1.4 million hectares = (1.4/1.75) × 100 = 80%; therefore, fire-related loss = 20%.
d) Forest loss decreased from 2002 to 2009, then it remained quite stable until the loss increased greatly in 2016. It then decreased in 2018 but increased again from 2018 to 2022.

Investigating the effect of variables on ecosystem stability

Mesocosms are experimental systems that can act as models to study natural environments under controlled conditions. They can range in size from a small plastic bottle to a large tank. Mesocosms can be closed systems, such as sealed glass vessels, allowing only energy exchange with the environment, or open systems, allowing exchange of materials as well. The advantages of a mesocosm are that many environmental conditions can be controlled and measured, the experiments can be replicated several times under the same conditions and it is easier to manipulate conditions than in a free environment. The disadvantage is that it does not represent the real environment exactly, so you could reach false conclusions. Aquatic or microbial ecosystems are likely to be more successful than terrestrial ones.

Nature of science

Care and maintenance of mesocosms should follow IB experimental guidelines. Any experimentation should not result in any pain or undue stress on any animal (vertebrate or invertebrate) or compromise its health in any way.

Role of keystone species in the stability of ecosystems

Keystone species are those that have a destabilizing disruptive impact on the ecosystem they live in. The keystone species has a disproportionate effect because even a few organisms affect the environment more than expected. They can be predators that influence the distribution, behaviour, and abundance of prey species in the ecosystem. They can also be herbivores that maintain the vegetation from overgrowing or plants that provide vital sources of food or shelter to other species. Examples of keystone species are the sea star *Pisaster*, sea otters, elephants, prairie dogs and mountain lions.

- A **keystone species** is one whose presence has a disproportionate impact on an ecosystem and its removal often leads to significant changes.

Assessing sustainability of resource harvesting from natural ecosystems

The term **sustainability** refers to the capability of a system to endure and maintain itself. The three main challenges for sustainable ecosystems are finding sources of both nutrients and energy that will not run out and avoiding the accumulation of toxic wastes that will become hard to store or have an impact on the environment. Because of the increase in our global population, it is becoming more important that we consider the sustainability of human activities. When harvesting resources from natural ecosystems, sustainability depends on the rate of harvesting being lower than the rate of replacement. Strategies for sustainable harvesting need to consider not only harvested species, but also other non-harvested species interacting with them in the same ecosystem.

- **Sustainability** in natural ecosystems depends on the rate of harvesting being lower than the rate of replacement.

Continuity and change

Example 6

The commonly exploited pike fish (*Esox lucius*) has sexual size dimorphism. Females generally grow faster and to a larger size than males, so size-selective harvesting is often sex selective. Simulations without harvest and harvest with lower size limit of 50 cm were run and the size distributions of fish recorded.

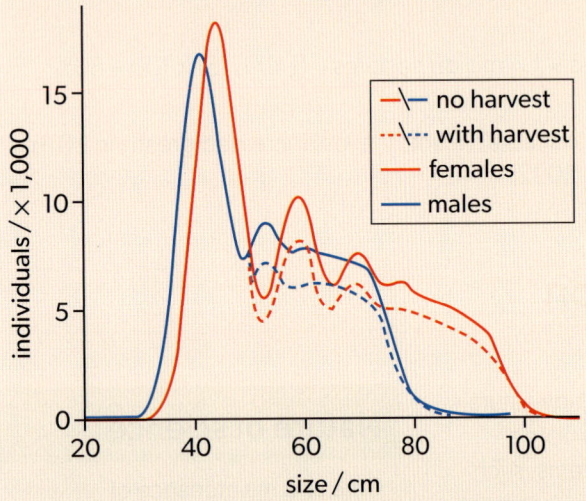

Source: Stubberud, M. W., et al. (2019). Effects of size- and sex-selective harvesting: An integral projection model approach. *Ecology and Evolution*, **9**, 12556–12570. https://doi.org/10.1002/ece3.5719 (CC BY 4.0)

a) Compare and contrast male and female size distribution with no harvest.
b) Describe the changes produced by harvesting in both males and females.
c) Suggest one method to prevent sex-selective harvesting of pikes.

Solution

a) Both start at 30 cm and numbers quickly peak. Males peak at 40 cm whereas females peak at 42 cm, as females are larger than males. There are no males larger than 90 cm whereas females reach up to 105 cm.

b) Harvesting decreases numbers of individuals in both males and females. Trends are nearly identical to no harvest but lower. Not much change in males larger than 75 cm and females larger than 95 cm.

c) Using slots, which means determining a range of sizes of pikes that can be caught, as mid-sized pikes include males and females. In this way they avoid catching only the larger females and smaller males.

Factors affecting the sustainability of agriculture

Our huge population growth has led to an increase in the demand for food, which, in turn, has led to intensive methods of agriculture. Intensive agriculture involves labour-intensive and costly practices to obtain high yields per unit of land. Issues associated with intensive crop farming methods are the use of toxic pesticides, which are often carcinogenic and bioaccumulate, harming non-target species, and the use of chemical fertilizers, which can leach into and cause eutrophication of waterways.

Sustainable agriculture can improve adaptive capacity, increase biodiversity and carbon storage in farmland soils and vegetation, and reduce greenhouse gas emissions. Improved management of cropland and grazing systems such as soil conservation, reduced soil erosion and reduction of fertilizer input will mitigate climate change. Agricultural intensification increases productivity per unit of agricultural area leading to free land for biodiversity conservation, but if not done sustainably the detrimental effects of intensification on the environment can outweigh the benefits of land sparing.

Eutrophication of aquatic and marine ecosystems due to leaching

Leaching of mineral nutrients, such as nitrogen and phosphate fertilizers, from agricultural land into rivers causes **eutrophication** and leads to increased **biochemical oxygen demand**.

- **Eutrophication** is the enrichment with nutrients of a body of water leading to the excessive growth of plants and algae, causing oxygen depletion.
- **Biochemical oxygen demand** is the amount of dissolved oxygen needed by aerobic organisms to break down organic material.

D4.2 Stability and change

Example 7

Stonefly nymphs (*Plecoptera*) are good indicators of unpolluted streams and rivers. Graph I shows the effect of a sewage outfall on oxygen concentration and biochemical oxygen demand (BOD) in a stream. Graph II shows the distribution curves of three different aquatic species living in the stream.

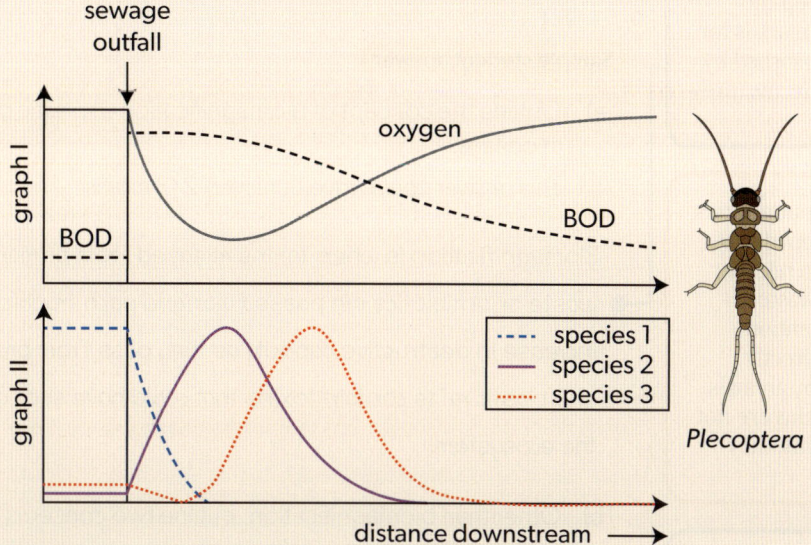

a) Identify which curve in graph II shows the distribution of stonefly nymphs.
b) Explain the effect of eutrophication on BOD.

Solution

a) Species 1 because stonefly nymphs do not survive in polluted waters.
b) BOD it is the amount of dissolved oxygen needed by aerobic organisms or bacteria. The higher the BOD, the more oxygen-lacking (anoxic) the environment. Sewage is rich in minerals and/or bacteria. Decomposition is carried out by aerobic organisms such as bacteria, lowering oxygen in the stream. There is an excessive growth of algae (algal bloom). Eutrophication leads to an increased BOD because bacteria decompose algae.

Biomagnification of pollutants and effects of plastic pollution

Pollutants become concentrated in the tissues of organisms at higher trophic levels by **biomagnification**. Examples of pollutants are pesticides, industrial wastes, heavy metals such as mercury, and plastics. Macroplastic and microplastic debris have accumulated in marine environments.

- **Biomagnification** is the increasing concentration of toxins in the tissues of organisms at successively higher levels in a food chain.

Example 8

Why are plastics persistent in the natural environment?
A. Because they are not toxic
B. Because they are large
C. Because of their nonbiodegradability
D. Because of their decomposition

Solution

The correct answer is **C**, as plastics do not biodegrade or decompose so they can last for many years in the environment. Plastics are toxic, microplastics are small and macroplastics are large and persist in the environment.

Nature of science

Scientists revealed the effects of plastics on the environment and popular media coverage reflected this worldwide, changing the public perception of the effects of plastic pollution on marine life. This drove citizens and governments to take measures to address this problem.

Dichlorodiphenyltrichloroethane (DDT) is a pesticide used to reduce pests that are disease vectors. DDT was widely used in the reduction of mosquitoes carrying malaria. This reduction of mosquitoes led to a reduction in malaria. Unfortunately, the rates of biomagnification in food chains caused a negative impact on health of the top predators, for example the thinning of eggshells and reduced reproductive success in birds of prey.

Sample student answer

a) Outline the concept of biomagnification of plastic pollution in oceans. [3]

This answer could have achieved 2/3 marks:

▲ This answer scored one mark for saying that toxins are concentrated in each successive level up the food chain. The second mark is for saying that toxins are absorbed by lower trophic level organisms (prey to predator).

> Biomagnification is when toxins released from primarily human general waste or antibiotics are passed along to each trophic level. The toxins increase in destructive capacity as they pass from prey to predator, and plastic rafts contain toxins that could have such an effect on the ecosystem.

▼ The question was about plastics, so it should have mentioned that plastics in the ocean can release toxins (not just human wastes). The answer does not mention that microplastics are directly ingested or consumed by animals such as birds. As the toxins are not metabolized by organisms, they accumulate in tissues.

b) Other than biomagnification, outline two concerns associated with mobility of plastic rafts and the communities they host. [2]

This answer could have achieved 2/2 marks:

▲ This answer scored one mark for saying that birds can choke on macroplastics and another for saying that plastics can worsen their health.

> The plastic on these rafts is hazardous to other animals that do not know what it is. Birds might consume the plastic as food. Ingesting macroplastic can choke the bird or worsen its health. Microplastics can be ingested by animals and the toxins can result in biomagnification. The fact that these islands are motile means they can affect a wide selection of animals.

▼ The answer did not mention that these islands can introduce pathogens into areas where the pathogen is not found. Another issue not mentioned is that the introduced species may become invasive.

Example 9

A study was carried out to investigate levels of mercury (Hg) contamination in the Rio Negro basin, Brazil. Mercury concentration was measured in fish of four different trophic levels and in water.

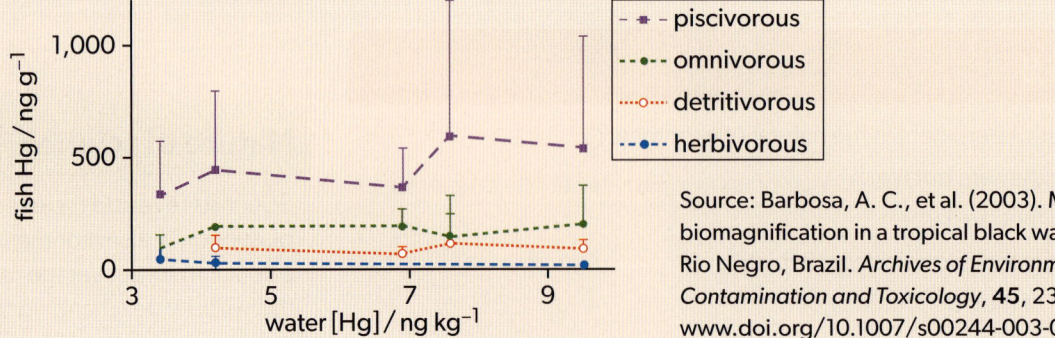

Source: Barbosa, A. C., et al. (2003). Mercury biomagnification in a tropical black water, Rio Negro, Brazil. *Archives of Environmental Contamination and Toxicology*, **45**, 235–246. www.doi.org/10.1007/s00244-003-0207-1

a) State the trophic level of fish that present the most mercury contamination.
b) Compare the levels of mercury found in omnivorous and detritivorous fish.
c) Explain biomagnification of mercury in fish.

> **Solution**
> a) Piscivores
> b) Omnivores and detritivores show a similar range of mercury concentrations, around 100–200 ng g^{-1}, although omnivores are slightly higher. A level of 200 ng g^{-1} is more common in omnivores, whereas 100 ng g^{-1} is more common in detritivores.
> c) Mercury bioaccumulation in aquatic systems varies considerably with food-chain structure and length. It can be simplified with producers at the base. Herbivorous fish are the primary consumer followed by omnivorous fish (secondary consumers) and piscivorous fish (tertiary consumers), where higher mercury concentrations are expected. Piscivorous fish feed at different trophic levels within the food chain. Piscivorous fish that feed on herbivores or omnivores will have lower levels of contamination, while piscivorous fish feeding on carnivores will have high levels. Piscivorous fish eat other fish. These animals include the top carnivores in the aquatic habitat, so they have the highest concentration of mercury.

Rewilding restores natural processes in ecosystems

Rewilding includes reintroduction of apex predators and other keystone species, re-establishment of connectivity of habitats over large areas and minimization of human impact by ecological management. An example of rewilding is the Hinewai Reserve in New Zealand. Botanist Hugh Wilson, in 1987, allowed the introduced "weed" gorse to grow as a nurse canopy to regenerate farmland into native forest, where birds and other wildlife are now abundant.

Stages of succession

Ecological successions are processes where the species structure of an ecological community changes over time. **Primary succession** occurs in environments where there was no soil or organisms, such as lava flows or glacier retreats. **Secondary succession** occurs in areas that previously supported organisms before an ecological disturbance such as a fire, flood or agricultural practices. Disturbance influences the structure and rate of change within ecosystems. During secondary succession there is an increase in size of plants, in amount of primary production, in species diversity, in complexity of food webs and in amount of nutrient cycling.

In some ecosystems, there is a cycle of communities rather than a single unchanging **climax community**. In this type of succession, there is a pattern of change where a small number of species tend to replace other species over a period of time. This usually occurs in the absence of large disturbances. For example, the interaction between herbivores of different size and predators can generate **cyclical succession** in a plant community by alternating periods of short vegetation dominated by high-quality plants, with periods of tall vegetation dominated by low-quality plants. When vegetation is tall, large herbivores promote the increase in density of medium and small herbivores by reducing plant quantity and increasing quality, eventually pushing larger herbivores out of the system. In time, the high density of small herbivores is reduced by small predators. As a result, vegetation becomes tall again and the population of the largest herbivore is restored, after which the cycle may start again.

Given any specific environmental conditions, ecological succession tends to lead to a particular type of climax community, but human influences can prevent this from developing. Grazing by farm livestock and drainage of wetlands are examples of **arrested succession**, with suppression of a return to the original structure.

- **Primary succession** occurs in areas that were not inhabited by living organisms.
- **Secondary succession** occurs in areas that were already inhabited but suffered disturbance.
- A **climax community** is when a succession has reached a steady state. The climax community is composed of species best adapted to average conditions in that area.
- In a **cyclic succession**, the community changes periodically, eventually returning to the original state.
- **Arrested succession** is an ecosystem permanently halted in an early stage of succession.

Continuity and change

Sample student answer

The diagram represents primary succession that occurs in an Arctic ecosystem, on a river floodplain in Alaska, USA. Permafrost is permanently frozen subsoil found in Arctic regions.

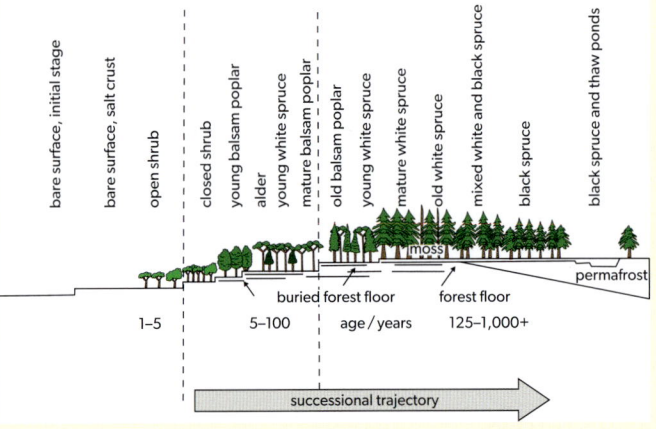

▼ The answer is too vague, as there is no description of how each factor affects the ecosystem. Very cold temperatures (with permafrost in inland soil) mean a short growth season or that trees would have to survive the cold or that the roots would have to be able to survive the permafrost temperatures. Low water availability would not allow plants to grow or the lack of soil does not let trees grow could have been another answer.

a) Outline primary succession. [1]
This answer could have achieved 1/1 marks:

> Colonization of areas that were not previously inhabited.

b) Describe **two** limiting factors on this ecosystem. [2]
This answer could have achieved 0/2 marks:

> Temperature and water.

▼ The sentence does not answer the question. Decomposers decay matter, increasing nutrients in soil. Organic material in humus helps increase and maintain moisture. Larger plants can grow and continue the cycle. The weathering of underlying rock by plant roots produces more and deeper soil.

c) Outline processes that must occur over time to produce deeper soil. [2]
This answer could have achieved 0/2 marks:

> The overpopulation of species and the permanent freezing of soil causes a drastic change.

- What is the distinction between artificial and natural processes?
- Over what time scales do things change in different biological systems?

D4.3 Climate change

You should know:

✔ climate change is caused mainly by human actions.
✔ there are positive feedback cycles in global warming.
✔ melting of landfast and sea ice is causing habitat change.
✔ there is a poleward and upslope shift in temperate species due to climate change.
✔ coral reefs are threatened by water temperature increase.

Additional higher level:

✔ phenology is research into timing of biological events.
✔ climate change disrupts the synchrony of phenological events.
✔ climate change is a driver of evolution.

You should be able to:

✔ describe how the change from net carbon accumulation to net loss in boreal forests is a tipping point.
✔ describe how changes in ocean currents affect nutrient upwelling.
✔ describe afforestation, forest regeneration and restoration of peat-forming wetlands as approaches to promote carbon sequestration.

Additional higher level:

✔ describe increases in insect life cycles within a year due to climate change.
✔ describe spring growth of the Arctic mouse-ear chickweed and arrival of migrating reindeer as an example of disruption of synchrony.

Anthropogenic causes of climate change

Scientific evidence for warming of the climate system is unequivocal. The global temperature of the Earth's surface has increased 1.8°C since 1880. Measurements of air trapped in polar ice showed atmospheric concentrations of methane and carbon dioxide remained relatively stable for many years until nearly 100 years ago when concentrations began to rise to present levels. Many scientific models have been used to explain these changes.

It is likely that this trend is partially due to **anthropogenic** increases in atmospheric concentrations of carbon dioxide and methane, as it is unlikely that these changes occurred only from natural internally generated variability of the climate system. Greenhouse gases include carbon dioxide (CO_2), methane (CH_4) and water vapour (H_2O). Carbon dioxide is produced mainly from respiration in organisms or by combustion of biomass or fossil fuels. Methane is produced by anaerobic respiration of biomass by methanogenic bacteria.

Methane is oxidized to carbon dioxide and water in the atmosphere, but this takes many years. Methane can affect climate directly through its interaction with long-wave infrared energy and indirectly through atmospheric oxidation reactions that produce carbon dioxide.

Short-wave radiations from the rays of the Sun penetrate these greenhouse gases in the atmosphere, reaching the surface of the Earth and thus delivering heat energy. However, when they reflect from the surface, they become long-wave radiation, which is absorbed by the gases and reflected back to the Earth or trapped in the atmosphere. This causes the Earth to become warmer. Long-wave radiation is unable to pass through the greenhouse gases. Short-wave radiation (as UV) passes through the greenhouse gases to the Earth's surface.

- An **anthropogenic** action is originated by human activity.

Nature of science

Data from Antarctic ice cores shows a positive relationship between global temperatures and atmospheric carbon dioxide concentrations over hundreds of thousands of years. This correlation does not prove that carbon dioxide in the atmosphere causes the increase in global temperatures, although other evidence confirms the causal link.

Example 10

The table shows recent estimates of the principal natural and anthropogenic global methane sources, in millions of metric tonnes per year.

natural		energy/refuse		agricultural	
wetlands	115	gas and oil	50	rice	60
oceans	15	coal	40	livestock	80
termites	20	charcoal	10	manure	10
burning	10	landfills	30	burning	5
		wastewater	25		
	160		155		155

Source: Johnson, K. A. & Johnson, D. E. (1995). Methane emissions from cattle. *Journal of Animal Science*, **73**(8), 2483–2492.

a) State the principal natural source of methane.
b) Calculate the percentage contribution of natural and of anthropogenic global methane sources.
c) Cattle can produce 200 to 500 litres of methane per day, around 70% of livestock contribution. Using the data provided, discuss the contribution of cattle to global warming.

Solution
a) Wetlands
b) Natural = 160/470 × 100 = 34%; anthropogenic (energy/refuse + agricultural) = 66%.
c) Livestock contribute 17% of the total methane, therefore cattle contribute around 12% of the total methane. Manure is produced by cattle, so they also contribute to that source.

Continuity and change

Example 11

The graph shows the global fossil fuel carbon dioxide annual emissions from 1850 to 2012. It also shows the yearly percent increase in emissions for different year periods.

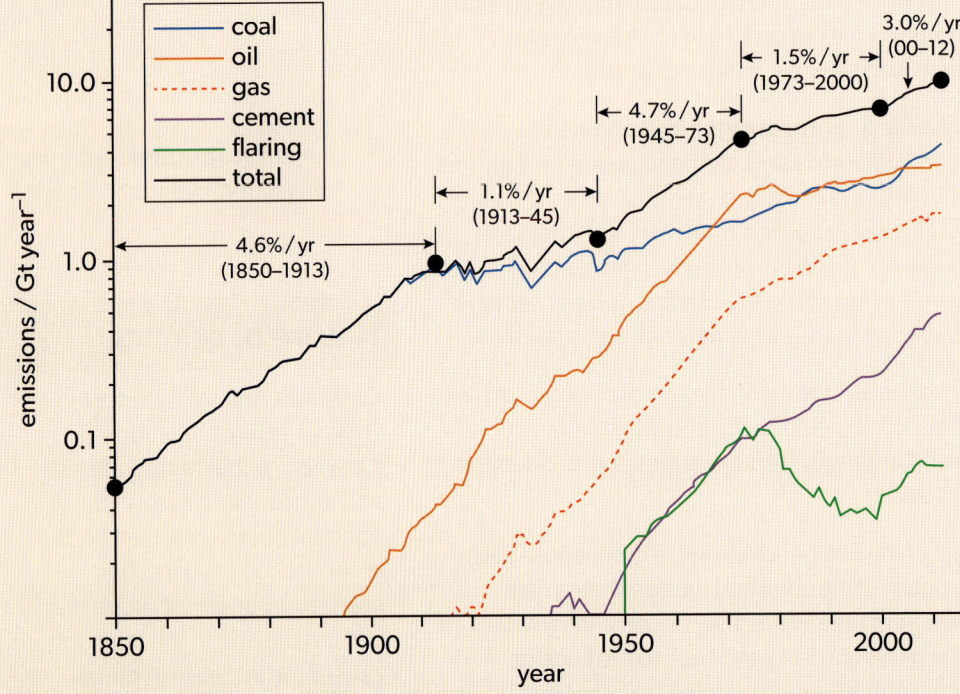

Source: Hansen, J., et al. (2013). Assessing "dangerous climate change": Required reduction of carbon emissions to protect young people, future generations and nature. *PloS ONE*, **8**(12), https://doi.org/10.1371/journal.pone.0081648 (CC0 1.0). Data from: BP Statistical Review of World Energy 2012, and Boden, T., G. Marland, and R. Andres, 2012: Global, regional, and national fossil-fuel CO2 emissions. Carbon Dioxide Information Analysis Center, Oak Ridge National Laboratory, US Department of Energy, Oak Ridge, Tenn., USA

a) State the emissions produced by burning coal in 1910.
b) Identify the major contributors to carbon dioxide emissions in 1980.
c) Identify the period of time when the emissions grew the most.
d) Suggest anthropogenic effects on climate change.

Solution

a) 1 Gt per year   b) Burning oil and coal   c) 1945–1973

d) Humans burn fossil fuels to obtain energy. In doing so, they increase the amounts of carbon dioxide and methane in the atmosphere. This increase in greenhouse gases causes an increase in global temperatures, causing climate change.

Positive feedback cycles in global warming

Example 12

The **enhanced greenhouse effect** has caused global warming. What is likely to occur in the Arctic if the Earth's surface temperature rises?

A. Increase in area of ice habitat
B. Decrease in numbers of pest species and pathogens
C. Decreased rates of decomposition of detritus
D. Increased number of predators from temperate regions

Solution

The correct answer is **D**, as all the other answers are the opposite of what happens. In **A**, there is a decrease in ice areas due to melting of icecaps. In **B**, there is an increase in pest species due to polewards migration of tropical and temperate species. In **C**, there is an increase in decomposition rates due to warmer temperatures accelerating metabolism of decomposers.

Warmer temperatures trigger carbon dioxide release from deep ocean back to the atmosphere. This can amplify climate change and reduce remaining emission budgets. At the same time, an increased absorption of solar radiation due to loss of reflective snow and ice will act as a positive feedback loop, increasing even more the atmospheric temperature. Warmer temperatures will accelerate the rates of decomposition of peat and previously undecomposed organic matter in permafrost. This will cause the release of methane from melting permafrost, increasing the amounts of greenhouse gases and therefore enhancing the greenhouse effect. Increases in droughts and forest fires have turned some regions from a carbon sink to a carbon source, contributing to the positive feedback cycles in global warming.

- The **enhanced greenhouse effect** is the process by which the Earth's surface is warmed due to entrapment of the radiation re-emitted by the Earth as long-wave radiation.

Climate change-related tipping points

The Intergovernmental Panel on Climate Change (IPCC) suggests that tipping points could be exceeded even between 1 and 2°C of warming in mean surface temperatures.

Example 13

Scientists explored feedback loops that could push towards a threshold, even when human emissions are reduced. Crossing the threshold would lead to continued climate warming, causing serious disruptions to ecosystems. The map shows global potential tipping cascades, colour-coded according to estimated thresholds in global average surface temperature (tipping points). Arrows show the potential interactions among the tipping elements.

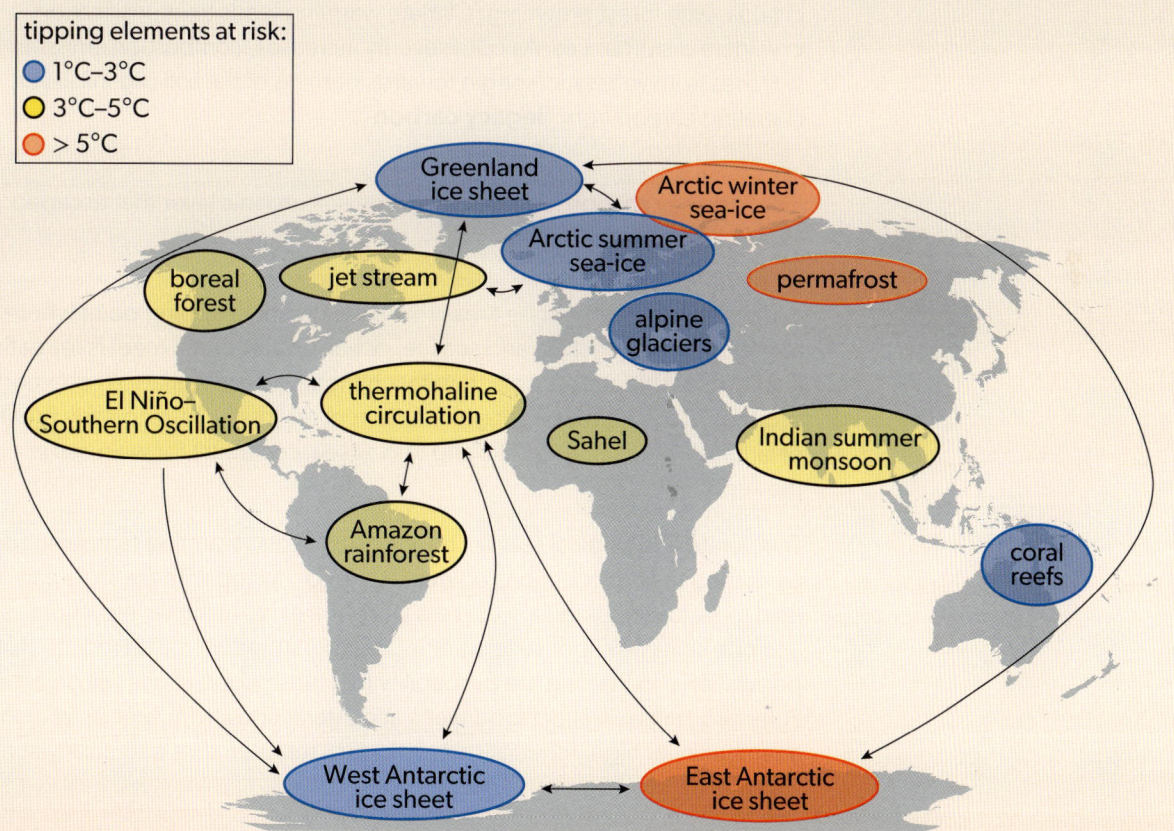

Source: Steffen, W., et al. (2018). Trajectories of the Earth system in the Anthropocene. Proceedings of the National Academy of Sciences, 115(33), 8252–8259. https://doi.org/10.1073/pnas.1810141115 (CC BY-NC-ND 4.0)

Use the diagram to exemplify one interaction of at least three tipping elements.

Solution

Melting of the Greenland ice sheet would cool nearby surface waters, with a significant effect on the ocean circulation. This could trigger a critical transition in the thermohaline circulation. Changes in ocean currents could alter the timing and extent of nutrient upwelling. Warmer surface water can prevent nutrient upwelling to the surface, decreasing ocean primary production and energy flow through marine food chains. Warmer surface water could also cause sea-level rise and Southern Ocean heat accumulation that would accelerate ice loss from the East Antarctic ice sheet on time scales of centuries.

Climate change can cause serious environmental issues, including:
- carbon loss in boreal forests
- melting of polar ice
- changes in ocean currents and nutrient upwelling
- **poleward** and **upslope shifts** in temperate habitats.

- **Poleward shift** is the distribution towards the poles in response to climate warming.
- **Upslope shift** is the distribution of populations to higher altitudes due to climate warming.

In boreal forests climate change has caused a switch from net carbon accumulation to net loss. Warmer temperatures and decreased winter snowfall lead to increased incidence of drought and reductions in primary production in taiga, with forest browning. Climate warming leads to intensification in wildfire disturbances. As burn severity increases, combustion emission switches from vegetation origin towards burning of the soil organic layer, liberating carbon from "**legacy carbon**".

- **Legacy carbon** is carbon from the soil organic layer that escaped combustion in previous fires.

In the poles there might be areas that have passed a tipping point where, due to melting of landfast ice and sea ice, ocean and bedrock meet is retreating irreversibly. This causes polar habitat change, with potential loss of breeding grounds of the emperor penguin (*Aptenodytes forsteri*) due to early breakout of landfast ice in the Antarctic and loss of sea ice habitat for walruses in the Arctic.

Temperate and tropical species have responded to warming temperatures by shifting their distributions poleward and/or upslope. These shifts can lead to serious ecological loss. Less land area exists at higher elevations, so upslope shifts in response to warming temperatures will cause reductions in populations, increasing the probability that these populations will go extinct. Poleward shifts will affect food webs and interactions, with species with low adaptability or highly specialised ecological requirements particularly at risk.

Example 14

Scientists studied shift rates of temperate-zone and tropical montane organisms in response to temperature increases. Changes in New Guinean bird species' upper elevational limits between historical (measured in the 1960s) and modern (measured in 2014) resurveys are plotted for Mt. Karimui (left) and Karkar Island (right). Points on the solid lines represent species with unchanged elevational limits.

D4.3 Climate change

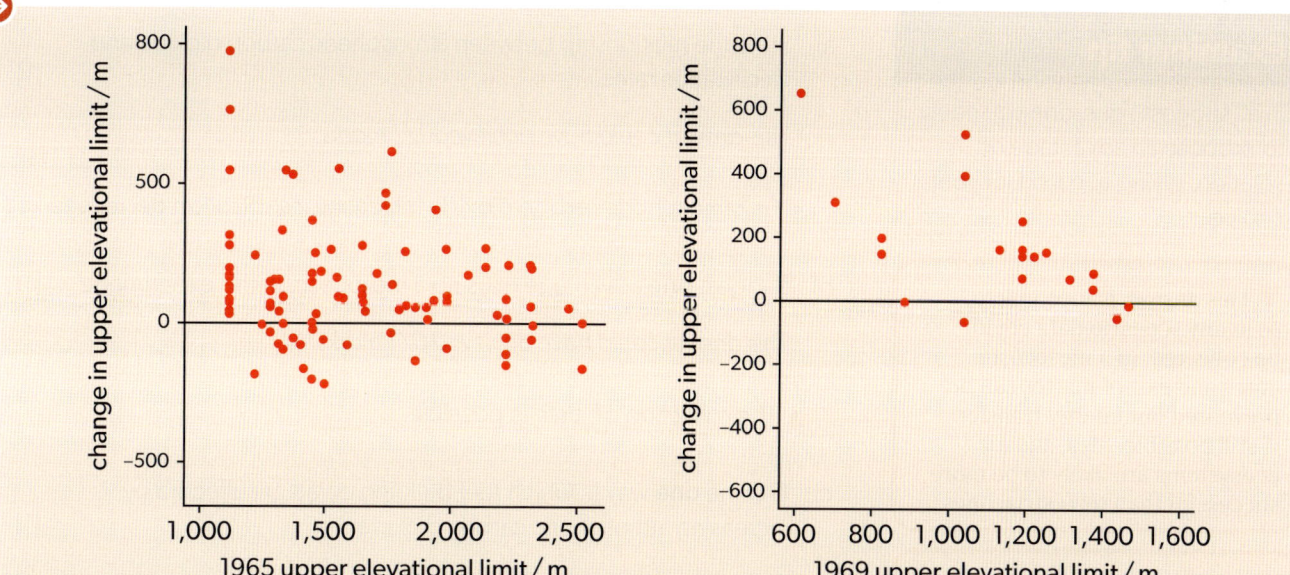

Source: Freeman, B. G. & Freeman, A. M. (2014). Rapid slope shifts in New Guinean birds illustrate strong distributional response of tropical montane species to global warming. *Proceedings of the National Academy of Sciences of the United States of America* 111 (12) 4490–4494. www.doi.org/10.1073/pnas.1318190111

a) Compare and contrast the trends in change in upper elevational limit in both locations.
b) Suggest effects of upslope shift on New Guinean bird species.

Solution

a) Both show a higher upslope shift than downslope shift. In Mt. Karimui there are more data points and there are more cases of downslope shift than in Karkar Island.
b) The upslope shift can cause changes in habitat and in food web relations.

Threats to coral reefs

Sample student answer

Increasing carbon dioxide concentration in the atmosphere leads to acidification of the ocean. This in turn reduces the amount of dissolved calcium carbonate. A study was undertaken to investigate the effect of increasing the concentration of atmospheric carbon dioxide on the calcification rate of marine organisms. Calcification is the uptake of calcium into the bodies and shells of marine organisms. The study was undertaken inside Biosphere 2, a large-scale closed mesocosm. The graph shows the results of the data collection.

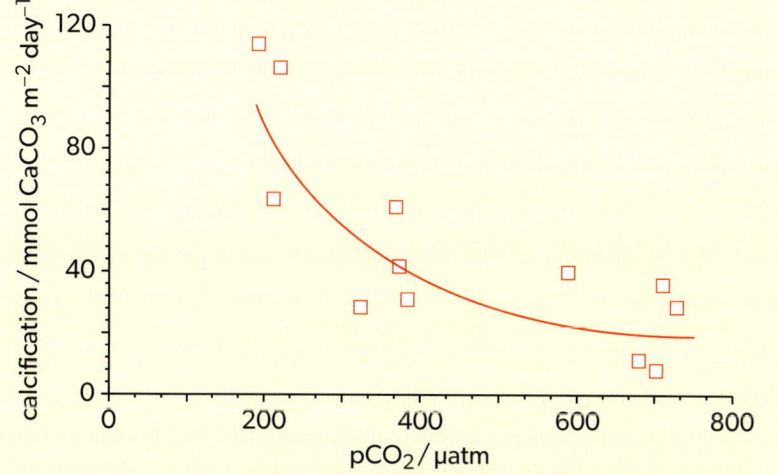

Source: Langdon, C., et al. (2000). Effect of calcium carbonate saturation state on the calcification rate of an experimental coral reef *Global Biogeochemical Cycles*, 14(2), 639–654. www.doi.org/10.1029/1999gb001195

> **Examiner tip**
>
> In a "suggest" question you need to propose an advantage or solution, giving a reason for your choice.

a) State the relationship between atmospheric carbon dioxide and calcification rates. [1]

This answer could have achieved 1/1 marks:

> As atmospheric carbon dioxide increases, calcification decreases.

b) Suggest **one** advantage of using a mesocosm in this experiment. [1]

This answer could have achieved 0/1 marks:

> The collection of data is possible.

▼ This answer is incomplete, as in any scenario data could be collected. In a mesocosm, entry and exit of matter can be prevented but energy transfer is still possible. Also, aquatic ecosystems are likely to be more successful than terrestrial ones.

c) Outline **one** way in which reef-building corals are affected by increasing atmospheric carbon dioxide. [1]

This answer could have achieved 0/1 marks:

> Reef-building corals are affected by increasing atmospheric carbon dioxide by reducing their building which can cause deterioration.

▼ Carbon dioxide in the atmosphere dissolves in the oceans, acidifying the water. This acid pH causes symbiotic *Zooxanthelle* algae to leave the corals (coral bleaching). Corals no longer obtain nutrients from these photosynthetic organisms.

Approaches to carbon sequestration

Carbon sequestration reduces the amount of carbon in the atmosphere. **Afforestation** with non-native tree species or rewilding with native species to regenerate forests promotes carbon sequestration.

- **Carbon sequestration** removes atmospheric carbon dioxide, transforming it into a form that cannot be easily used.
- **Afforestation** is forestation in areas where there was no previous forest.

Peat-forming wetlands are ecosystems consisting of partially decomposed plant remnants and organic matter that accumulate over thousands of years to form carbon-rich soils (carbon sinks). They occur naturally in waterlogged soils in temperate and boreal zones and are rapidly formed in some tropical ecosystems. Degradation of peatlands through land-use change and drainage is responsible for global annual anthropogenic carbon dioxide emissions. Therefore, restoration of these peat-forming wetlands is an approach to carbon sequestration.

> **Nature of science**
>
> Scientists debate over whether plantations of non-native tree species or rewilding with native species offers the best approach to carbon sequestration. Non-native plants grow faster compared to native plants as they do not have competition, but they decompose faster and accelerate the release of carbon dioxide from the soil.

Phenology and disruption of phenological events

Phenology is the study of seasonal plant and animal activity driven by environmental factors. Phenological studies include photoperiod and temperature patterns on the timing of biological events such as flowering and budding in deciduous trees and migration and nesting in birds. Phenological synchrony of related species can affect interspecific interactions, resource allocation, species survival, range shift and evolutionary trajectory. Climate change will lead to changes in synchrony of phenological events.

- **Phenology** studies the timing in biological cycles and how they are influenced by seasonal or interannual variations.

▲ **Figure 2** (a) Reindeer (*Rangifer tarandus*) and (b) Arctic mouse-ear chickweed (*Cerastium arcticum*)

Migrating reindeer (*Rangifer tarandus*) are the most abundant large terrestrial herbivores across the North Pole, playing a key role in northern ecosystems through grazing on plant communities and by supporting predator populations. Reindeer must forage effectively to recover from the nutritional stress of the previous winter. In reindeer, the timing of migration, where calves are born, is cued by daylength (photoperiod). The calving season in migrating reindeer is highly synchronized with the emergence of the plant species they forage. Arctic mouse-ear chickweed (*Cerastium arcticum*) is a flowering plant found in polar areas. The onset of the plant growing season is cued by local temperatures. If plant phenology advances in response to climatic warming, there could be a mismatch between the peak of resource demands by reproducing herbivores and the peak of resource availability in plant species, decreasing the calves' survival chances.

Climate change affects insect life cycles

The spruce bark beetles *Ips typographus* and *Dendroctonus micans* are major pests in northern European coniferous forests. Initiation of a new generation is cued by temperature. Climate warming may produce a second generation in southern Scandinavia and even a third generation in lowland parts of central Europe, causing even greater damage to the spruce trees.

Example 15

Scientists have published models of phenological shifts, indicating that the life stages of 203 plant and animal species have advanced by about 2.8 days per decade.

The diagram shows phenological shifts within some taxonomic groups. Each circle represents the phenological change of one species as it shifts a life-cycle stage to earlier or later by a number of days per decade. Circles appear as overlapping when two or more species in the same taxonomic group shift at similar rates.

Continuity and change

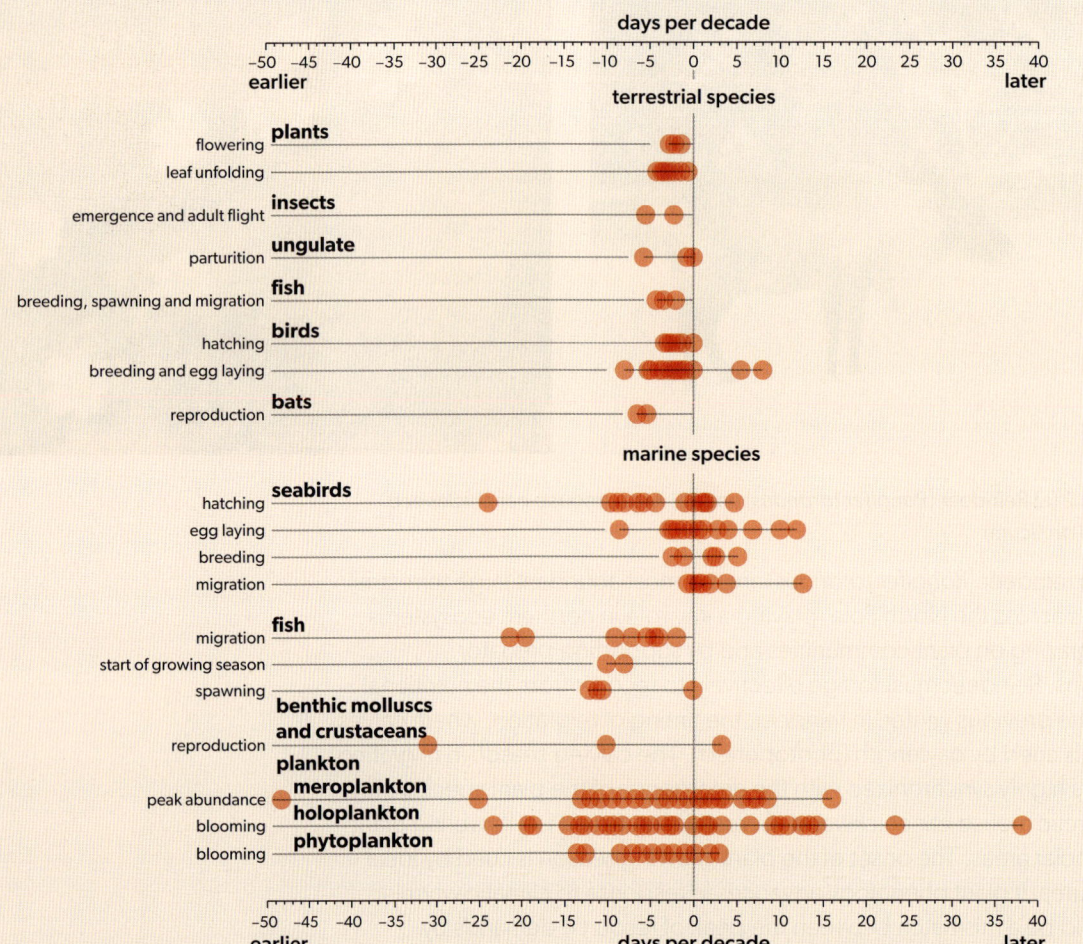

Source: Visser, M. E. (2022). *Phenology: Climate change is shifting the rhythm of nature*. In Frontiers 2022: Noise, Blazes and Mismatches: Emerging Issues of Environmental Concern (2022 ed., Vol. Frontiers report, pp. 41-58). United Nations. https://wedocs.unep.org/bitstream/handle/20.500.11822/38062/Frontiers_2022CH3.pdf

a) State the maximum number of days seabirds have shifted in their hatching.
b) Outline the phenological shift in phytoplankton blooming.
c) Suggest a possible effect of a phenological mismatch between insect emergence and egg hatching in insectivorous birds.

Solution

a) −25 days (25 days earlier)
b) Most species studied bloom earlier, as only two species bloom later and one remains the same.
c) Insectivorous birds depend on insects for food. If insects emerge earlier, the source of food will be earlier, and chicks might starve.

Evolution due to climate change

Example 16

Climate change can lead to evolution. Changes in snow cover have determined changes in the fitness of colour variants of the tawny owl (*Strix aluco*). A cryptic plumage allows tawny owls to hide from predators. The graph shows the least square means with 95% confidence intervals of detection probability for both colour morphs in landscapes with no snow and with snow.

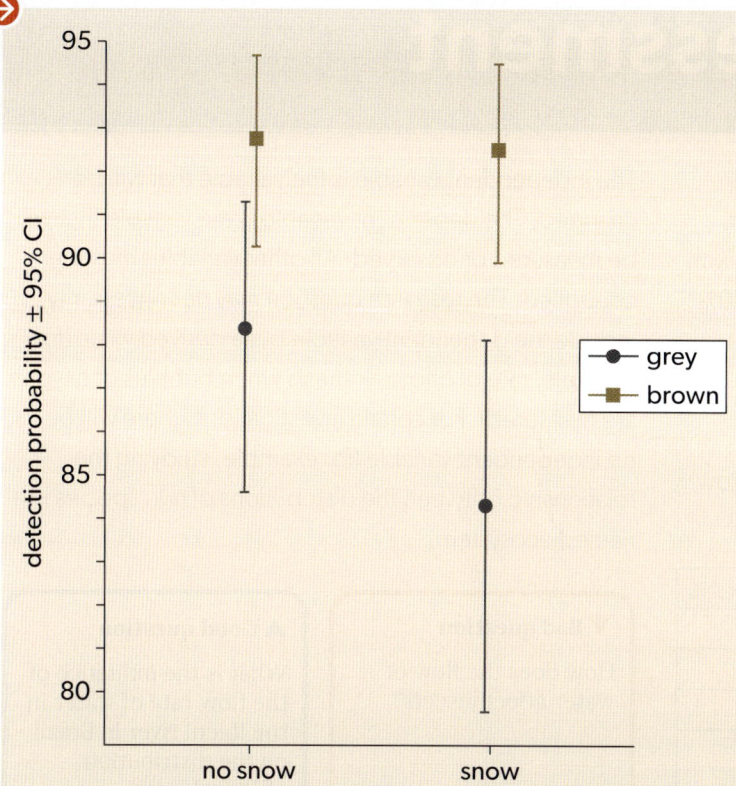

Source: Koskenpato, K., et al. (2020). Gray plumage color is more cryptic than brown in snowy landscapes in a resident color polymorphic bird. *Ecology and Evolution*, **10**, 1751–1761. www.doi.org/10.1002/ece3.5914 (CC BY 4.0)

a) State the mean percentage detection probability for grey and brown owls in snow.
b) Calculate the difference in mean detection in grey owls in no snow compared to snow landscapes.
c) Deduce whether the grey tawny owl morph could benefit from its coloration via increased crypsis in snowy landscapes.

Solution
a) Grey: 84%, brown: 92.5%
b) (88.5% − 84%) = 2.5%
c) Detectin probability is always lower for a grey than for a brown tawny owl morph, but it is even lower in snowy landscapes. There is a strong selection pressure against the brown owls during snowy winters, showing a benefit of crypsis for grey owls in snowy landscapes.

Practice problems

1. Explain how diversity changes through selection.
2. What Hardy–Weinberg condition needs to be maintained for a population to be in genetic equilibrium?
   A. No migration
   B. Large genetic flow
   C. High number of mutations
   D. Small population size
3. Outline how sustainability can be assessed in a terrestrial plant species.
4. Discuss the processes in the carbon cycle that affect concentrations of carbon dioxide and methane in the atmosphere and the consequences for climate change.

- What are the impacts of climate change at each level of biological organization?
- What processes determine the distribution of organisms on Earth?

Internal Assessment

For your internal assessment (IA), you will be asked to carry out an individual investigation and write a report. This report should be between 6 and 12 pages long, with a maximum word count of 3,000 words. It will form 20% of your final IB Biology grade.

Assessment criteria

Your investigation is assessed using the six criteria below, out of a total of 24 marks.

Criterion	Maximum marks	Percentage of mark / %
Research design	6	25
Data analysis	6	25
Conclusion	6	25
Evaluation	6	25
Total	24	100

▲ Table 1 IA assessment criteria

Choice of research question

It is important that your investigation yields good data. You can be adventurous in the choice of your investigation, but you must make sure it will provide you with enough data for analysis. Choosing an interesting research question that yields little data, or insufficiently useful data, will score poorly.

Assessment tip

Research and try out your ideas to make sure you can generate sufficient useful data before you embark on a full investigation.

Not all IAs are based on data obtained from experiments (primary data). You can use data generated from online databases. In cases where you work with a group of candidates collaborating to generate a database, you must generate your own unique research question.

Assessment tip

Collect sufficient data over a meaningful range, with adequate repeats, and use the most precise apparatus you have access to.

The independent variable is the variable that will be changed. The dependent variable is the variable that will be measured or observed. All other variables must be controlled. The research question may not necessarily include the dependent variable but a derived value (for example, rate of photosynthesis when bubbles of oxygen are measured). For certain investigations, there will be no independent variable (for example, studying the relationship between the distributions of two species in a named ecosystem).

▼ Bad question	▲ Good question
How does the flow of water affect insects?	What is the influence of the flow rate of water in the Ibicui river in Brazil on the distribution of *Paracloedes leptobranchus* nymphs?

▲ Table 2 Good and bad research questions

Assessment tip

Make sure your research question is focused and precisely expressed. For example, "Measuring the effect of light on photosynthesis" is too vague as it does not indicate what the dependent and independent variables are. The independent variable could be light intensity, wavelength, photons, type of lamp, etc., whereas the dependent variable could be rate of photosynthesis, presence of starch, and many others.

When formulating your hypothesis, consider:
- the independent variable or the two variables being correlated
- the range of the independent variable (try to keep it limited to a reasonable value)
- the dependent variable (or derived dependent variable) and how you are going to measure it
- the scientific name of the organism used
- a prediction of what you think is going to happen.

Once you have an investigative hypothesis, then you can add the statistical "null" and "alternative" test hypotheses:
- The null hypothesis will state a lack of correlation/association or impact between variables or groups.
- The alternative hypothesis will state a significant correlation/association or difference between variables or groups.

Criterion 1: Research design

This criterion includes the choice of the research question, description and justification of the methodology. This criterion is about establishing a scientific context, stating a clear and focused research question and using appropriate concepts and techniques in your method. In addition, it assesses your awareness of safety, environmental and ethical considerations.

For this criterion, make sure that you:
- communicate your interest in your investigation
- pose a question to which you do not already know the answer
- do not use a known method without making any modifications to it—examiners value originality in the design of your method
- explain why you are using the method you have chosen. Ensure that the method works, and if it does not, modify it and explain how and why you modified it in your report.

To gain the research design marks, include the following sections:
- your research question
- your hypothesis
- background information
- a statement of the variables
- apparatus
- method
- safety considerations.

Assessment tip

If you use data obtained from other researchers (secondary data) or data from models and simulations instead of collecting primary data in the lab, you will need more than one source of data so that the variance between sources can be evaluated. This will make your error analysis and evaluation more successful.

After stating your research question you need to include suitable background information, which should focus on the specifics of your chosen research question and methodology. Describe the context for your question and existing knowledge.

Assessment tip

Remember, the background information provided for the investigation should be relevant and should enhance the understanding of the context of the investigation.

You will need to include a statement of the variables and how you will control them. You may find it helpful to use a table while planning your investigation to make sure you do not miss any out variables when writing your statement.

Include a detailed list of the apparatus and materials used in your investigation. Choose apparatus that will enable you to collect reliable data, for example, by using graduated pipettes and measuring cylinders to measure volume rather than beakers.

▼ Bad naming of apparatus	▲ Good naming of apparatus
measuring cylinders	$3 \times 25.0 \, cm^3 \pm 0.5 \, cm^3$ measuring cylinders

▲ Table 3 Good and bad descriptions in an apparatus section

Make sure your method is written in sufficient detail for any reader to repeat the investigation and so that an idea of the associated uncertainties can be gained. It is a good idea to describe in a paragraph how you chose and developed your methodology. This narrative will help explain the amount of data you are collecting and give insight into your decision-making. You should plan how many measurements you will need to take for a valid conclusion, and your method must allow sufficient repeats to ensure each measurement is valid.

After the method section, you should add in a short section about the safety, ethical and environmental issues relevant to your methodology. This might be quite basic, such as noting the need for gloves and safety glasses, but could also include the application of the IB animal experimentation policy, use of consent forms, disposal of waste and impact on field sites. If you cannot identify any safety, ethics or environmental problems, a statement to this effect will show that you have at least considered them.

Internal Assessment

Criterion 2: Data analysis

The data analysis criterion includes the communication of the recording and processing of the data, considering the uncertainties. The communication includes the appropriate choice of graphs, tables, etc. and the correct use of units. The processing of data (manipulation of data into useful information) needs to be accurate.

In your analysis, you should include:

- raw data
- sample calculations for processed data
- qualitative and quantitative data
- statistical analysis of your data
- associated uncertainty table
- graphical analyses.

> **Assessment tip**
>
> Remember that bar charts are of limited analytical use and are used only if you have discontinuous variables.

You need to record enough data about the independent and dependent variables to enable meaningful processing and interpretation. You should also record qualitative observations. For example, fieldwork should always have some form of site description, which could take the form of maps, sketches or photographs. However, if you use photos make sure that these are supporting and not replacing written qualitative data. Also, be careful not to include unnecessary photos, for example of the set-up of a respirometer under different conditions, where one picture will be enough.

▼ Bad graph	▲ Good graph
Units missing on the y-axis	Graph has a title
Line of best fit is straight, but the data indicates a curve would be a better fit	Axes have labels
	Data points are clearly marked
	Scale of graph is appropriate

▲ Table 4 Good and bad aspects of a graph

Once you have collected the data, include the associated uncertainty in a table. The uncertainty is how the estimate might differ from the actual value. You need to interpret and evaluate the uncertainty in the data and the size of any discrepancy between your results and those in the literature (if they exist). This gives an indication of how accurate your data is.

If you are familiar with statistical analyses, you can consider the impact of measuring uncertainty on your results using the following techniques:

- propagate the errors through numerical calculations
- calculate the standard deviation (if you have a sufficiently large data set)
- draw a line of best fit on your graph
- consider including error bars and/or the maximum or minimum slopes that can be drawn with the data you have, based on the range of values at each data point.

Statistical analysis of your data is expected. The layout of this will depend upon the tools used. Percentages, means, medians, standard deviations or other statistics can be presented at the end of the column or row of data they represent. For more complex processing of data using spreadsheets, such as correlation coefficient, t-test and χ^2 (chi-squared) test, screenshots are acceptable. For other types of processing, such as analysis of variance (ANOVA) tests, a worked example may be necessary.

- The **mean** is the sum of the values divided by the number of values; i.e. the average.
- The **median** value is found exactly in the middle of a data set.
- Statistical tests are to prove that any difference between sets of data is not random. The result of a test is **statistically significant** when a test shows that your data sets are reliably different, and therefore not different by chance.
- **Standard deviation** is a measure that is used to quantify the amount of variation or dispersion of a set of data values.
- **Correlation coefficient (r)** is a numerical measure of some type of correlation. It shows whether there is a statistical relationship between two variables.
- **Coefficient of determination (R^2)** is a statistic that gives some information about the goodness of fit of a model. An R^2 of 1 indicates that the regression predictions perfectly fit the data.
- A ***t*-test** is a statistical test used to determine whether there is a significant difference between the means of two groups of data. A t-test is commonly applied when the data sets follow a normal distribution.

- A χ^2 **(chi-squared) test** is a statistical test used to determine whether there is a significant difference between the expected frequencies and the observed frequencies in one or more sets of data.
- An **ANOVA (analysis of variance) test** is a collection of statistical models that helps you to calculate how much of the total variance comes from variance between groups of data and from variance within groups of data. ANOVA is useful for comparing three or more groups of data. It will show if there are any statistically significant differences between the means of three or more independent groups of data.

Assessment tip

When presenting your data, check you have considered:
- degrees of precision of instruments used
- variation in the material used
- standard deviations
- standard errors
- ranges (maximum–minimum)
- outlier data.

Outliers need to be recorded but marked as such and should not be used in the statistical analysis. You should check whether any data is really an outlier by repeating the experiment. You should not exclude outliers from processing just because they do not "fit well" in the general trend of the data. Their exclusion requires a justification.

Completing several trials will make it easier for you to decide which results are inconsistent. If you can, compare the data from different data sources (secondary data examples) to evaluate reproducibility. If you are using data from a database, you need to show you have done some research into the uncertainties associated with database data and taken these uncertainties into consideration.

Assessment tip

In this section, you need to keep an eye on your use of both significant figures and decimal places—they need to be consistent.

- A **random error** is a statistical fluctuation (in either direction) in the measured data due to the precision limitations of the measurement device.
- **Systematic error** is a problem that persists throughout the experiment that can be solved through a better experimental design.
- The **uncertainty** of a measurement is as accurate as the tool being used to make this measurement. It is one half of the smallest measurement possible with the device.
- An **outlier** is any value that is numerically distant from most of the other data points in a set of data.
- **Accuracy** refers to the closeness of a measured value to a standard or known value.
- **Precision** refers to the closeness of two or more measurements to each other.
- **Reliability** is the degree of consistency of a measurement.

Assessment tip

Using error bars and lines of best fit on graphs will help you to portray the data as realistically and honestly as possible.

Criterion 3: Conclusion

The conclusion should reflect how well the data has been interpreted and how well it answers the research question in an appropriate scientific context. Drawing a conclusion that is consistent with the data is straightforward, but correctly describing or justifying your conclusion through relevant comparisons to the accepted scientific context is more difficult. However, you can do this by comparing your experimentally determined data with other available data and/or checking whether any trends and relationships that you identify are in line with accepted theory.

- A **correlation** indicates a predictive relationship between two variables.
- A **trend** is a pattern or general tendency of a series of data points.
- A **causal link** indicates that one event is the result of the occurrence of the other event, i.e. the independent variable affects the dependent variable.

Criterion 4: Evaluation

This criterion assesses your evaluation of your investigation and your results in relation to the research question and accepted scientific context. It also assesses your ability to identify limitations and suggest improvements.

In the evaluation you must identify limitations and suggest improvements to your investigation. Try to identify and address any systematic and random errors in detail. Include a discussion of the likely effect of systematic errors. For example, if you measured the volume of an enzyme with a pipette and left a drop behind each time you will have a smaller volume than recorded. Do not be tempted to suggest things that are unnecessary, such as more repeats if you had sufficient results that agreed. You can also calculate the percentage error of your results. Lastly, if you can see a good extension to your investigation, suggest it here.

> **Assessment tip**
> - Make sure you have a basis to your conclusion that you can then compare with secondary data or a scientific theory.
> - Identify whether errors are random or systematic.
> - Suggest improvements to address the errors identified.

Report checklist

Complete this checklist before you submit your report.

Tasks
The report is within the 3,000-word limit
The title page has the candidate name and number on it
There is a contents page and all pages are numbered
The research question is specific
All background information included is relevant to the research question
The apparatus is listed precisely
The methods are described clearly (and are repeatable)
All illustrations have figure numbers
Raw data has been included
All graphs/tables have units and all graphs have labelled axes
Sample calculations have been included for processed data
Data is shown honestly, e.g. using lines of best fit or error bars
Safety, ethical and environmental issues have been included
All sources are correctly referenced
There is a conclusion
The evaluation makes recommendations for improvements

Practice Exam Papers

At this point, you will have re-familiarized yourself with the content from the topics of the IB Biology syllabus. Additionally, you will have picked up some key techniques and skills to refine your exam approach. It is now time to put these skills to the test; in this section you will find practice examination papers 1A, 1B and 2, with the same structure as the external assessment you will complete at the end of the DP course. Answers to these papers are available at www.oxfordsecondary.com/ib-prepared-2e.

Paper 1

SL: 1 hour 30 minutes – 55 marks
HL: 2 hours – 75 marks
Weighting: 36%
Paper 1 is presented in two separate booklets.

Paper 1A: Multiple-choice questions. No marks are deducted for incorrect answers.
Paper 1B: Four data-based questions related to experimental work and the syllabus.

Paper 1A

Answer all the questions. For each question, choose the answer you consider to be the best.

The maximum mark for the SL examination paper is 30 marks.

The maximum mark for the HL examination paper is 40 marks.

SL candidates: answer questions 1—30 only.
HL candidates: answer all questions.

1. What is a consequence of cohesion between water molecules?
    A. Turbidity of lakes
    B. Transport in xylem
    C. Dissolving of nutrients
    D. Low melting point of ice

2. What structure can only be found in prokaryotic and not in eukaryotic cells?
    A. Ribosome
    B. Naked DNA in a loop
    C. Nucleus
    D. Cytoplasm

3. What could have caused the current biodiversity crisis?
    I   Hunting
    II Spread of pests
    III Spread of invasive species by global transport
    A. I only
    B. I and II only
    C. I and III only
    D. I, II and III

4. Beak shape in Darwin's finches (*Geospiza*) is emblematic of natural selection. Representatives of four distinct beak morphologies of finches that feed on different parts of cacti are shown.

    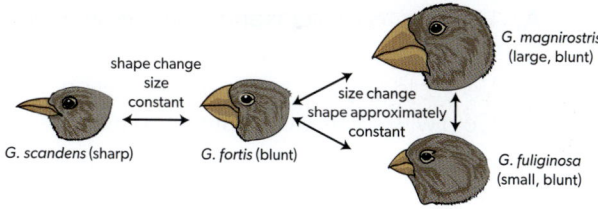

    What is this an example of?
    A. Speciation
    B. Competition
    C. Convergent evolution
    D. Selective breeding

5. The diagram shows the structure of different biological molecules. Which diagram shows the structure of alpha-glucose?

    A. [amino acid structure with H₂N—C(H)(R)—C(=O)OH]
    B. [HOCH₂ pentose ring with OH groups, HO and OH at bottom]
    C. [CH₂OH hexose ring with OH groups: HO, OH, OH]
    D. [straight-chain structure: H—C—C—C—C—C—C—C—C—C(=O)—O—H with H substituents]

251

6. Which is amphipathic?
   A. Phospholipid with hydrophobic and hydrophilic regions
   B. Steroid with a hydrophobic region
   C. Hydrophobic amino acid
   D. Hydrophilic phosphate group

7. Where can peripheral proteins be found in membranes?
   A. Embedded in one lipid layer
   B. Embedded in both lipid layers
   C. Across from one side to the other
   D. Attached to the outer surface of the bilayer

8. What is an advantage of compartmentalization in the cytoplasm of cells?
   A. Concentration of metabolites
   B. Faster diffusion of molecules
   C. Increase in surface area-to-volume ratio
   D. Availability of enzymes in all cytoplasm

9. What is a similarity between most veins and capillaries?
   A. Neither have valves
   B. Neither have smooth muscle cells
   C. They both have elastic tissue
   D. They both carry blood at low blood pressure

10. Which adaptation enhances gas exchange in leaves?
    A. Thick waxy cuticle
    B. Few air spaces in spongy mesophyll
    C. Few stomata
    D. Thin epidermis

11. What defines a tundra biome?
    A. Hot days, cold nights and scarce vegetation
    B. Rainy, hot and many trees
    C. Very cold, dry and hardly any vegetation
    D. Cold with some precipitation and trees

12. In which environment could an obligate anaerobe thrive?
    A. An airtight can
    B. A jar without a lid
    C. A frozen lake
    D. A freshwater river

13. How is energy transferred to ATP (adenosine triphosphate) in respiration?
    A. Energy from the breakdown of glucose phosphorylates ADP
    B. The breakdown of glucose hydrolyses the ADP to form ATP
    C. The formation of glucose phosphorylates one ATP molecule
    D. The breakdown of carbon compounds joins phosphate to ATP

14. The graphs show the absorption spectra of three organisms.

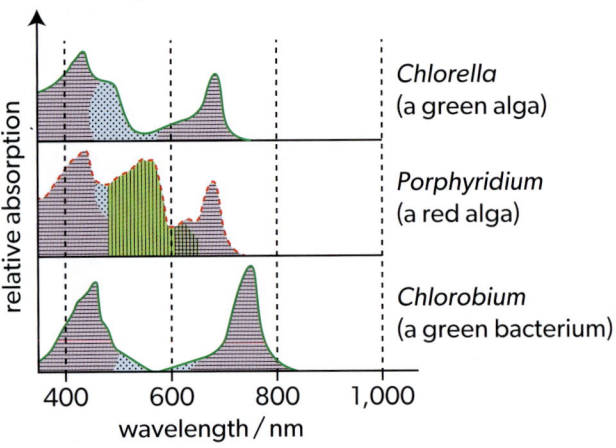

Source: Stanier, R., Doudoroff, M., & Adelberg, E. (n.d.). General Microbiology. Macmillan.

What can be concluded from this data?
   A. Green algae photosynthesize less than green bacteria at all wavelengths
   B. Green algae photosynthesize more in green light than red algae
   C. Red algae absorb more red light than green light
   D. Red algae absorb more green light than green bacteria

15. The graph shows the percentage activity of the enzymes X and Y at 130°C for 60 minutes.

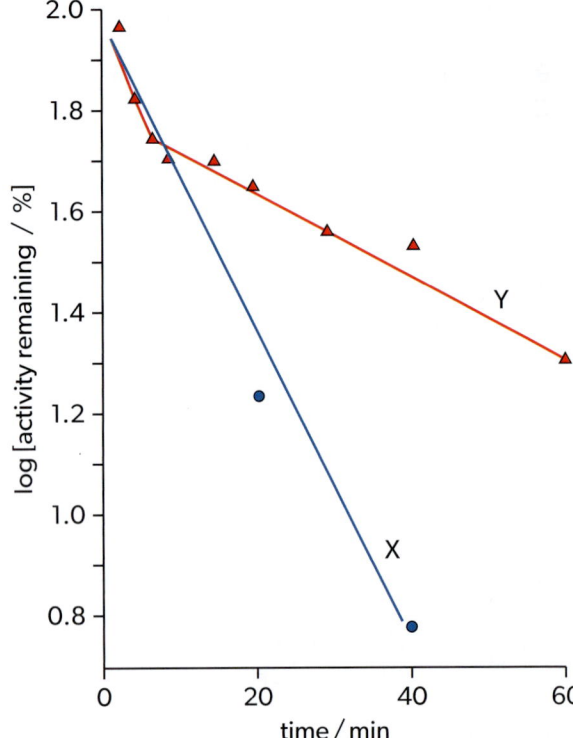

Source: Daniel, R. M., et al. (1996). The denaturation and degradation of stable enzymes at high temperatures. *Biochemistry Journal*, **317**, 1–11. www.doi.org/10.1042/bj3170001

What can be concluded from the graph?
   A. X is not denatured by temperatures above 130°C
   B. Y is more resistant to high temperatures than X

C. X has fewer collisions with the substrate than Y
D. Both X and Y are fully denatured after 10 minutes

16. The graph shows the myelin thickness in nerve fibres of rats of different ages.

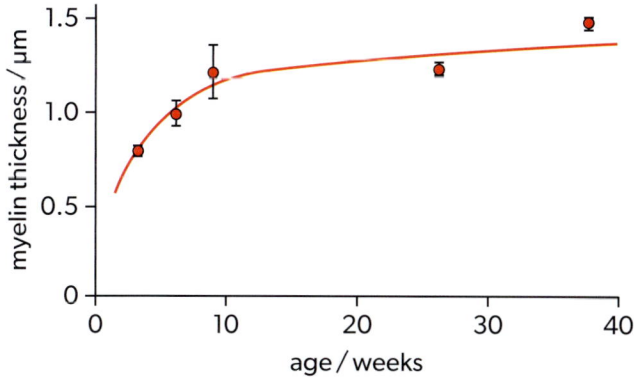

Source: Ellis, T. J., et al. (1980). Automated measurement of peripheral nerve fibres in transverse section. *Journal of Biomedical Engineering*, 2(4), 272–280. www.doi.org/10.1016/0141-5425(80)90120-x

What describes the change with aging in rat nerve fibres?
A. Myelin at 30 weeks is 100% thicker than at 3 weeks old
B. Myelin sheath nearly duplicates every 5 weeks
C. Nerve impulses will travel faster as they age until week 20 where it plateaus
D. There is no change in the transmission of impulses with aging

17. What is the direct role of calcium ions in nerve impulse transmission?
A. To signal acetylcholine vesicle fusion with the membrane in the presynaptic neuron
B. To open sodium channels in the presynaptic neuron
C. To inhibit sodium channels in the postsynaptic neuron
D. To close potassium channels in the postsynaptic neuron

18. How do the hormonal and nervous systems differ?
A. Only the nervous system has signalling
B. Hormonal signals travel through blood and nervous through fibres
C. Only hormonal signals are chemical
D. Hormonal signalling is faster than nervous signalling

19. The diagram shows the production of antibodies.

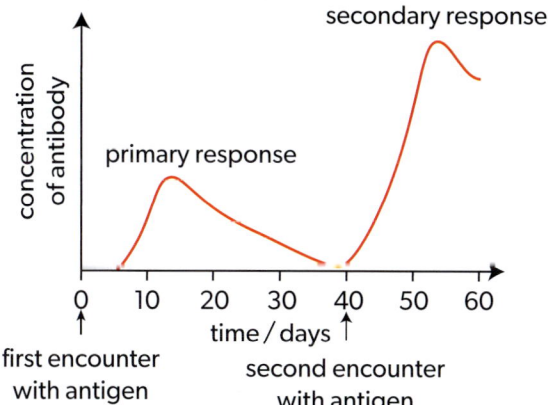

What describes the production of antibodies when an antigen is encountered?
A. In the second encounter, the antibodies are larger in size than in the first encounter
B. In the first encounter, plasma cells produce fewer antibodies than in the second
C. In the first encounter, there are no B-lymphocytes, only T-lymphocytes
D. In the second encounter, the amount of T-lymphocytes is greater than in the first

20. How is peristalsis controlled in the human digestive system?
A. Voluntary control by the enteric nervous system only
B. Voluntary control by the central nervous system only
C. Involuntary control by the central nervous system and voluntary control by the enteric nervous system
D. Involuntary control by the enteric nervous system and voluntary control by the central nervous system

21. In a field trip, a student used a quadrat to count the presence of two species (1 and 2) in an ecosystem. To test whether there is association between the presence of the two species, a chi-squared test was performed.

Species	Number of organisms observed	Expected result if there is no association	$\frac{(\text{Observed} - \text{expected})^2}{\text{expected}}$
1 only	3	2.5	0.1
2 only	1	2.5	0.9
1 and 2	4	2.5	0.9
none	2	2.5	0.1
total	10	10	2.0

Degrees of freedom	Probability 0.05
1	3.841
2	5.991
3	7.815
4	9.488

What does the information show?
A. The alternative hypothesis is rejected as the calculated value is smaller than the critical value
B. The null hypothesis is rejected as the critical value is smaller than the calculated value
C. The null hypothesis is accepted as the critical value is larger than the probability
D. There is no association between the presence of the two species as there are four degrees of freedom

22. Orchids have life stages that depend on fungi for carbon and other essential resources, and orchids provide carbohydrates through photosynthesis. The relationship between density of showy orchid (*Galearis spectabilis*) plants and DNA of mycorrhizal fungus *Ceratobasidium* in the soil close to the orchids was monitored.

    What does this graph show?
    A. The competitive relationship is intraspecific
    B. The mutualistic relationship is interspecific and density related
    C. The amount of mycorrhizal fungal DNA is not related to the number of plants
    D. The relationship is not density related and the mycorrhizal fungus is a parasite

    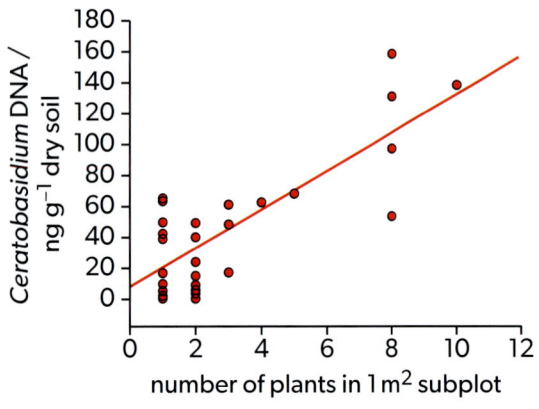

    Source: McCormick, et al. (2018). Mycorrhizal fungi affect orchid distribution and population dynamics *New Phytologist*, **219**, 1207–1215. www.doi.org/10.1111/nph.15223

23. The table shows the first, second and third positions of the three bases of the messenger RNA (mRNA) codons in the genetic code.

		Second base				
		U	C	A	G	Third base
First base	U	UUU UUC } Phe UUA UUG } Leu	UCU UCC UCA UCG } Ser	UAU UAC } Tyr UAA Stop UAG Stop	UGU UGC } Cys UGA Stop UGG Trp	U C A G
	C	CUU CUC CUA CUG } Leu	CCU CCC CCA CCG } Pro	CAU CAC } His CAA CAG } Gln	CGU CGC CGA CGG } Arg	U C A G
	A	AUU AUC AUA } Ile AUG Met	ACU ACC ACA ACG } Thr	AAU AAC } Asn AAA AAG } Lys	AGU AGC } Ser AGA AGG } Arg	U C A G
	G	GUU GUC GUA GUG } Val	GCU GCC GCA GCG } Ala	GAU GAC } Asp GAA GAG } Glu	GGU GGC GGA GGG } Gly	U C A G

    A sequence of mRNA nucleotides is: UGUCUAAAUGUAUAAAGA.
    What sequence of amino acids will the sequence of mRNA translate to?
    A. Met Cys Leu Asn Val Arg
    B. Thr Asp Gly His Ile Ser
    C. Cys Leu Asn Val Arg
    D. Cys Leu Asn Val

24. When is a gene expressed?
    A. If DNA is transcribed to mRNA and translated to protein
    B. If chromatids are replicated in meiosis
    C. Always after it is synthesized by DNA replication
    D. Only after cytokinesis of the parent cell

25. What is a cause of Down syndrome?
    A. Point mutation in chromosome 2
    B. Non-disjunction of a chromosome in meiosis
    C. Deletion in chromosome 9
    D. Substitution in chromosome 21

26. What is an example of unequal cytokinesis?
    A. Formation of monkey male gametes by meiosis
    B. Formation of sperm in humans by meiosis
    C. Growth of a xylem cell by mitosis
    D. Budding in yeast cells by mitosis

27. How are hormones used in in vitro fertilization (IVF)?
    A. Normal secretion of hormones is enhanced with vitamins
    B. Artificial hormones are injected to induce ovulations
    C. The menstrual cycle is arrested
    D. Hormones are used to join sperm with eggs

28. How can cross-pollination be promoted?
    A. Pollen matures at a different time than the stigma in the same flower
    B. Flowers contain both male and female reproductive systems
    C. Male and female plants are grown near each other
    D. By ensuring self-pollination

29. What is phenotypic plasticity?
    A. A change in phenotype due to different DNA composition
    B. The difference between genotype and phenotype
    C. The change in a trait due to varying gene expression patterns
    D. Adaptation to the environment by convergent evolution

30. The diagram shows the gross forest gain and loss in Africa and Asia in the year 2020.

    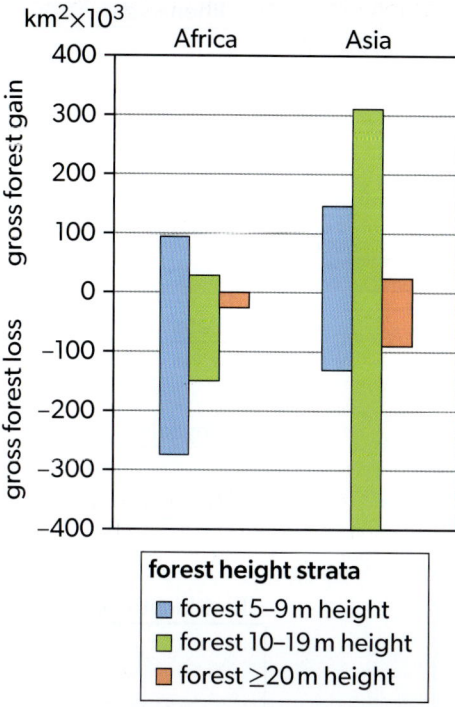

    Source: Potapov, P., et al. (2022). The global 2000–2020 land cover and land use change dataset derived from the Landsat archive: first results. *Frontiers in Remote Sensing*, 3. www.doi.org/10.3389/frsen.2022.856903 (CC BY 4.0)

    What can be concluded from this graph?
    A. Total forest area is larger in Asia than in Africa
    B. Total forest area will decrease in Africa
    C. Forests are taller in Asia than in Africa
    D. The percent change in tall forest is larger in Africa than in Asia

31. What allowed the retention of extraplanetary water on Earth?
    A. Adhesion of water to rocks only
    B. Gravity and condensation of water
    C. Temperatures cold enough for freezing only
    D. Temperatures warm enough for water evaporation

32. What occurs only during the lysogenic cycle of a lambda bacteriophage?
    A. The host is destroyed
    B. New bacteriophages are assembled
    C. Bacteriophage DNA is destroyed in the host
    D. DNA of the bacteriophage fuses with the host's chromosome

33. Which property of some viruses explains the lack of a secondary immune response in humans?
    A. Viruses can have a high mutation rate
    B. B-cells do not attack viruses
    C. Antibodies do not attack viruses
    D. Viruses do not induce a primary response

34. The cladogram shows evolutionary relationships between fish orders.

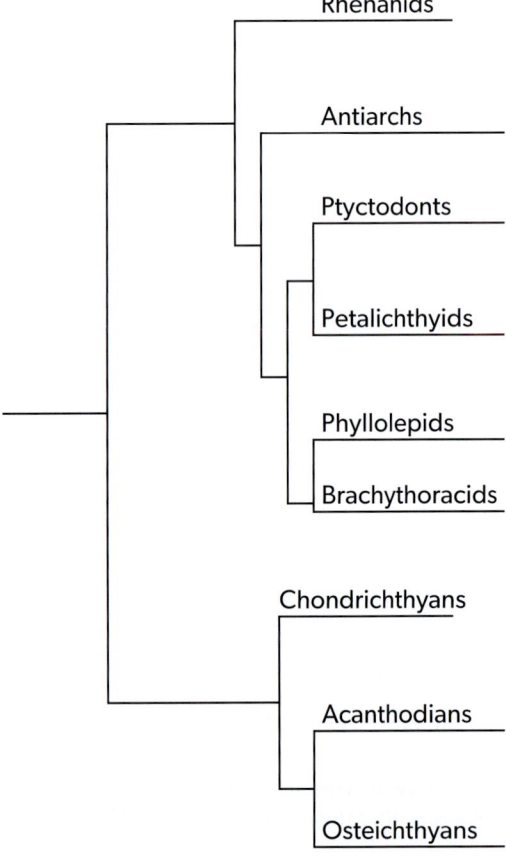

Source: Carr, R. K. & Jackson, G. L. (2008). Guide to the geology and paleontology of the Cleveland Member of the Ohio Shale. *Ohio Geological Survey Guidebook 22*, Chapter 5.

What can be deduced from this cladogram?

A. Rhenanids and Antiarchs are more closely related than Ptyctodonts and Petalichthyids
B. Chondrichthyans are the closest relatives to Brachythoracids
C. Acanthodians and Osteichthyans have a different ancestor
D. Petalichthyids and Phyllolepids have a common ancestor

35. Which is a transmembrane protein?

A. Clathrin
B. Potassium ion channel
C. Neurotransmitter
D. Myelin

36. What can cause the Bohr shift?

A. An increase in water in tissues
B. A decrease in carbon dioxide in blood
C. An increase in acidity of blood due to carbon dioxide
D. A decrease in temperature

37. What are photosystems?

A. Arrays of pigments that generate and emit electrons
B. Chloroplasts found in plants and blue-green algae
C. Chlorophyll and other pigments in the stroma of the chloroplast
D. The site where photolysis of water occurs

38. The diagram shows a pair of homologous chromosomes during meiosis in a cell in the human testis. The positions of the alleles of some genes are indicated.

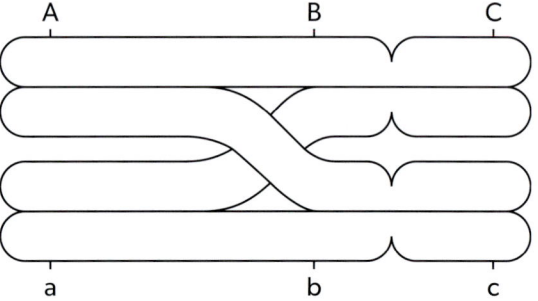

Which alleles are possible recombinants?

A. AABBCC
B. AABbCC
C. AaBbCc
D. AaBBCC

39. What contributes to a high water potential inside a plant cell?

A. Lower absolute solute potential inside the cell
B. Higher osmotic pressure on both sides of the membrane
C. Lower pressure potential outside the cell
D. Positive solute potential outside the cell

40. The electron micrograph shows cells in the proximal convoluted tubule in the kidney. The brush border of these cells increases surface area.

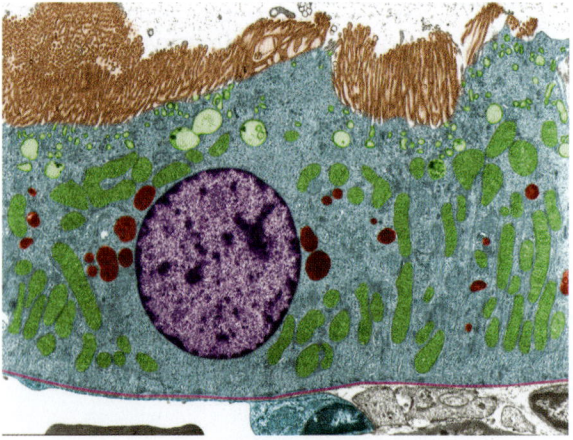

What is the function of these cells?

A. Reabsorption of material from the glomerular filtrate
B. Reabsorption of water into blood
C. Secretion of aquaporins
D. Ultrafiltration

Paper 1B

SL: 25 marks
HL: 35 marks

1. The electron micrograph shows a rat liver cell ×11,000.

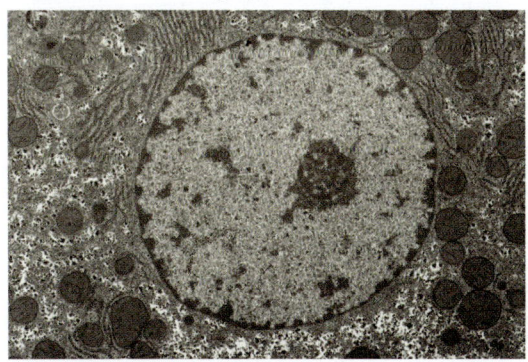

   (a) On the micrograph, label a Golgi apparatus and a mitochondrion. [2]
   (b) Suggest a reason for the number of mitochondria. [1]
   (c) Calculate the actual size of the diameter of the nucleus. [1]
   (d) Describe the function of the Golgi apparatus. [2]
   (e) Suggest **one** way liver cells could have differentiated to perform a specific function. [1]

2. The altitudinal ranges in New Guinea of two closely related nectar-drinking parrots, *Vini placentis* and *V. rubronotata*, on Mount Nibo, the Sepik Mountains and Mt. Karimui change with the presence or absence of competitors. In the graph, the horizontal line under each mountain's name indicates the summit elevation.

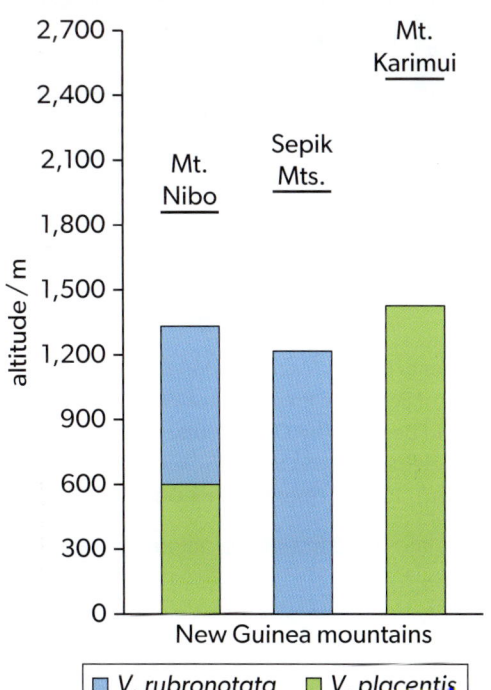

Source: Diamond, J. M. (1978). Niche shifts and the rediscovery of interspecific competition: Why did field biologists so long overlook the widespread evidence for interspecific competition that had already impressed Darwin? *American Scientist*, 66(3), 322–331. www.jstor.org/stable/27848643

   (a) Deduce the effect of competition on the altitudinal niche distribution of the parrots. [3]
   (b) Describe a method for quantifying the number of nectar-drinking parrots in this ecosystem. [3]

3. In a study to see the effects of progesterone on the endometrium, female mammals were injected with either progesterone or corn oil (control with no progesterone) for 28 days. The height of the endometrium was measured (A) and an electrophoretic analysis (Western blot) of uterine secretions from mammals of both treatments was performed (B).

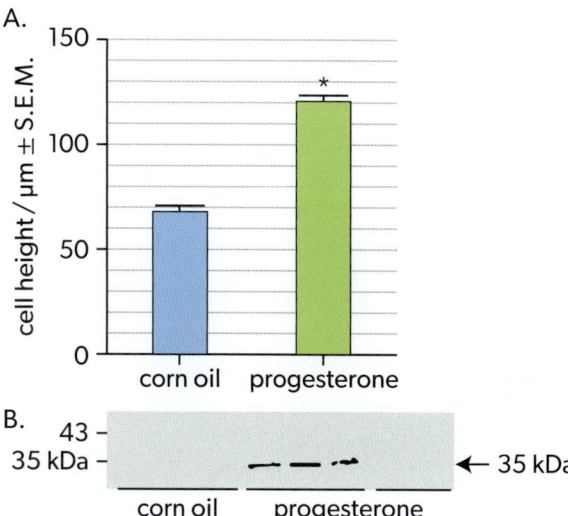

Source: Bailey, D. W., et al. (2010). Effects of long-term progesterone on developmental and functional aspects of porcine uterine epithelia and vasculature: progesterone alone does not support development of uterine glands comparable to that of pregnancy. *Reproduction*, 140, 583–594. www.doi.org/10.1530/rep-10-0170

   (a) State the independent variable in this experiment. [1]
   (b) Describe the electrophoresis performed to separate proteins. [3]
   (c) (i) Calculate the percentage increase in cell height in progesterone-treated endometrium. [1]
       (ii) Explain the effect of progesterone on endometrium cell height. [2]
   (d) A 32 kDa protein appeared in the progesterone-treated mammals. Suggest **one** reason for the appearance of a new protein in the treatment group. [1]
   (e) Explain the control of pregnancy and childbirth by progesterone. [2]

4. The table shows the approximate production of organisms in an ecosystem.

Organism	Trophic level	Production / $kJ\,m^{-2}\,y^{-1}$
birds	tertiary consumer	15
crabs	secondary consumer	120
spiders	secondary consumer	180
insects	primary consumer	1,200
alga	producer	8,000
grass	producer	16,000

(a) Construct a pyramid of energy showing the trophic levels. [3]
(b) Explain the change in energy between trophic levels. [3]
(c) Sketch a food chain of four organisms in this ecosystem. [2]
(d) Describe the changes occurring during a primary succession in a named habitat. [4]

Paper 2

SL: 1 hour 30 minutes – 50 marks
HL: 2 hours 30 minutes – 80 marks
Weighting: 44%

Section A

SL: 34 marks
HL: 48 marks

Data-based and short-answer questions

1. The Colorado potato beetle (*Leptinotarsa undecemlineata*) is an important economic pest of potatoes (genus *Solanum*) in North America, Europe and Asia.

The number of mating females in a population of Colorado potato beetles was recorded per day in Veracruz, Mexico. The days were separated into four periods, each of 20 days, starting on day 2.

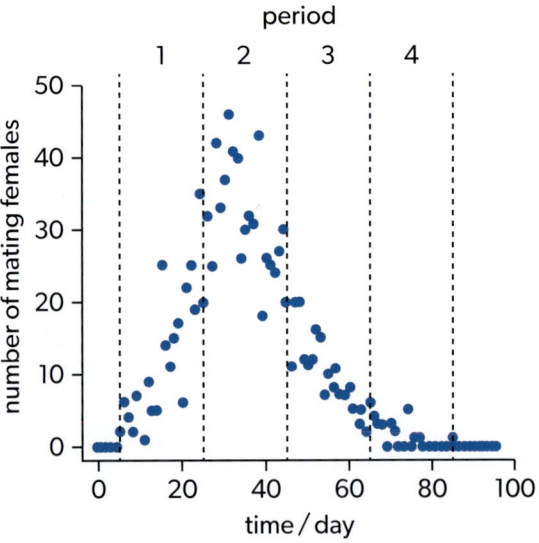

Source: Baena, M. L. & Macías-Ordóñez, R. (2012). Phenology of scramble polygyny in a wild population of chrysolemid beetles: The opportunity for and the strength of sexual selection. *PLoS ONE*, 7(6), e38315. www.doi.org/10.1371/journal.pone.0038315 (CC BY 4.0)

(a) State the period where most mating was observed in Colorado potato beetles. [1]
(b) On the diagram, sketch the trendline for mating females per day. [3]

The number of Colorado potato beetle females (F) and males (M) per plant were observed over the four periods of the mating season.

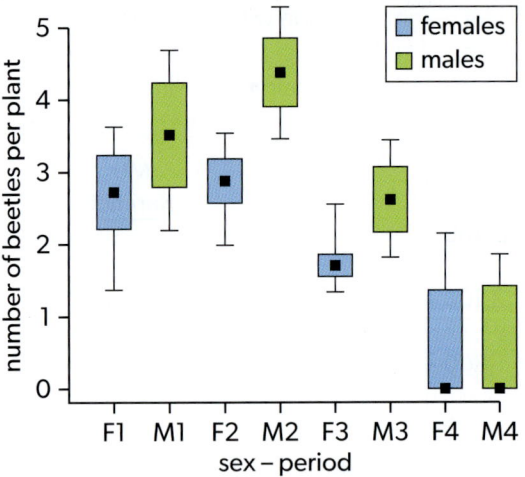

Source: Baena, M. L. & Macías-Ordóñez, R. (2012). Phenology of Scramble Polygyny in a Wild Population of Chrysolemid Beetles: The Opportunity for and the Strength of Sexual Selection. PLoS ONE, 7(6), e38315. www.doi.org/10.1371/journal.pone.0038315 (CC BY 4.0)

(c) Calculate the percentage difference in means between males and females in period 3. [1]
(d) Identify the range of females per plant in the third quartile in period 1. [1]
(e) Describe the trend for numbers of males in all four periods. [2]
(f) Female mating success depends on male density. Comment on this statement. [2]

The Colorado potato beetle feeds and completes its life cycle only on *Solanum lanceolatum*, and although *S. myriacanthum* grows sympatrically in Veracruz, it is never used by this beetle.

Under nutritional stress, female Colorado potato beetles reabsorb the egg cells, thus causing an interruption in their reproductive activity. The effect of the non-host plant, *S. myriacanthum*, on egg cell maturity was studied under laboratory conditions. Females fed on the natural host, *S. lanceolatum*, were compared to females fed on *S. myriacanthum*.

The percentages of mature and immature egg cells in the ovaries of the female Colorado potato beetles were recorded at four different ages when they were fed on *S. lanceolatum* or on *S. myriacanthum*. Measurements were taken for each diet at 14, 19, 34 and 54 days.

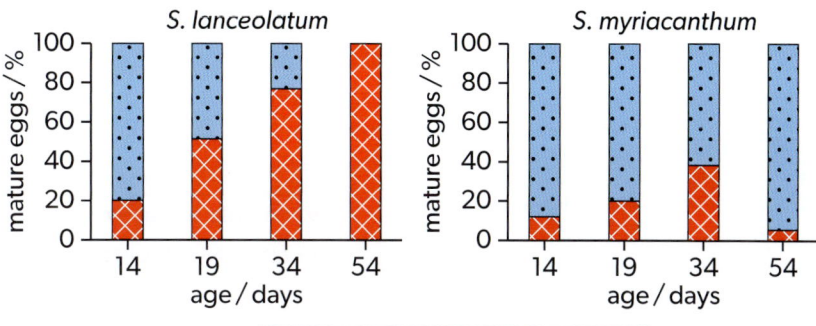

Source: López-Carretero, A., et al. (2005). Phenotypic plasticity of the reproductive system of female *Leptinotarsa undecimlineata*. The Netherlands Entomological Society Entomologia Experimentalis et Applicata, 115, 27–31. www.doi.org/10.1111/j.1570-7458.2005.00279.x

(g) State the percentage of immature eggs at 14 days in Colorado potato beetles fed on *S. lanceolatum*. [1]

(h) Compare and contrast the percentage of mature eggs in Colorado potato beetles fed on *S. lanceolatum* and those fed on *S. myriacanthum*. [3]

(i) To record mating success in Colorado potato beetles, scientists fed them on *S. lanceolatum*. Suggest **one** reason for not feeding them on *S. myriacanthum*. [1]

2. Carbon dioxide absorption has been studied in the Pacific Ocean near Hawaii by measuring carbon dioxide concentrations in the atmosphere and in surface water every month from October 1988 onwards. The graph shows the carbon dioxide concentration expressed as partial pressures (pCO_2).

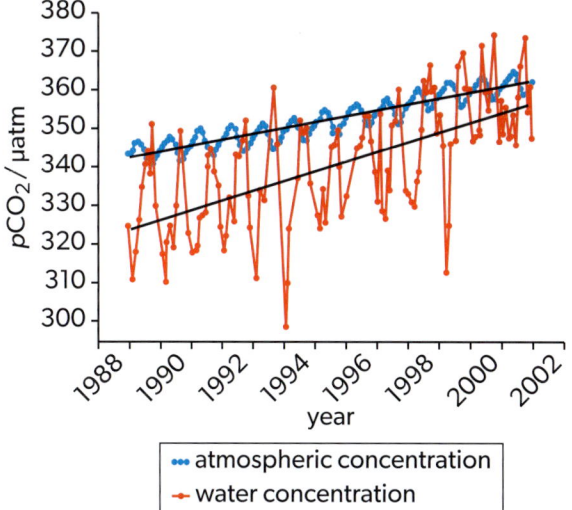

Source: Dore, J. E., et al. (2003). Climate-driven changes to the atmospheric CO_2 sink in the subtropical North Pacific Ocean. *Nature*, 424, 754–756. www.doi.org/10.1038/nature01885

(a) Suggest **two** reasons for the trends in carbon dioxide. [2]

(b) (i) Diffusion of carbon dioxide only occurs when there is a concentration gradient. Deduce the pattern of carbon dioxide diffusion between water and atmosphere, from 1988 to 2002. [2]

(ii) The graph provides evidence for the hypothesis that there will be no net diffusion of carbon dioxide between water and atmosphere by 2020. Explain this evidence. [1]

3. Cerebral toxoplasmosis is a leading cause of the central nervous system disorders in acquired immune deficiency syndrome (AIDS). It is a severe disease, caused by *Toxoplasma gondii*, an opportunistic intracellular parasite that can infect and destroy any nuclear cells. A study included 90 HIV-infected patients with cerebral toxoplasmosis and a control group of 225 HIV-infected patients without toxoplasmosis. HIV infection was confirmed with a polymerase chain reaction (PCR) test. The results were classified according to HIV viral load: below 50 copies/ml (small viral load), above 50 copies/ml (large viral load).

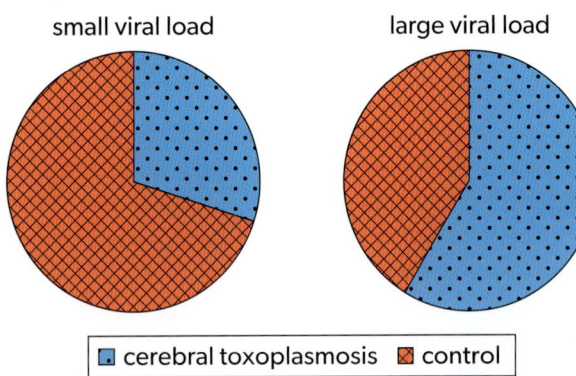

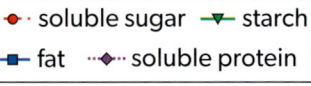

Source: Azovtseva, O. V., et al. (2020). Cerebral toxoplasmosis in HIV-infected patients over 2015–2018 (a case study of Russia). *Epidemiology and Infection*, **148**, e142, 1–6. www.doi.org/10.1017/s0950268820000928 (CC BY-NC-ND 4.0)

(a) Suggest the relationship between viral load and toxoplasmosis. [2]
(b) Describe the PCR process performed. [3]

4. The graph shows the percentage change in seed content reserve for *Chloris virgata* and *Lespedeza hedysaroides* at various germination stages (relative to imbibed seeds): (1) imbibition; (2) 1% germination; (3) 50% germination; (4) highest germination; (5) early seedling.

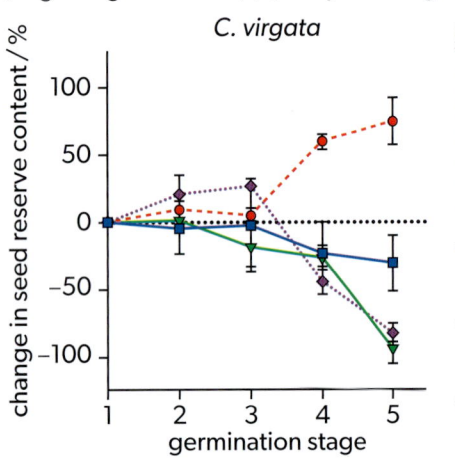
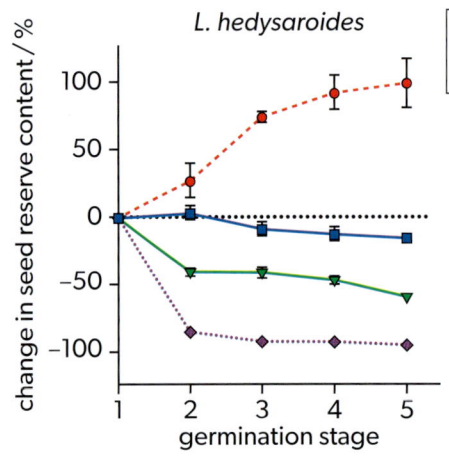

Source: Zhao, M., et al. (2018). Mobilization and Role of Starch, Protein, and Fat Reserves during Seed Germination of Six Wild Grassland Species. *Frontiers in Plant Science*, 9(234). https://doi.org/10.3389/fpls.2018.00234 (CC BY 4.0)

(a) Describe the imbibition stage of seeds. [1]
(b) Describe the use of different reserve sources in *C. virgata* and *L. hedysaroides*. [1]
(c) Explain the change from starch to soluble sugars. [2]

5. The graph shows the trophic strategies of 3,020 marine organisms as a function of their length.

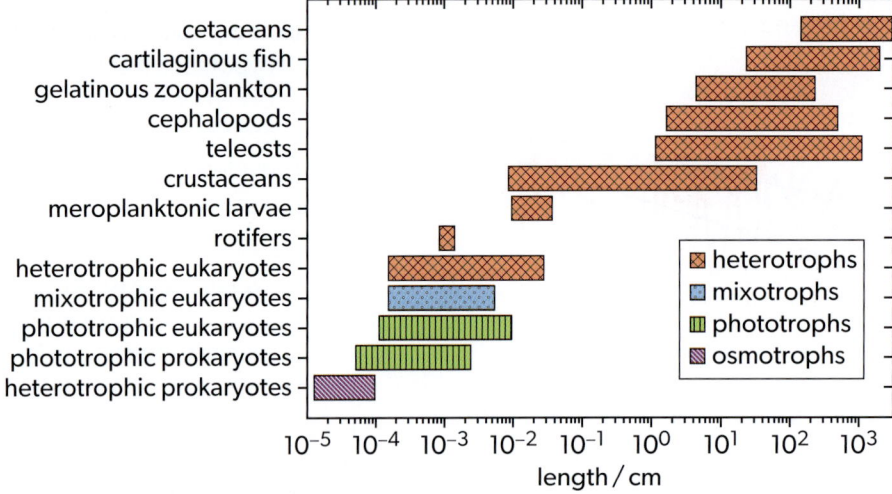

Source: Andersen, K. H., et al. (2016). Characteristic sizes of life in the oceans, from bacteria to whales. *Annual Review of Marine Science*, **8**, 217–241. www.doi.org/10.1146/annurev-marine-122414-034144

(a) Define mixotrophs. [1]
(b) Describe the mode of nutrition of phototrophs. [1]
(c) Identify the mode of nutrition of the largest organisms. [1]

6. The mechanisms of control regulating coastal benthic food webs can be bottom-up (BU) or top-down (TD) interactions.

   Identify the effect involved by completing the table with a positive sign (+) when abundance at the receiving trophic level is enhanced and a negative sign (−) when there is a reduction in abundance at the receiving trophic level. [4]

	Mechanism of control	Effect on	BU	TD
1	Producer biomasses increase by photosynthesis	herbivores		
2	Herbivores graze on producers	predators		
3	Producers provide habitat and shelter from predation	herbivores		
4	Herbivores graze on producers	producers		
5	Predators consume herbivores	producers		
6	Predators consume other predators	herbivores		
7	Deposition of nutrients in sediment of lake	producers		
8	Human disturbances (overfishing, excess nutrients)	predators		

7. The taxonomic correction process retests and updates existing species classification based on new evidence. Scientists studied taxonomic correction in the *Checklist of American Birds* over the last 130 years. They identified 142 lumps and 95 splits across 63 versions of the *Checklist*.

   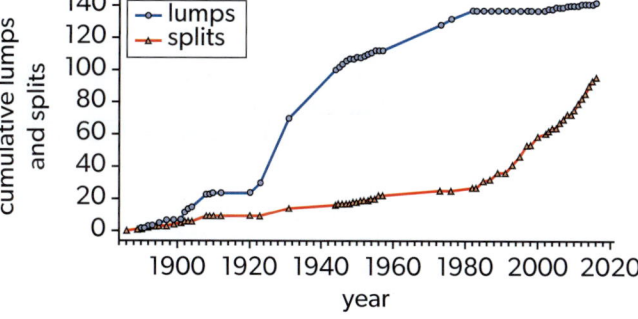

   Source: Vaidya, G., et al. (2018). The tempo and mode of the taxonomic correction process: How taxonomists have corrected and recorrected North American bird species over the last 127 years. *PLoS ONE*, 13(4), e0195736. www.doi.org/10.1371/journal.pone.0195736 (CC BY 4.0)

   (a) Comment on the trends for lumps and splits throughout the years. [3]

   Scientists investigated the uptake of RNA or protein into membrane vesicles. Either radioactive uracil (U^{15}) or radiolabelled albumin protein (BSA) was present during membrane formation. They then measured the fluorescence intensity inside and outside the vesicles. The histograms show the number of vesicles with increasing inside/outside ratios for U^{15} and BSA.

   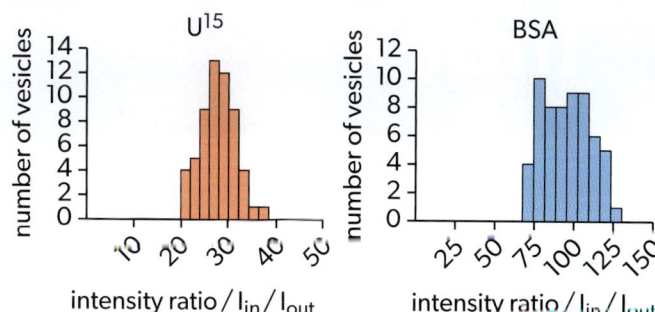

   Source: Cakmak, F. P., et al. (2021). Phospholipid membrane formation templated by coacervate droplets. *Langmuir*, 37(34), 10366–10375. www.doi.org/10.1021/acs.langmuir.1c01562 (CC-BY-NC-ND 4.0)

   (b) Suggest **one** reason for the use of U^{15} in this experiment. [1]
   (c) Outline RNA as the presumed first genetic material. [3]
   (d) Suggest **one** advantage of membrane-bound compartmentalization in cells. [1]
   (e) Explain how the structure of membranes allows for the formation of vesicles. [4]

Section B

SL: 16 marks
Answer one of the two extended-response questions.
HL: 32 marks
Answer two of the three extended-response questions.
An additional mark is available for the construction of your answer.

8. Organisms can adapt to their living environment.
   (a) Describe **two** abiotic factors that will affect the transpiration rate in plants. [4]
   (b) List **four** adaptations of plants to a desert biome. [4]
   (c) Explain, using a named example, how a response to an environmental change could lead to the evolution of a species. [7]

9. Different molecules have different functions.
   (a) Describe **two** properties of water. [4]
   (b) List **four** functions of lipids. [4]
   (c) Explain how enzymes are adapted to their function. [7]

10. Organisms perform different processes in life. Movement is one of these processes.
    (a) List **four** reasons for locomotion. [4]
    (b) Describe contraction in a sarcomere. [4]
    (c) An action potential on muscle cell leads to contraction due to the release of calcium ions. Explain the action potential in neurons. [7]

Index

Key terms are in **bold**.

abduction 93
ABO blood groups 59, 93, 154–5, 185, 213
absorption spectra 124–5
acetylcholine receptor signalling 61, 134–5, 141, 143–4
acquired immune deficiency syndrome (AIDS) *see* human immunodeficiency virus
acrosome reaction 210
action potentials 90, 138, 139, 142–3
action spectra 124, 125
activation energy, enzyme effects 109, 112, 113
active sites 109–10, 114–15
active transport, across membranes 59
adaptations
    food relationships in habitats 104–5
    increasing surface area-to-volume ratios of cells 70
    natural selection 225
    pneumocytes in alveoli 71
    sperm and egg cells 71
    striated muscle cells 71
    to environment 96–101
adaptive immune system 152–5
adaptive radiation 38–9
adduction 93
adenosine triphosphate (ATP)
    active transport 59, 62, 88
    muscle contractions 90
    photosynthesis 123, 128–9
    respiration 64–6, 116, 118–21
ADH *see* antidiuretic hormone
adhesion
    cell to cell 59, 63
    water molecules 3, 84, 88
adrenaline *see* epinephrine
aerobic respiration 4, 116, 118–19
afforestation 242
AIDS *see* human immunodeficiency virus
alleles 212
allelopathy 165
allopatric reproductive isolation 38
allosteric effects 61, 79, 112, 113
alveoli 71, 73
amino acids
    essential 53
    recycling by proteasomes 183
    structure 52, 55
    translation of RNA 53, 178–9
amphipathic molecules 51, 57, 58
anabolic reactions 109
anaerobic respiration 116
analogous structures 35
anaphase of mitosis 189
animal cells 19
antagonistic muscles 93
anthropogenic causes of climate change 237–9
anthropogenic species extinction 41
antibiotics/antibiotic resistance 30, 114, 155–6, 165, 171
antibodies 153–5
antidiuretic hormone (ADH) 148, 222
antigens 154–6, 185

anucleate cells 189
aquaporins 58, 222
aquatic environments 5–6, 232–3
archaea 26, 30, 33, 103, 153, 187
arrested succession 235
artificial selection 229
asexual reproduction 29, 204
ATP *see* **adenosine triphosphate**
autotrophs 103
auxin 150–2

bacteria 15, 29–30, 33, 132–3, 154
    see also antibiotics
bacteriophages 10–12, 22–3
bases (in nucleic acids) *see* **complementary base pairing**; genetic code; **nitrogenous bases**
benign tumours 194
bilayers, membranes 15, 51, 57–60
binomial system of classification 27
biochemical oxygen demand 232
biodiversity 40
    adaptive radiation 38–9
    threats 41–4
    see also diversity of organisms
biological species concept 27, 29–30
biomagnification of pollutants 233–5
biomes 100–1
blood cells *see* red blood cells; white blood cells
blood groups 59, 93, 154–5, 185, 213
blood supply to organs 222–3
blood vessels 81–3
B-lymphocytes 154–5
Bohr shift 79
Bowman's capsule 220, 221
box-and-whisker plots 215–16
brain 144, 146, 149
buoyancy 5

calcium ion signalling 133, 134
Calvin cycle 129–31
cancer 184–6, 194
carbohydrates 46–9, 59, 121
carbon atoms, chemical properties 46–7
carbon compounds on early Earth 13, 14
carbon cycle 169–71
carbon dioxide in the atmosphere
    anthropogenic increases 237–8
    enrichment effects on plant biomass 127
    Keeling curve
    ocean acidification 241–2
carbon fixation 129–30
carbon sequestration 242
carbon sinks and carbon fluxes 169, 171
cardiac cycle 86–7
cardiac muscle cell adaptations 71
carrying capacity of habitats 159
catabolic reactions 109
cell-cell adhesion 59, 63
cell-cell recognition 49, 59
cell cycle 192–4
cell and nuclear division 188–94
cells
    differentiation 69

    effects of osmotic water movement 200–1
    origins on earth 12–15
    respiration 115–21
    structure 16–21
cell signalling pathway regulation 137
centromeres 190
cerebellum 146
cervix 205
chemical mutagens 184
chemical signalling 132–7
chemiosmosis 120–1, 129
chemoreceptors 149
chiasma 191
chi-squared test 163–4
chlorophyll 122, 124–5, 128
chloroplasts 19, 21, 65–6, 129–30
cholesterol 59, 60, 82–3, 112
cholinergic synapses 143
    see also acetylcholine receptor signalling
chromatids 189–90
chromatin 189, 196, 197
chromosomes
    linkage of genes on same chromosome 218
    loci of genes 218
    in nuclear division 189
    number 28–30
    segregation and independent assortment in meiosis 216
circadian rhythms 147, 157
circulatory system 81–3, 86–7
CK *see* cytokinin
clades/cladistics 31–3
cladograms 32–3, 37
classification of organisms 27, 31–3
    see also **species**
clathrin 67
climate change 236–45
climax community 235
coalescence 15
codominant alleles 213
cohesion, water molecules 3, 84
collecting duct 220, 221
common ancestors
    of clades 31–3
    evidence for universal common ancestor 8
    homologous structures 35, 36
    last universal common ancestor 15, 30
    most recent common ancestors of modern humans 28
    unicellular ancestor of eukaryotes 21
communities 161
compartmentalization in eukaryotic cells 63–8
competition 105–6, 161, 162
competitive exclusion 106, 163
competitive inhibitors, enzymes 112, 113
complementary base pairing 8–11, 172–3, 177–8
condensation reactions 47, 52
conjugated proteins 55, 56
consciousness 144
conservation of biodiversity 40–4
conserved sequences, DNA 187
consumers in ecosystems 169

Index

continuous variation 212, 215
convergent evolution 23–4, 33, 35–6, 100
convoluted tubules (proximal and distal) 220, 221
cooperation 161
coral reefs 99, 241
coronary artery occlusion 82–3
cortical reaction 210
cotransporters 62–3
covalent bonds 2, 47
COVID-19 pandemic 22, 156–7
crenation 200–1
CRISPR–Cas9 complex 187
crossing over, DNA in sister chromatids 190–1, 218, 225
cross-pollination 207
cyclic succession 235
cytokine signalling 133, 134
cytokinesis 189
cytokinin (CK) 152

decomposers 103, 169, 233, 239
defence against disease 152–8
denaturation of proteins 54, 110
density-dependent factors 159, 164, 166
density-independent factors 164, 225, 226
dentition, diet relationship 103–4
depolarization, synaptic membranes 67, 138, 139, 142–3
desert biomes 100, 101
diabetes 219
dichotomous keys 30
diet
    dentition relationship 103
    essential amino acids 53
differentiation of cells 21, 68
diffusion, across membranes 58, 59
diploid nuclei 28, 190, 192, 206, 209–10, 212
directionality of DNA polymerases 175–6
directionality of transcription and translation 180
directional selection 228
disaccharides 47
discrete variation 215
disease
    defences 152–8
    pathogenicity 161
disruptive selection 228
divergent evolution 25, 30, 35
diversity of organisms 27–31
    see also biodiversity
DNA 7–11
    conserved sequences 187
    methylation 196
    protein synthesis 176–80
    replication 172–6, 189
    structure 8–10
DNA barcoding 30–1
DNA ligase 175
DNA polymerases 172–6
DNA primase 175
domains, classification of organisms 33
dominant alleles 212, 214, 216
dopamine 144

early Earth 6, 13–14
early hominids 103–4
ecosystem diversity 40
ecosystems
    causes of loss 43

    energy and matter transfer 167–70
    niches 102–6
    stability and change 229–36
    tipping points 230–1
EDGE of Existence programme 44
egg cells 68, 71, 197, 204, 206, 209, 210
electron microscopy 16–18, 20
electron transport chain 120
endangered species 41–2, 44
endocrine system
    control 147–8
    see also hormones
endocytosis 61, 67
endosymbiosis, theory of multicellularity 21
end-product inhibition, of metabolic pathways 113
energy
    in ecosystems 167–9
    storage and distribution in body 47, 116
    see also adenosine triphosphate
enhanced greenhouse effect 239
enteric nervous system (ENS) 149
enzymes 108–15
    activation energy 109, 112, 113
    active sites 109–10
    allosteric sites 112
    DNA replication 172–3
    induced-fit binding 109
    measurements of reactions 111
    rate of activity 110
epididymis 204, 205
epigenesis/epigenetics 21, 68, 195–7, 212
epinephrine 147–8
erythrocytes 19, 59, 70, 79, 155, 179, 185, 200
essential amino acids 53
estrogen see **oestradiol**
ethylene as a phytohormone 152
eukaryotes
    atypical cells 19
    cell structure 63–8
    common unicellular ancestor 21
eutrophication 165, 232–3
evolution 34–9
    antibiotic resistance 229
    cladistic relationships 31–3, 37
    climate change effects 244–5
    conserved DNA sequences 187
    eukaryotic cells and multicellularity 21
    evidence 35–6
    natural selection 224–9
    phylogenetic trees 28
    theory 35
    viruses 23–6
evolutionary distinctness, endangered species 44
excitatory neurotransmitters 144
excretion 220
exocytosis 61
exogenous chemicals, synaptic transmission effects 143
exons in mRNA 181
expiratory reserve of lungs 73
extinction 36, 41, 43, 44, 240

facilitated diffusion, across membranes 59
FASTA DNA sequencing 32, 34
fatty acids 49–50, 60
feedback loops
    cell signalling pathway regulation 137

    gene expression regulation 198
    see also negative feedback; positive feedback
fertilization 206, 207, 210
fibrous proteins 56
fitness, natural selection 225
flowering plants, reproduction 207–8
follicle stimulating hormone (FSH) 205–6
food chains and food webs 167–8, 235, 240
forest fires 239
FSH see **follicle stimulating hormone**
fundamental niches 105
fungi 19, 103, 155, 161, 165

gas exchange 72–80
    animals 73–6
    plants 75–8
gated ion channels 61–2, 139–40, 142–3
gel electrophoresis 173
gene expression 195–8
    bacterial quorum sensing 132
    differentiation of cells 21, 68
    external factors 197–8
    intracellular receptors 136
    methylation/epigenetics 196–7
    regulation 66, 136, 180, 195–6
    transcription 177
gene knockout (KO) technology 186
gene mutations 184–6
gene pools, natural selection 227–8
genetic code 8, 15, 23, 177–9, 184
genetic crosses 212
genetic disorders 179, 187, 214
genetic diversity (variation) 29, 33, 40, 186, 191, 207, 230
genomes 29, 195
genotypes 195, 212, 213, 218–19
germ cells
    epigenetic tags 197
    fertilization 207, 210
    formation 209–10
    mutations 186
    see also egg cells; sperm cells
globular proteins 56
glomerulus 220, 221
glucose
    functions 47
    photosynthesis 124
    regulation in blood 136, 182, 219
    respiration 116, 118–19
    structure 46
    transport across membranes 62–3, 136
glycolipids 59
glycolysis 112, 118–19, 120
glycoproteins 49, 59, 154
Goldilocks zone 6
Golgi apparatus 61, 64, 66–7
G protein-coupled receptor signalling 135, 137
grassland biomes 43, 100
greenhouse gasses 169, 232, 237–9
growth curves 159–61

habitats
    carrying capacity 159
    definition 96
    loss/destruction 41, 44
haemoglobin 56, 79–80
haploid nuclei 28, 190–1, 204, 209–10, 212
Hardy–Weinberg equation 228

heart 82–3, 86–7, 147, 149
heat capacity of water 5
helicase 172
herbivory/herbivores 103, 104, 161, 167, 235
herd immunity 156, 157
hermaphroditic plants 207
heterotrophs 103
heterozygous genotype 212
histones 10, 189, 196, 197
HIV see human immunodeficiency virus
holozoic nutrition 103
homeostasis 219–23
    see also negative feedback
homologous chromosomes 28, 190–2, 216, 218
homologous structures 35, 36
homozygous genotype 212
horizontal gene transfer 29–30
hormone replacement therapy (HRT) 211
hormones
    affecting gene expression 136–7
    changes at puberty 209
    endocrine glands 148
    epinephrine 147–8
    menstrual cycle 205–6
    steroids crossing phospholipid bilayers 51
    see also phytohormones
human immunodeficiency virus (HIV) 22, 25, 26, 155, 158
hybridization 29, 30, 39, 197
hydrogen bonding
    DNA strands 8, 172–4, 177
    protein structure 55–6
    water 2–3, 5, 84, 199
hydrolysis 47
hydrophilic/hydrophobic groups 3, 51, 55, 56, 57, 58
hyperpolarization of neurons 143, 144
hypertonic solutions 199–201, 220
hypothalamus 136, 146, 148, 209, 220, 222
hypotonic solutions 199–201

immune system 49, 152–8
    see also human immunodeficiency virus
immunization (vaccines) 25, 156
imprinted genes 197
incomplete dominance 213
independent assortment in meiosis 216
independent segregation of unlinked genes 218
induced-fit binding, at enzyme active sites 109
infections 153
    see also antibiotics; immune system; viruses
inheritance 211–19
inhibitors, enzymes 112–15
inhibitory neurotransmitters 144
innate immune system 152–3
insecticides 143
insect life cycles, climate change effects 243
inspiratory reserve 73
insulin 56, 136, 147, 148, 182–3, 219
integral proteins, in membranes 57
integration of body systems 145–52
intercostal muscles 73, 74, 93–4, 149
interphase 192–3
interspecific competition 106, 161–2

intracellular receptors 134, 136–7
intraspecific competition 106, 161–2
introns in mRNA 180
invasive species 98, 162–3
in-vitro fertilization (IVF) 206
irreversible binding of inhibitors 114–15
isotonic solutions 199–201
IUCN Red List categories of endangered species 41–2
IVF see in-vitro fertilization

joints 92–3

karyograms/karyotyping 28–9
Keeling curve 169, 170
keystone species 231, 235
kidneys 220–2
Krebs cycle 112, 119–20

large vacuoles 19, 20, 151
last universal common ancestor (LUCA) 15, 30
leaves 75–8, 84, 101, 104
legacy carbon 239
ligaments 92
ligands 61, 132, 135
light-dependent reactions in photosynthesis 122–3, 129
light-independent reactions in photosynthesis 122–3, 129–30
Lincoln index 159
linkage of genes 218
link reaction in cell respiration 118
lipids 49–51, 121
loci of genes 218
locomotion 93, 94
loop of Henle 220, 221
LUCA see last universal common ancestor
lumpers, taxonomy 40
lungs 71, 73–5, 93
luteinizing hormone (LH) 205
lymphatic system 85–6
lymphocytes 153–5
lysogenic cycle, of viruses 22–3
lytic cycle, of viruses 22–3

magnification 17–19
malignant tumours 194
MCRAs see most recent common ancestors
mechanism-based inhibition, enzymes 114–15
meiosis 190–2, 204
melatonin 147
membrane potentials
    neurons 138, 139
    see also depolarization
membranes 57–63
    active transport 61–2
    fatty acid content 60
    fluid mosaic model 59–60
    ion channels and pumps 61–2
    origins of lipid bilayers 15
    osmosis of water 58, 199–201
    permeability to substances 51, 57–9
    phospholipid bilayers 15, 51, 57–60
    proteins 57, 128
    temperature effects 60
memory cells 154
menstrual cycle 205–6
metabolism, enzymes 108–15
metaphase of mitosis 189
methane, in atmosphere 237–8
methylation, epigenetic tags 196–7

microscopy 16–20
Miller, Stanley 14
mitochondria 20, 64–5, 116, 119–21
mitochondrial DNA 29, 33
mitosis 189–90
mitotic index 194
mixotrophs 103
modes of nutrition 102–3
molecular clock 31
monosaccharides 46, 47
most recent common ancestors (MCRAs) 28
motility of animals 90–5
motor units, in skeletal muscle 91–2
multicellularity 21
multipotent stem cells 68
muscles
    antagonistic 93
    blood supply 222
    cell adaptations 71
    intercostal 73, 74, 93–4, 149
    mechanism of contraction 90
    motor units 91–2
    structure 91
mutagens 184–5
mutation rates 31–2
mutations 179, 184–6, 194
    single-nucleotide polymorphisms 29, 184, 185, 213
mutualism 161
myelination of neurons (myelin sheath) 140, 142, 146

NADH see reduced NAD
NADPH see **reduced NADP**
natural selection 8, 15, 224–9
    see also evolution
negative feedback 219
    cell signalling pathway regulation 137
    control of population size 159
    end-product inhibition of enzymes 113
    see also homeostasis
Neo-Darwinism 227
nerves
    impulses 138, 140, 142, 144
    integration of body systems 146, 149
    in muscle 90–1
nervous system 137–45
neurons 67, 138, 139
neurotransmitters 133–5
    excitatory and inhibitory 144
    gated ion channels 61
    neural synapses 67, 138, 141, 143–4
niches 102–6
nitrogenous bases 7–9, 116
    see also adenosine triphosphate; complementary base pairing; genetic code
non-coding sequences in DNA 180
non-competitive inhibitors, enzymes 112, 113
non-polar molecules, steroids 51
nucleic acids 7–11
    directionality 9
    protein synthesis 176–82
    structure 7–9
    see also DNA; RNA
nucleosomes 10, 196–7
nucleotides 7
nucleotide substitution rates 33
nucleus, separation from cytoplasm 64, 65
nutrition see diet; modes of nutrition

oestradiol 205–6
oncogenes 194
oogenesis 209
optimum conditions for enzymes 110–11
organelles 63–8
osmoregulation 220–2
osmosis 58, 89, 199–201
    see also **chemiosmosis**
ovaries 148, 205, 209
oviducts 205
oxidative phosphorylation 116
oxygen
    biochemical oxygen demand 232
    by-product of photosynthesis 123
    electron acceptor in respiration 121
    gas-exchange surfaces 73
oxygen dissociation curves, haemoglobin 79–80

pacemakers
    heart 87
    intestinal muscles 149
pain perception 144
pain reflex arcs 146–7
paradigm shift 225
parasitism 161
    see also **viruses**
passive transport, across membranes 59
pathogenicity 161
PCR see polymerase chain reaction
peat 170, 239, 242
pedigree charts 214
penis 205
peptide bonds 52–3
peripheral proteins, on membranes 57
peristalsis control 149
pH
    blood 149
    effect on proteins 54, 110–11
phagocytes 153, 154
phenology 242–4
phenotype 195, 212–13
phenylketonuria (PKU) 212
phloem 84–5, 88–9
phospholipid bilayers 15, 51, 57–60
photolysis 123
photosynthesis 122–31
    adaptations of plants to harvest light 105
    chlorophyll 122, 124–5, 128
    chloroplasts 19, 21, 65–6, 129–30
    gas exchange relationship 75–6
    limiting factors 126
photosynthetic pigments 123
    see also chlorophyll
photosystems 128
phylogenetic trees 28
phytohormones 150–2
pituitary gland 148
plant cells 19
plants
    adaptation for harvesting light 105
    adaptations to abiotic environment 98, 101
    adaptations to resist herbivory 104
    allelopathy 165
    gas exchange in leaves 75–8
    genetic crosses 212
    phytohormones 150–2
    polyploidy 39
    root and stem structures 85, 89
    sexual reproduction 207–8

transport of water and materials 83–5, 88–9
    tropic responses 149–50
plasma cells 154
plasmodesmata 88, 89
plasmolysis 200–1
pluripotent stem cells 68
pneumocytes in alveoli 71
poleward shift, of habitats 240
pollination 207
pollutants
    biomagnification 233–5
    gene expression effects 197
polymerase chain reaction (PCR) 173–4
polypeptides
    amino acid sequences 53
    formation 52
    modification 182–3
    translation of RNA 177–8
polyploidy 39
polysaccharides 48
polyspermy prevention 210
populations 159–61, 164–5
positive feedback
    cell signalling pathway regulation 137
    global warming effects 238–9
post-transcription modifications 181, 182
post-translational modification of polypeptides 182–3
predation 105, 161
pregnancy and childbirth 210–11
pressure potential 201
primary production 169
primary structure of proteins 55, 56
primary succession 235
primary tumours 194
producers in ecosystems 169
progesterone 205–6
prokaryotes
    cell structure 18
    endosymbiotic theory of eukaryotic evolution 21
    see also archaea; bacteria
prokaryotic cells 64
promoters in gene expression 196
prophase of mitosis 189
proteasomes 183
proteins 52–6
    denaturation 54, 110
    diversity 53, 55
    formation of polypeptides 52, 53
    glycoproteins 49
    levels of structure 55–6
    modification into their functional state 182–3
    mutations 179
    pH and temperature effects 54
    synthesis 176–83
    see also enzymes
proteome 195
proto-oncogenes 194
psychoactive drugs 144
puberty 209
pulse rate 82
purines 9–11
pyrimidines 9–11

quaternary structure of proteins 55, 56
quorum sensing 132–3

radiation exposure 183–4
rates of reaction, enzymes 110
realized niches 105
receptor tyrosine kinases (RTK) 135–6

recessive alleles 212, 216
red blood cells 19, 59, 70, 79, 155, 179, 185, 200
redox reactions 118, 120
reduced NAD (NADH) 116, 118–19, 120
reduced NADP (NAPDH) 129
reflex arcs 146–7
regulation of blood glucose 136, 182, 219
regulation of cell signalling pathways 137
regulators of gene expression 66, 136, 180, 195–8
replication of DNA 172–6
repolarization, synaptic membranes 138, 142
reproduction 204–11
    human 204–5, 209–11
    inheritance of genes 212
    plants 207–8
    sexual versus asexual 204
reproductive isolation 29, 38
resolution, of microscopes 17
respiration 115–21
    aerobic versus anaerobic 116
    rate of cell respiration 117–18
resting potential, neurons 138
rewilding 235
R-groups, of amino acids 52, 55
ribosomes 18–19, 66
ribulose-1,5-bisphosphate carboxylase oxygenase (Rubisco) 129–30
RNA 7, 9
    first genetic material 15
    transcription from DNA 177–80
    translation into polypeptides 177–82
root pressure 88
RTK see **receptor tyrosine kinases**
Rubisco see ribulose-1,5-bisphosphate carboxylase oxygenase

saprotrophs 103
    see also **decomposers**
saturated fatty acids 49, 50
saturated or trans fats 83
scrotum 204
secondary production 169
secondary structure of proteins 55, 56
secondary succession 235
secondary tumours 194
second messengers 135
seed dispersal 207
seed germination 208
segregation in meiosis 216
self-pollination 207, 208
seminal vesicles 205
sex chromosomes 213–14, 218
sex determination in zygotes 213
sex linkage 214, 218
sexual reproduction 204–11
sexual selection 226
sigmoid growth curve 159–61
signalling
    chemical 132–7
    neural 137–45
signal transduction pathways 132
single-nucleotide polymorphisms (SNPs) 29, 184, 185, 213
sinks, phloem transport materials 84
sister chromatids 190–1
skeleton 92
SNPs see **single-nucleotide polymorphisms**
sodium cotransport 62
sodium–potassium pumps 62, 138, 142

solute potential 201
solvents, water 3, 199
sources, phloem transport materials 84
speciation 27
    adaptive radiation 38
    barriers to hybridization 39
    evolution relationship 36
    reproductive isolation 29, 38
species
    anthropogenic extinction 41
    biological species concept 27, 29–30
    environmental effects on distribution 98–9
    identification by DNA barcoding 30–1
species diversity 40
species evenness 40
species richness 40
spermatogenesis 209–10
sperm cells 68, 71, 148, 197, 204, 206, 213
sperm ducts 204
spinal cord 146
spindle fibres 189
splitters, taxonomy 40
spontaneous generation theory of life 13
stabilizing selection 228
steroids
    cholesterol 59, 60, 82–3, 112
    crossing phospholipid bilayers 51
    intracellular receptors affecting gene expression 136–7
    sex hormones 209–11
stomata 75–8
striated muscle cells 68, 71
stroke volume, heart 149
stroma, in chloroplasts 65, 66, 129–30
substrate, of enzymes 109
succession stages 235–6
surface area-to-volume ratio 70–1, 73
sustainability 231–2
sympatric reproductive isolation 38
synapses/synaptic transmission 67, 138, 141, 143–4
synovial fluid/synovial joints 92–3

taiga biomes 100
taxonomy, splitters and lumpers 40
teeth, dentition/diet relationship 103
telophase of mitosis 189, 190
temperature
    climate change 236–45
    thermoregulation 220
temperature effects
    membrane fluidity 60
    photosynthesis 126, 131
    proteins 54, 110, 111
    water oxygen concentration 4
tendons 92
tertiary structure of proteins 55, 56
testes 148, 204, 209
thermal conductivity 5
thermoregulation 220
thylakoids (thylakoid membrane/thylakoid space) 65, 66, 128, 129
tidal volume of lungs 74–5
tipping points 230–1, 239–40
tissue fluid 85–6
T-lymphocytes 154–5
top-down and bottom-up control of populations 164–5
totipotent cells 68
transcription 177, 181
transcription factors 195
transcriptome 195
transduction pathways 132
translation 177–82
translocation 88, 89
transmembrane proteins 57
transmembrane receptors 134–6
transpiration 3, 76, 78, 84, 88, 201–2
transport
    in humans 81–3, 85–7
    in plants 83–5, 88–9
triglycerides 50
tropical rainforest biomes 100
tropic responses in plants 149–50
tumours/cancer 184–6, 194
tundra biomes 100
type I/type II pneumocytes 71

Type I/Type II pneumocytes 71
tyrosine kinase receptors 135–6

unicellular organisms 19, 21, 200
universal common ancestor 8, 15, 30
unsaturated fatty acids 49
upslope shift, of habitats due to global warming 240–1
Urey, Harold 14
uterus 205

vaccines 25, 156
vacuoles 19, 20, 64, 151, 154, 200
vagina 205
variation, natural selection 245
ventilation rate of lungs 74–5
vesicles 15, 61, 67
viruses 22–6
    origins 23–4
    see also bacteriophages
viscosity 5
vital capacity of lungs 73
voltage-gated ion channels 61–2, 139–40, 142–3
vulva 205

water 2–6
    cellular uptake/loss 200–1
    origins on Earth 6
    properties 2–5
    as a solvent 3, 199
    transport in plants 3, 78, 83–4, 88
water potential 199, 201–3
white blood cells 153–5
whole genome sequencing 29

X chromosomes 213–14
xylem 83–5, 88–9

Y chromosomes 213

zoonoses 156
zygotes 68, 212, 213